SAS/GRAPH® 9.2
Reference
Second Edition

SAS® Documentation

The correct bibliographic citation for this manual is as follows: SAS Institute Inc. 2010. *SAS/GRAPH® 9.2 Reference, Second Edition*. Cary, NC: SAS Institute Inc.

SAS/GRAPH® 9.2 Reference, Second Edition

Copyright © 2010, SAS Institute Inc., Cary, NC, USA

978-1-60764-449-1

All rights reserved. Produced in the United States of America.

For a hard-copy book: No part of this publication may be reproduced, stored in a retrieval system, or transmitted, in any form or by any means, electronic, mechanical, photocopying, or otherwise, without the prior written permission of the publisher, SAS Institute Inc.

For a Web download or e-book: Your use of this publication shall be governed by the terms established by the vendor at the time you acquire this publication.

U.S. Government Restricted Rights Notice. Use, duplication, or disclosure of this software and related documentation by the U.S. government is subject to the Agreement with SAS Institute and the restrictions set forth in FAR 52.227-19 Commercial Computer Software-Restricted Rights (June 1987).

SAS Institute Inc., SAS Campus Drive, Cary, North Carolina 27513.

1st electronic book, May 2010
1st printing, May 2010

SAS® Publishing provides a complete selection of books and electronic products to help customers use SAS software to its fullest potential. For more information about our e-books, e-learning products, CDs, and hard-copy books, visit the SAS Publishing Web site at **support.sas.com/publishing** or call 1-800-727-3228.

SAS® and all other SAS Institute Inc. product or service names are registered trademarks or trademarks of SAS Institute Inc. in the USA and other countries. ® indicates USA registration.

Other brand and product names are registered trademarks or trademarks of their respective companies.

Contents

What's New **xv**
Overview **xv**
The SAS/GRAPH Statistical Graphics Suite **xv**
The SAS/GRAPH Network Visualization Workshop **xvi**
Support for Multiple Open ODS Destinations **xvii**
Support for ODS Styles **xvii**
Devices **xvii**
Colors **xviii**
Fonts and Font Rendering **xviii**
Changing the Appearance of Output to Match That of Earlier SAS Releases **xix**
Procedures **xix**
Global Statements **xxiii**
Graphics Options **xxiv**
Transparent Overlays **xxiv**
ActiveX Control **xxiv**
Java Map Applet **xxiv**
Java Tilechart Applet **xxiv**
The Annotate Facility **xxv**
New Map Data Sets **xxv**
Updated Map Data Sets **xxv**
Map Data Set Descriptions **xxx**
New Data Set for Military ZIP Codes **xxx**
Changes in SAS/GRAPH Documentation **xxx**

PART 1 SAS/GRAPH Concepts 1

Chapter 1 △ Introduction to SAS/GRAPH Software 3
Overview 4
Components of SAS/GRAPH Software 4
Device-Based Graphics and Template-Based Graphics 6
Graph Types 7
About this Document 24
Conventions Used in This Document 25
Information You Should Know 28

Chapter 2 △ Elements of a SAS/GRAPH Program 31
Overview 31
A Typical SAS/GRAPH Program 31

Chapter 3 △ Getting Started With SAS/GRAPH 39
Introduction 39
Introduction to ODS Destinations and Styles 40

Generating Output With SAS/GRAPH Procedures 43
Controlling the Graphics Output Format With the DEVICE= Option 48
Summary of Default Destinations, Styles, and Devices 49
Sending Output To Multiple Open Destinations 51
Related Topics 52

Chapter 4 △ SAS/GRAPH Processing 53
Running SAS/GRAPH Programs 53
SAS Data Sets 54
Specifying an Input Data Set 54
Using Engines with SAS/GRAPH Software 56
RUN-Group Processing 56

Chapter 5 △ The Graphics Output Environment 59
Overview 59
The Graphics Output and Device Display Areas 59
Controlling Dimensions 60
Controlling Display Area Size and Image Resolution 61
Units 62
Maintaining the Quality of Your Image Across Devices 65
How Graphic Elements are Placed in the Graphics Output Area 65
How Errors in Sizing Are Handled 66

Chapter 6 △ Using Graphics Devices 67
Overview 67
What Is a SAS/GRAPH Device? 68
Commonly Used Devices 68
Default Devices For ODS Destinations 69
Viewing The List Of All Available Devices 70
Deciding Which Device To Use 71
Overriding the Default Device 72
Device Categories And Modifying Default Output Attributes 72
Using Universal Printer Shortcut Devices 75
Using Scalable Vector Graphics Devices 77
Viewing and Modifying Device Entries 85
Creating a Custom Device 86
Related Topics 86

Chapter 7 △ SAS/GRAPH Output 87
About SAS/GRAPH Output 88
Specifying the Graphics Output File Type for Your Graph 91
The SAS/GRAPH Output Process 93
Setting the Size of Your Graph 94
Setting the Resolution of Your Graph 95
Controlling Where Your Output is Stored 97

Replacing an Existing Graphics Output File Using the GSFMODE= Graphics Option **104**
Storing Multiple Graphs in a Single Graphics Output File **104**
Replaying Your SAS/GRAPH Output **106**
Previewing Output **109**
Printing Your Graph **110**
Exporting Your Output **111**

Chapter 8 △ Exporting Your Graphs to Microsoft Office Products 113
What to Consider When Choosing an Output Format **113**
Comparison of the Graphics Output **116**
Enhancing Your Graphs **120**
Importing Your Graphs into Microsoft Office **120**

Chapter 9 △ Writing Your Graphs to a PDF File 123
About Writing Your Graphs to a PDF File **123**
Changing the Page Layout **124**
Adding Metadata to Your PDF File **124**
Adding Bookmarks for Your Graphs **124**
Changing the Default Compression Level for Your PDF File **125**
Examples **125**

Chapter 10 △ Controlling The Appearance of Your Graphs 133
Overview **133**
Style Attributes Versus Device Entry Parameters **134**
About Style Templates **135**
Specifying a Style **139**
Overriding Style Attributes With SAS/GRAPH Statement Options **140**
Precedence of Appearance Option Specifications **141**
Viewing the List of Styles Provided by SAS **141**
Modifying a Style **142**
Graphical Style Element Reference for Device-Based Graphics **144**
Turning Off Styles **153**
Changing the Appearance of Output to Match That of Earlier SAS Releases **154**

Chapter 11 △ Specifying Fonts in SAS/GRAPH Programs 155
Introduction: Specifying Fonts in SAS/GRAPH Programs **155**
SAS/GRAPH, System, and Device-Resident Fonts **155**
TrueType Fonts That Are Supplied by SAS **156**
Determining What Fonts Are Available **157**
Default Fonts **157**
Viewing Font Specifications in the SAS Registry **158**
Specifying a Font **159**
Methods For Specifying Fonts **163**

Chapter 12 △ SAS/GRAPH Colors and Images 167
Using SAS/GRAPH Colors and Images **167**

Specifying Colors in SAS/GRAPH Programs 168
Specifying Images in SAS/GRAPH Programs 181

Chapter 13 △ Managing Your Graphics With ODS 191
Introduction 191
Managing ODS Destinations 191
Specifying a Destination 192
ODS Destination Statement Options 192
ODS and Procedures that Support RUN-Group Processing 194
Controlling Titles and Footnotes with Java and ActiveX Devices in HTML Output 194

Chapter 14 △ SAS/GRAPH Statements 197
Overview 197
Example 1. Ordering Axis Tick Marks with SAS Date Values 294
Example 2. Specifying Logarithmic Axes 297
Example 3. Rotating Plot Symbols Through the Color List 299
Example 4. Creating and Modifying Box Plots 302
Example 5. Filling the Area between Plot Lines 304
Example 6. Enhancing Titles 307
Example 7. Using BY-group Processing to Generate a Series of Charts 309
Example 8. Creating a Simple Web Page with the ODS HTML Statement 313
Example 9. Combining Graphs and Reports in a Web Page 315
Example 10. Creating a Bar Chart with Drill-Down Functionality for the Web 321

Chapter 15 △ Graphics Options and Device Parameters Dictionary 327
Introduction 327
Specifying Graphics Options and Device Parameters 327
Dictionary of Graphics Options and Device Parameters 328

PART 2 Bringing SAS/GRAPH Output to the Web 437

Chapter 16 △ Introducing SAS/GRAPH Output for the Web 439
Which Device Driver or Macro Do I Use? 439
Types of Web Presentations Available 440
Selecting a Type of Web Presentation 447
Generating Web Presentations 451

Chapter 17 △ Creating Interactive Output for ActiveX 453
Overview 453
When to Use the ACTIVEX Device 454
Installing the SAS/GRAPH ActiveX Control 455
Generating Output for ActiveX 457
About Languages in ACTIVEX 458
About Special Fonts and Symbols in ACTIVEX 459
SAS Formats Supported by ACTIVEX 459
Configuring Drill-Down Links with ACTIVEX 460

ActiveX Examples 461

Chapter 18 △ Creating Interactive Output for Java 469
Overview 469
When to Use the JAVA Device 470
Generating Output for Java 470
Configuring Drill-Down Links for Java 475
Examples of Interactive Java Output 475

Chapter 19 △ Attributes and Parameters for Java and ActiveX 485
Specifying Parameters and Attributes for Java and ActiveX 485
Parameter Reference for Java and ActiveX 488

Chapter 20 △ Generating Static Graphics 503
What is a Static Graphic? 503
Creating a Static Graphic 504
ACTXIMG and JAVAIMG Devices Compared to GIF, JPEG, SVG, and PNG Devices 506
Developing Web Presentations with the GIF, JPEG, SVG, and PNG Devices 508
Developing Web Presentations with the JAVAIMG and ACTXIMG Devices 510
Adding Drill-Down Links to Web Presentations Generated with a Static-Graphic Device 511
Sample Programs for Static Images 512

Chapter 21 △ Generating Web Animation with GIFANIM 519
Developing Web Presentations with the GIFANIM Device 519
When to Use the GIFANIM Device 519
Creating an Animated Sequence 520
GOPTIONS for Controlling GIFANIM Presentations 521
Sample Programs: GIFANIM 522

Chapter 22 △ Generating Interactive Metagraphics Output 531
Developing Web Presentations for the Metaview Applet 531
Advantages of Using the JAVAMETA Device 532
Using ODS With the JAVAMETA Device 532
Enhancing Web Presentations for the Metaview Applet 533
Specifying Non-English Resource Files and Fonts 533
Metaview Applet Parameters 534
Example: Generating Metacode Output With the JAVAMETA Driver 536

Chapter 23 △ Generating Web Output with the Annotate Facility 539
Overview of Generating Web Output with the Annotate Facility 539
Generating Web Output with the Annotate Facility 539
Examples 541

Chapter 24 △ Creating Interactive Treeview Diagrams 543
Creating Treeview Diagrams 543
Enhancing Presentations for the Treeview Applet 546

DS2TREE Macro Arguments 547
Sample Programs: Treeview Macro 547

Chapter 25 △ Creating Interactive Constellation Diagrams 553
Creating Constellation Diagrams 553
Enhancing Presentations for the Constellation Applet 559
DS2CONST Macro Arguments 560
Sample Programs: Constellation Macro 560

Chapter 26 △ Macro Arguments for the DS2CONST and DS2TREE Macros 569
Macro Arguments 569

Chapter 27 △ Enhancing Web Presentations with Chart Descriptions, Data Tips, and Drill-Down Functionality 595
Overview of Enhancing Web Presentations 596
Chart Descriptions for Web Presentations 596
Data Tips for Web Presentations 598
Adding Links with the HTML= and HTML_LEGEND= Options 601
Controlling Drill-Down Behavior For ActiveX and Java Using Parameters 608
Example: Creating Bar Charts with Drill-Down for the Web 618

Chapter 28 △ Troubleshooting Web Output 633
Troubleshooting Web Output 633
Checking Browser Permissions 636
Using HTML Character Entities 636
Connecting to Web Servers that Require Authentication 637
Removing CLASSPATH Environment Variables 637
Setting the SAS_ALT_DISPLAY Variable for X Window Systems on UNIX 637
Correcting Text Fonts 638
Resolving Differences Between Graphs Generated with Different Technologies 638

PART 3 The Annotate Facility 639

Chapter 29 △ Using Annotate Data Sets 641
Overview 641
About the Annotate Data Set 643
About Annotate Graphics 649
Creating an Annotate Data Set 654
Producing Graphics Output from Annotate Data Sets 655
Annotate Processing Details 656
Examples 658

Chapter 30 △ Annotate Dictionary 667
Annotate Dictionary Overview 669
Annotate Functions 669
Annotate Variables 700
Annotate Internal Coordinates 737

Annotate Macros   738
Using Annotate Macros   759
Annotate Error Messages   761

PART 4 The Data Step Graphics Interface   767

Chapter 31 △ The DATA Step Graphics Interface   769
Overview   770
Applications of the DATA Step Graphics Interface   773
Using the DATA Step Graphics Interface   774
DSGI Graphics Summary   776

Chapter 32 △ DATA Step Graphics Interface Dictionary   813
Overview   813
GASK Routines   816
GDRAW Functions   855
GRAPH Functions   866
GSET Functions   870
Return Codes for DSGI Routines and Functions   908
See Also   909
References   910

PART 5 SAS/GRAPH Procedures   911

Chapter 33 △ The GANNO Procedure   913
Overview   913
Procedure Syntax   914
Examples   916

Chapter 34 △ The GAREABAR Procedure   931
Overview   931
Concepts   932
Procedure Syntax   933
Examples   937

Chapter 35 △ The GBARLINE Procedure   947
Overview   947
Concepts   949
Procedure Syntax   958
Examples   981

Chapter 36 △ The GCHART Procedure   989
Overview   990
Concepts   996
Procedure Syntax   1003
Examples   1066

References 1093

Chapter 37 △ The GCONTOUR Procedure 1095
Overview 1095
Concepts 1097
Procedure Syntax 1098
Examples 1115
References 1123

Chapter 38 △ The GDEVICE Procedure 1125
Overview 1126
Concepts 1126
Procedure Syntax 1128
Using the GDEVICE Procedure 1136
Examples 1143

Chapter 39 △ The GEOCODE Procedure 1147
Overview of the GEOCODE Procedure 1147
Concepts 1149
Procedure Syntax 1154
Street Geocoding 1162
Examples 1167

Chapter 40 △ The GFONT Procedure 1175
Overview 1175
Concepts 1176
Procedure Syntax 1178
Creating Fonts 1187
Examples 1199

Chapter 41 △ The GINSIDE Procedure 1205
Overview 1205
Procedure Syntax 1205
Examples 1207

Chapter 42 △ The GKPI Procedure 1213
Overview 1213
Concepts 1216
Procedure Syntax 1225
Examples 1230

Chapter 43 △ The GMAP Procedure 1239
Overview 1240
Concepts 1244
Procedure Syntax 1251
Using FIPS Codes and Province Codes 1289
Using Formats for Map Variables 1291

Using SAS/GRAPH Map Data Sets   **1294**
Examples   **1301**

Chapter 44 △ The GOPTIONS Procedure   **1319**
Overview   **1319**
Procedure Syntax   **1320**
Examples   **1322**

Chapter 45 △ The GPLOT Procedure   **1325**
Overview   **1325**
Concepts   **1329**
Procedure Syntax   **1332**
Examples   **1366**

Chapter 46 △ The GPROJECT Procedure   **1395**
Overview   **1395**
Concepts   **1397**
Procedure Syntax   **1402**
Using the GPROJECT Procedure   **1407**
Examples   **1409**
References   **1418**

Chapter 47 △ The GRADAR Procedure   **1419**
Overview   **1419**
Calculating Weighted Statistics   **1420**
Procedure Syntax   **1421**
Examples   **1435**

Chapter 48 △ The GREDUCE Procedure   **1447**
Overview   **1447**
Concepts   **1449**
Procedure Syntax   **1450**
Using the GREDUCE Procedure   **1452**
Examples   **1454**
References   **1457**

Chapter 49 △ The GREMOVE Procedure   **1459**
Overview   **1459**
Concepts   **1460**
Procedure Syntax   **1462**
Examples   **1465**

Chapter 50 △ The GREPLAY Procedure   **1473**
Overview   **1474**
Concepts   **1475**
Procedure Syntax   **1477**
Using the GREPLAY Procedure Windows   **1500**

Running the GREPLAY Procedure Using Code-based Statements   1504
Replaying Catalog Entries   1505
Creating Custom Templates   1506
Replaying Graphics Output in a Template   1506
Creating Color Maps   1507
Examples   1508

Chapter 51 △ The GSLIDE Procedure   1517
Overview   1517
Procedure Syntax   1518
Examples   1522

Chapter 52 △ The GTILE Procedure   1527
Overview   1527
Concepts   1527
Procedure Syntax   1529
Examples   1536

Chapter 53 △ The G3D Procedure   1541
Overview   1541
Concepts   1543
Procedure Syntax   1546
Examples   1560
References   1570

Chapter 54 △ The G3GRID Procedure   1571
Overview   1571
Concepts   1573
Procedure Syntax   1576
Examples   1581
References   1590

Chapter 55 △ The MAPIMPORT Procedure   1593
Overview   1593
Procedure Syntax   1594
Examples   1597

PART 6   Appendixes   1599

Appendix 1 △ Summary of ActiveX and Java Support   1601
Introduction   1602
Global Statements   1602
PROC GAREABAR   1612
PROC GBARLINE   1613
PROC GCHART   1615
PROC GCONTOUR   1620
PROC GMAP   1622

PROC GPLOT **1625**
PROC GRADAR **1630**
PROC GTILE **1633**
PROC G3D **1633**
Annotate Functions **1635**

Appendix 2 △ Using SAS/GRAPH Fonts **1643**
Introduction **1643**
Rendering Bitstream Fonts **1643**
Listing or Displaying SAS/GRAPH Fonts on Your System **1644**
SAS/GRAPH Font Lists **1644**
The SIMULATE Font **1652**
Font Locations And the Default Search Path **1653**

Appendix 3 △ Using Device-Resident Fonts **1655**
Introduction **1655**
Default Device-Resident Fonts **1655**
Specifying the Full Font Name **1657**
Specifying Alternative Device-Resident Fonts **1657**

Appendix 4 △ Transporting and Converting Graphics Output **1659**
About Transporting and Converting Graphics Output **1659**
Transporting Catalogs across Operating Environments **1659**
Converting Catalogs to a Different Version of SAS **1662**

Appendix 5 △ GREPLAY Procedure Template Code **1663**
Overview **1663**
H2: One Box Left and One Box Right **1663**
H2S: One Box Left and One Box Right with Space **1664**
H3: Three Boxes Across **1664**
H3S: Three Boxes Across with Space **1665**
H4: Four Boxes Across **1665**
H4S: Four Boxes Across with Space **1666**
L1R2: One Box Left and Two Boxes Right **1666**
L1R2S: One Box Left and Two Boxes Right with Space **1667**
L2R1: Two Boxes Left and One Box Right **1667**
L2R1S: Two Boxes Left and One Box Right with Space **1668**
L2R2: Two Boxes Left and Two Boxes Right **1668**
L2R2S: Two Boxes Left and Two Boxes Right with Space **1669**
U1D2: One Box Up and Two Boxes Down **1670**
U1D2S: One Box Up and One Box Down with Space **1670**
U2D1: Two Boxes Up and One Box Down **1671**
U2D1S: Two Boxes Up and One Box Down with Space **1671**
V2: One Box Up and One Box Down **1672**
V2S: One Box Up and One Box Down with Space **1672**
V3: Three Boxes Vertically **1672**

V3S: Three Boxes Vertically with Space   **1673**
Whole: Entire Screen Template   **1673**

Appendix 6 △ Recommended Reading   1675

Recommended Reading   **1675**

Glossary   1677

Index   1693

What's New

Overview

The changes and enhancements for SAS/GRAPH 9.2 are very extensive. Highlights include the following:

- The new SAS/GRAPH statistical graphics suite provides a new set of procedures, a new language, and a graph editor specifically designed for creating and editing statistical graphics.
- All SAS/GRAPH procedures now support ODS styles for all devices.
- SAS/GRAPH now automatically selects an appropriate device and style for all open destinations.
- SAS/GRAPH now provides TrueColor support, which allows over 16 million colors in a single image.
- The new Network Visualization Workshop enables you to visualize and investigate the patterns and relationships hidden in network data (node-link data).
- The new GKPI procedure generates several key performance indicators.
- The new GTILE procedure generates tile charts.
- The new GEOCODE procedure enables you to add geographic coordinates to data sets that contain location information such as mailing addresses or to perform geolocation with non-address location data.
- The new GINSIDE procedure determines which polygon in a map data set contains the geographic coordinates in your input data set.
- All procedures now support graphics output filenames up to 256 characters long.
- Many procedures have significant enhancements and new options. See "Procedures" on page xix for a complete list.
- The new Scalable Vector Graphics devices enable you to generate SVG output.
- Several new map data sets, as well as new feature data sets, have been added to the MAPS library. Several existing map data sets have been updated.

The SAS/GRAPH Statistical Graphics Suite

ODS Statistical Graphics (referred to as ODS Graphics for short) is major new functionality for creating statistical graphics that is available in a number of SAS software products, including SAS/STAT, SAS/ETS, SAS/QC, and SAS/GRAPH. Many statistical procedures have been enabled to use this functionality, and these procedures now produce graphs as automatically as they produce tables. In addition, the new statistical graphics (SG) family of SAS/GRAPH procedures use this functionality to produce plots for exploratory data analysis and customized statistical displays.

ODS Graphics includes the new SAS/GRAPH statistical graphics suite. This suite provides the following new features:

SAS/GRAPH statistical graphics procedures
: provide a simple syntax for creating graphics commonly used in exploratory data analysis and for creating customized statistical displays. These new procedures include the SGPANEL, SGPLOT, and SGSCATTER procedures. In addition, the SGRENDER procedure provides a SAS procedure interface to the new Graph Template Language. For more information, including changes and enhancements for SAS 9.2 Phase 2, see *SAS/GRAPH: Statistical Graphics Procedures Guide*.

Graph Template Language (GTL)
: is the underlying language for the default templates that are provided by SAS for procedures that use ODS Statistical Graphics. You can use the GTL either to modify these templates or to create your own highly customized graphs. Templates written with the GTL are built with the TEMPLATE procedure. For more information about Graph Template Language, see the *SAS/GRAPH: Graph Template Language Reference* and the *SAS/GRAPH: Graph Template Language User's Guide*.

ODS Graphics Editor
: is an interactive editor that enables you to edit and enhance graphs that are produced by procedures that use ODS Statistical Graphics. You can use the ODS Graphics Editor to modify the existing elements of a graph such as titles and labels, or to add features such as text annotation for data points. For more information, including changes and enhancements for SAS 9.2 Phase 2, see *SAS/GRAPH: ODS Graphics Editor User's Guide*.

ODS Graphics Designer
: provides a point-and-click interface for creating ODS graphics. Using the ODS Graphics Designer does not require knowledge of ODS templates or the Graph Template Language. With the ODS Graphics Designer, you can easily create multi-cell graphs, classification panels, scatter plot matrices, and more. You can save your output as an image file or as an ODS Graphics Designer file (SGD file) that you can edit later.

    The ODS Graphics Designer is available beginning with SAS 9.2 Phase 2. For more information, including information about changes and enhancements for the third maintenance release of SAS 9.2, see *SAS/GRAPH: ODS Graphics Designer User's Guide*.

Note: For additional information on the ODS Statistical Graphics functionality, see *SAS Output Delivery System: User's Guide* and *SAS/STAT User's Guide*. △

The SAS/GRAPH Network Visualization Workshop

The Network Visualization (NV) Workshop application enables you to visualize and investigate the patterns and relationships hidden in network data (node-link data). Some common applications that use network data include supply chains, communication networks, Web sites, database schema, and software module dependencies. NV Workshop is designed for visualizing large networks. Using a combination of data tables, statistical graphs, and network graphs, NV Workshop enables you to extract information that would otherwise remain hidden. Help is available from the menu within the product. To start NV Workshop, select **Start ▶ Programs ▶ SAS ▶ SAS GRAPH NV Workshop 2.1**.

For more information, including changes and enhancements for SAS 9.2 Phase 2, see *SAS/GRAPH: Network Visualization Workshop User's Guide*.

Support for Multiple Open ODS Destinations

If you have multiple ODS destinations open, SAS/GRAPH automatically selects the appropriate device for each destination. In addition, each graph uses the ODS style associated with each destination. You do not need to specify a device or style to get optimal results. For example, if you do not specify a device, then SAS/GRAPH automatically selects the PNG device for the HTML destination if it is open and the SASEMF device for the RTF destination.

Also, if you have multiple ODS destinations open and you are using a device other than the Java or ActiveX devices (ACTIVEX, JAVA, ACTXIMG, or JAVAIMG), a different GRSEG is created for each open destination. The GRSEGs for the first destination are stored in WORK.GSEG. The GRSEGs for any other open destinations are stored in catalogs named according to the destinations, for example, WORK.HTML.

Support for ODS Styles

All SAS/GRAPH procedures and devices now support ODS styles. By default, all colors, fonts, symbols, and graph sizes are derived from the current style. Procedure statement options and SAS/GRAPH GOPTIONS override individual elements of the style, so you can easily customize the appearance of any graph.

Additionally, the colors used by the styles have been updated to enhance the appearance of your graphics output.

The use of ODS styles by default is controlled by the GSTYLE system option. For information on the GSTYLE option, refer to *SAS Language Reference: Dictionary*.

Devices

- The new Scalable Vector Graphics devices enable you to create SVG graphs. The SVG devices (SVG, SVGZ, SVGView, and SVGT) are supported for the LISTING, HTML, and PRINTER destinations.
- The default device for the ODS HTML destination has changed from GIF to PNG, which provides TrueColor support. Using the PNG device might result in graphs that have spacing or size differences, such as slightly narrower bars in bar charts.
- Data tips are now supported by the JAVAIMG device.
- Several devices have been added for compatibility with previous releases of SAS/GRAPH. These devices are named Z*device*, where device is the name of the device in previous releases.
    - The following devices ignore the FONTRENDERING= system option and force host font rendering (see "Fonts and Font Rendering" on page xviii): ZGIF, ZGIF733, ZGIFANIM, ZJPEG, ZPNG, ZSASBMP, ZTIFFB, ZTIFFBII, ZTIFFBMM, ZTIFFG3, ZTIFFG4, and ZTIFFP.

- The following devices support printer-resident fonts only: ZPCL5, ZPDF, ZPDFC, ZPSCOLOR, ZPSEPSFC, ZPSL, and ZPSLEPSF. They will not work well with ODS styles (see "Support for ODS Styles" on page xvii) because they do not support TrueType fonts, which are used by the styles.
- Several Universal Printing shortcut devices have been added. The UPCL5, UPCL5E and UPCL5C devices have been added for printing support. The UPDF and UPDFC devices have been added for PDF support. The UPSL and UPSLC devices have been added for PostScript support. See also "Using Universal Printer Shortcut Devices" on page 75.

Colors

- SAS/GRAPH now provides TrueColor support, which allows over 16 million colors in a single image.
- The number of colors in the default color list has been increased to 38.

Fonts and Font Rendering

- The following fonts are now obsolete: DAVID, NHIRA, NKATA.
- Some of the characters in the Hebrew font are mapped differently to the Roman character set than they were previously.
- Fonts are now rendered using the FreeType engine. This new font rendering might result in fonts appearing larger than they did in previous versions of SAS/GRAPH. See also "Changing the Appearance of Output to Match That of Earlier SAS Releases" on page xix.
- Many new TrueType fonts have been added. These new fonts are listed in Table 0.1 on page xviii.

Table 0.1 TrueType Fonts Supplied by SAS

Albany AMT*	Thorndale Duospace WT SC	GungsuhChe
Cumberland AMT*	Thorndale Duospace WT TC	Dotum
Thorndale AMT*	Arial Symbol*	DotumChe
Symbol MT	Times New Roman Symbol*	Gulim
Monotype Sorts	MS PMincho	GulimChe
Monotype Sans WT J	MS Mincho	NSimSun
Monotype Sans WT K	MS PGothic	SimHei
Monotype Sans WT SC	MS UI Gothic	SimSun
Monotype Sans WT TC	Batang	PMingLiU
Thorndale Duospace WT J	BatangChe	MingLiU
Thorndale Duospace WT K	Gungsuh	HeiT

* Albany AMT, Cumberland AMT, Thorndale AMT, Arial Symbol, and Times New Roman Symbol are font families. Normal, bold, italic, and bold italic versions of these fonts are provided.

Changing the Appearance of Output to Match That of Earlier SAS Releases

SAS/GRAPH 9.2 introduces many new features that significantly change the default appearance of your SAS/GRAPH output. To produce output that looks as if it was produced with previous versions of SAS/GRAPH, do the following:

- Specify the NOGSTYLE system option. This option turns off the use of ODS styles. See "Turning Off Styles" on page 153.
- Specify the FONTRENDERING=HOST_PIXELS system option. This option specifies whether devices that are based on the SASGDGIF, SASGDTIF, and SASGDIMG modules render fonts by using the operating system or by using the FreeType engine. This option applies to certain native SAS/GRAPH devices (see "Device Categories And Modifying Default Output Attributes" on page 72). For example, this option works for GIF, TIFFP, JPEG, and ZPNG devices, but it is not applicable to PNG, SVG, or SASPRT* devices.
- Specify DEVICE=ZGIF on the GOPTIONS statement when you are sending output to the HTML destination.
- In other cases where your application specifies a device, specify a compatible Z device driver, if applicable. See "Devices" on page xvii for more information.

Procedures

Support for Long Filenames

The NAME= option for each procedure has been enhanced to allow you to specify filenames up to 256 characters long for graphics output files (PNG files, GIF files, and so on). See the documentation for the specific SAS/GRAPH procedures for more information.

GAREABAR Procedure

The GAREABAR procedure has the following new options and enhancements:

- The GAREABAR procedure now supports the BY and LEGEND statements.
- The CONTINUOUS option enables you to display a range of numeric values along the width axis.
- The DESCRIPTION= option specifies the description of the catalog entry for the plot.
- The LEGEND= option assigns the specified LEGEND definition to the legend generated by the SUBGROUP= option.
- The NOLEGEND option suppresses the legend automatically generated by the SUBGROUP= option.

GBARLINE Procedure

The GBARLINE procedure has the following new options and enhancements:

- The PLOT statement supports the creation of multiple plot lines on a single bar chart.

- The SUBGROUP= option divides the bar into segments according to the values of the SUBGROUP variable values.
- The HTML= option on the PLOT statement supports data tips and drill-down links on the markers of the line plot.
- The HTML_LEGEND= option supports data tips and drill-down legend links.
- The IMAGEMAP= option enables you to generate an image map with drill-down functionality in an HTML file.
- The LEGEND= option enables you to generate both BAR and PLOT legends.
- The LEVELS=ALL option has been enhanced to display any number of midpoints.
- The ASCENDING and DESCENDING options now join plot points from left-to-right by default when the bars are reordered.
- The PLOT statement now supports several options for references lines on the plot (right) response axis.
    - The AUTOREF option draws a reference line at each major tick mark.
    - The REF= option draws reference lines at the specified positions.
    - The CREF=, LREF=, and WREF= options enable you to specify the color, line style, and width of user-defined reference lines.
    - The CAUTOREF=, LAUTOREF=, and WAUTOREF= options enable you to specify the color, line style, and width of AUTOREF lines.
- The WREF= and WAUTOREF= options on the BAR statement enable you to specify the width of reference lines on the bar (left) response axis.
- The PLOT statement now supports the following options:

CAXIS=	specifies a color for the tick marks and the axis area frame
CTEXT=	specifies a color for all text on the plot response axis and legend
NOAXIS	suppresses the right PLOT response axis

GCHART Procedure

The GCHART procedure has the following new options and enhancements:
- The COUTLINE= option has been enhanced to include outlines on cylinder-shaped bars.
- The GAXIS= option is now supported by the ACTIVEX, ACTXIMG, JAVA, and JAVAIMG devices.
- The MAXIS= option is now supported by the ACTIVEX, ACTXIMG, JAVA, and JAVAIMG devices.
- The NOPLANE option enables you to remove walls from three-dimensional bar charts.
- The PCTSUM option in the HBAR statement displays a column of percentages for the sum variable values.
- The new PCTSUMLABEL= option enables you to specify the text for the column label for the PCTSUM statistic in the table of statistics.
- The PLABEL= option enables you to specify the font, height, and color of pie slice labels.
- The NOZERO option on the BAR statement is now supported by the JAVA and JAVAIMG devices.
- The RADIUS= option on the PIE statement enables you to specify the radius of the pie chart.

- The RAXIS= option is now supported by the ACTIVEX, ACTXIMG, JAVA, and JAVAIMG devices.
- The SHAPE= option on BLOCK statement is now supported by the ACTIVEX, ACTXIMG, JAVA, and JAVAIMG devices.
- The WREF= and WAUTOREF= options enable you to specify the width of reference lines.
- The pie and bar name variable now support up to 256 characters.

GCONTOUR Procedure

The GCONTOUR procedure has the following changes and enhancements.
- When used with the Java and ActiveX devices, the LJOIN option displays filled contour areas with separated by contour lines.
- When used with the Java and ActiveX devices, the SMOOTH option produces smooth gradient areas between levels.
- The WAUTOHREF= and WAUTOVREF= options specify the line width for reference lines generated with the AUTOHREF and AUTOVREF options, respectively.
- The WHREF= and WVREF= options specify the line width for reference lines generated with the HREF= and VREF= options, respectively.

GEOCODE Procedure

The new GEOCODE procedure enables you to add geographic coordinates (latitude and longitude) to data sets that contain location information such as mailing addresses. You can also perform geolocation, which is adding geographic coordinates to non-address locations such as sale territories.

For SAS 9.2 Phase 2 and later, the new RANGE geocoding method enables you to perform geolocation for IP addresses.

For the third maintenance release of SAS 9.2, the new STREET geocoding method enables you to perform geolocation for street addresses.

GINSIDE Procedure

The new GINSIDE procedure determines which polygon in a map data set contains the X and Y coordinates in your input data set. For example, if your input data set contains coordinates within Canada, you can use the GINSIDE procedure to identify the province for each data point.

GKPI Procedure

The new GKPI procedure generates key performance indicators, including sliders, bullet graphs, speedometers, dials, and traffic lights. This GKPI procedure is supported by the JAVAIMG device only.

GMAP Procedure

The GMAP procedure has the following new features:
- The AREA statement enables you to control the appearance of regions in block maps and prism maps.

- The CDEFAULT= option specifies the color for empty map areas.
- The DENSITY= option enables you to reduce the number of map points that are drawn.
- The RELZERO= option specifies that the heights of bars and regions are relative to zero, rather than the minimum value.
- The STATISTIC= option specifies a statistic to use for the response variable.
- The STRETCH option stretches the extents of a map to fill the output device.
- The UNIFORM option specifies that each map that is created when you use the BY statement uses the same colors and legend.
- The WOUTLINE= option on the BLOCK and CHORO statements is now supported by the JAVA and JAVAIMG devices.

GPLOT Procedure

The GPLOT procedure has the following new options and enhancements:
- The BFILL= option enables you to generate gradient, solid-filled bubble plots.
- The FRONTREF= option specifies that reference lines are drawn in front of filled areas.
- The OVERLAY option is no longer required to display a legend when the PLOT (or PLOT2)statement specifies only one plot.
- The WAUTOHREF= and WAUTOVREF= options specify the line width for reference lines generated with the AUTOHREF and AUTOVREF options, respectively.
- The WHREF= and WVREF= options specify the line width for reference lines generated with the HREF= and VREF= options, respectively.
- Enhanced features in box plots enable you to click on the interior of the boxes for simple drill-down functionality. Previously, you could click only on visible box elements. Now, you can click anywhere inside the box to drill down to more detailed data.

GPROJECT Procedure

The NODATELINE option enables contiguous projections when projecting maps that cross the line between 180 degrees and –180 degrees longitude.

The following options for the GPROJECT procedure have been renamed:

Old Name	New Name
DEGREE	DEGREES
PARALEL1	PARALLEL1
PARALEL2	PARALLEL2

GRADAR Procedure

The GRADAR procedure has the following new options and enhancements:
- The CALENDAR option produces a chart showing twelve equal-sized segments, one for each month of the year.

- The NLEVELS= option specifies the number of colors to use in calendar charts.
- The NOLEGEND option turns off the automatically generated legend.
- The SPOKESCALE= option specifies whether every spoke is drawn to the same scale or each spoke is drawn to a different scale.
- The WINDROSE option produces a windrose chart, which is a type of histogram.
- The FREQ= option now supports only non-zero integers. Zero and negative values are dropped. Decimal values are truncated to integers.
- The WEIGHT= option is no longer supported.
- The GRADAR procedure now draws missing overlay values to the center. Previously, missing values were drawn to zero.

GREMOVE Procedure

The GREMOVE procedure has the following new options:
- The FUZZ= option specifies an error tolerance for the point matching algorithm.
- The NODECYCLE option enables some types of polygons to be closed properly.

GTILE Procedure

The new GTILE procedure enables you to create and display tile charts using the Java or ActiveX device drivers. Tile charts are designed for visualizing a large quantity of hierarchical-type data and are sometimes referred to as rectangular tree maps. Tile charts display rectangles of varying sizes and colors based on the magnitude of the variables specified and provides drill-down links to more detailed data.

MAPIMPORT Procedure

The ID statement for the MAPIMPORT procedure enables you to group related polygons.

Global Statements

- The REPEAT= option on the LEGEND statement enables you to specify the number of times a plot symbol is displayed in a single legend item in the legend.
- The VALUE=EMPTY option on the PATTERN statement is now supported by three-dimensional bar charts.
- The STAGGER option offsets the axis values on a horizontal axis.
- The TICK= suboption on the VALUE= option of the LEGEND statement is now supported by the Java Map Applet.
- The ROWMAJOR and COLMAJOR options on the LEGEND statement enable you to control whether legend entries are listed by row or by column.

Graphics Options

- The ACCESSIBLE graphics option generates descriptive text and the summary statistics that are represented by the graph. This option is valid for the Java and ActiveX devices only.
- The ALTDESC option enables you to specify whether the text specified in the DESCRIPTION= option is used as the data tip text.
- The TRANSPARENCY option is supported by the ACTIVEX and ACTXIMG devices when the output is used in a PowerPoint presentation.

Transparent Overlays

Transparent overlays from GIF files are now supported in SAS/GRAPH output. You can use transparent GIFs with the IMAGE function in the Annotate facility and with the IBACK and IFRAME graphics options.

ActiveX Control

The following are enhancements for the ActiveX Control:
- The ActiveX control now displays calendar and windrose charts generated by the GRADAR procedure.
- The control also displays tile charts created by the new GTILE procedure.
- Support for UNICODE fonts has been added.
- A new field in the user interface enables you to provide interactive graphs in Microsoft Powerpoint slideshows.
- The user interface now enables you to specify the properties of scroll bars in your graph.
- Data tips are supported for scatter plots generated with the GCONTOUR procedure.
- Enhanced support of the Annotate Facility listed under "The Annotate Facility" on page xxv.

Java Map Applet

The Java Map Applet user interface enables you to change block sizes.
Support has been added for the MENUREMOVE parameter, which enables you to remove menu items from the applet user interface.

Java Tilechart Applet

The new Java Tile Chart applet creates and displays tile charts. Tile charts are designed for visualizing a large quantity of hierarchical-type data and are sometimes referred to as rectangular tree maps. They display rectangles of varying sizes and colors based on the magnitude of the variables specified and provide drill-down links to more detailed data. You can generate the applet with the GTILE procedure and the JAVA device.

The Annotate Facility

The following new features are available for the Annotate facility:

- The ANGLE, CBORDER, CBOX, LINE, and ROTATE variables are now supported by the ACTIVEX and ACTXIMG devices.
- The ARROW function and %ARROW macro enable you to draw arrows.
- A new value for the HSYS= option, 'D', specifies points as the unit of measurement for font sizes.
- The IMAGE function is now supported by the JAVA and JAVAIMG devices.
- The WIDTH variable for the PIE function specifies the thickness of the outline around the pie slice.

New Map Data Sets

New map data sets are provided for Antarctica (ANTARCTI, ANTARCT2), Montenegro (MONTENEG, MONTENE2), Romania (ROMANIA, ROMANIA2), Rwanda (RWANDA, RWANDA2), and Serbia (SERBIA, SERBIA2).

The continent map data sets now have corresponding feature data sets (ANTARCT2, AFRICA2, EUROPE2, OCEANIA2, NAMERIC2, SAMERIC2).

Note: Antarctica uses the new continent code 97. △

Updated Map Data Sets

Some of the map data sets in the MAPS library have been updated. Table 0.2 on page xxv contains a list of the changes.

Table 0.2 Changes to the Map Data Sets

Data Set(s)	Changes
Continent data sets (ASIA, AFRICA, EUROPE, NAMERICA, OCEANIA, SAMERICA)	updated to include new geographic features. Each data set includes a new DENSITY variable.
	Brunei, Indonesia, and the Philippines have moved from OCEANIA to ASIA. The continent code for these countries has changed from 96 to 95.
	OCEANIA replaces SPACIFIC as continent 96. Tasmania has been added to the OCEANIA data set.
CHINA, CHINA2	updated with new province names and ID numbers. The new OLDID and OLDIDNAME variables in the CHINA2 data set contain the old ID numbers and province names.
	Because the ID numbers and province names have changed, you might need to change your response data in any existing SAS programs that use these data sets.

Data Set(s)	Changes
GERMANY, GERMANY2	updated with new districts and states. The following new variables have been added: ☐ AREA ☐ DISTNAME ☐ DISTRICT ☐ ID2 The IDNAME variable contains the values that were previously in IDNAME2. The IDNAME2 variable has been removed.
INDIA, INDIA2	updated with new states and new ID numbers. The new OLDID variable in the INDIA2 data set contains the old ID numbers. Additionally, the IDNAME2 variable in the INDIA2 data set contains alternate spellings for the state names. The INDIA data set contains a new DENSITY variable. Because the ID numbers have changed, you might need to change your response data in any existing SAS programs that use these data sets.
ITALY, ITALY2	updated with new provinces and ID numbers. The new OLDID variable in the ITALY2 data set contains the old ID numbers. The ITALY data set contains new DENSITY, NUTS, and REGNAME2 variables. Because the ID numbers have changed, you might need to change your response data in any existing SAS programs that use these data sets.
JAOSAKA, JAOSAKA2	updated with new ID values. The new TYPE variable in JAOSAKA2 contains feature types.
JATOKYO, JATOKYO2	updated with new ID values. The new TYPE variable in JATOKYO2 contains feature types.
LUXEMBOU, LUXEMBO2	updated with more detail and new variables. The LUXEMBOU data set has a new DENSITY variable. The LUXEMBO2 data set has the following new variables: ☐ DISTNAME ☐ DISTRICT ☐ IDNAME2 ☐ NUTS4

Data Set(s)	Changes
NAMES (feature table for the WORLD data set)	contains three new variables: ID2 for territories, specifies the ID values for the countries that the territory is associated with. For example, Greenland has an ID2 value of **315** because it is a territory of Denmark. If a territory is claimed by more than one country, its ID2 value might consist of several three–digit ID values to identify each country. _REGION_ specifies a geographic region for each country or territory. For example, Panama belongs to the **Central America** region. TERRITORY for territories, describes the association between the territory and the country or countries that are identified by ID2. For example, Togo is described as **Overseas territory of France**.
PHILIPPI, PHILIPP2	updated with more detail and new variables. The PHILIPPI data set has a new DENSITY variable. The PHILIPP2 data set has the following new variables: - ISLANDG - ISLAND_GROUP - OLDID - PROVINCE - PSGC_PROV - PSGC_REG - REGION - REGNAME - REGNAME2 The ID numbers for these data sets have changed. You might need to change your response data in any existing SAS programs that use these data sets.
POLAND, POLAND2	updated with new values and variables. The POLAND data set has a new DENSITY variable. The POLAND2 data set has new PROVNAME and PROVNAME2 variables.
SPACIFIC	renamed to OCEANIA.
SPAIN, SPAIN2	updated with values and new variables. The SPAIN data set contains a new DENSITY variable. The new OLDID variable in the SPAIN2 data set contains the old ID numbers. The new REGION and REGNAME variables identify regions. The new IDNAME2 and REGNAME2 variables contain alternate spellings for the province and region names. Because the ID numbers have changed, you might need to change your response data in any existing SAS programs that use these data sets.

Data Set(s)	Changes
SWEDEN, SWEDEN2	updated with new provinces and ID numbers. The SWEDEN data set contains a new DENSITY variable. The new OLDID variable in the SWEDEN2 data set contains the old ID numbers. The new REGNAME variable contains region names.
	Because the ID numbers have changed, you might need to change your response data in any existing SAS programs that use these data sets.
SWITZERL, SWITZER2	updated with new province names and ID numbers, and new variables. The new OLDID and OLDNAME variables in the SWITZER2 data set contains the old names and ID numbers. The SWITZERL data set contains new DENSITY and LAKE variables.
	Because the province names and ID numbers have changed, you might need to change your response data in any existing SAS programs that use these data sets.
THAILAND, THAILAN2	updated with more detail and new variables. The THAILAND data set has a new DENSITY variable. The THAILAN2 data set has the following new variables:
	□ IDNAME2
	□ OLDID
	□ REGION
	□ REGNAME
	The provinces have new ID numbers. The new OLDID variable in the THAILAN2 data set contains the old ID numbers.
	Because the ID numbers have changed, you might need to change your response data in any existing SAS programs that use these data sets.
UKRAINE, UKRAINE2	updated with more detail and new variables. The UKRAINE data set has a new DENSITY variable. The UKRAINE2 data set has the following new variables:
	□ IDNAME2
	□ OLDID
	□ OLDIDNAME
	The provinces have new names and ID numbers. The new OLDIDNAME and OLDID variables in the UKRAINE2 data set contain the old province names and ID numbers.
	Because the province names and ID numbers have changed, you might need to change your response data in any existing SAS programs that use these data sets.

Data Set(s)	Changes
US, USCENTER, USCITY	Puerto Rico added as state 72. The new STATECODE variable in the US and USCITY data sets contains two-letter state abbreviations.
	The USCITY data set has new cities, and some city names have been standardized. The PLACE variable now includes the state FIPS code as the first two digits.
	Note: The projected X and Y values might be different due to the need to re-project the data sets with the addition of more cities in USCITY. △
VIETNAM, VIETNAM2	updated with more detail and new variables. The VIETNAM data set has a new DENSITY variable. The VIETNAM2 data set has the following new variables:
	☐ PROVINCE
	☐ REGION
	☐ REGNAME
	The ID numbers for these data sets have changed. You might need to change your response data in any existing SAS programs that use these data sets.

Data Set(s)	Changes
WORLD	simplified to use fewer observations. In addition, the following changes have been made:
	☐ The values are now projected using the CYLINDRI algorithm.
	☐ Continent 96 has been renamed from South Pacific to Oceania.
	☐ Antarctica has been added as continent 97.
	☐ Brunei, Indonesia, and the Philippines have been reassigned from continent 96 to continent 95.
	☐ French Southern Territories and Heard & McDonald Islands have been reassigned from continent 96 to continent 97.
	☐ St. Helena has been reassigned from continent 91 to continent 94.
	☐ The former country of Yugoslavia has been split into Serbia and Montenegro.
	☐ Newfoundland has been added to Canada (ID 260).
	☐ Tasmania has been added to Australia (ID 160).
	☐ More data points are included for Cuba (ID 300).
	☐ The Galapagos Islands have been added to Ecuador (ID 325).
	☐ Hong Kong is now included as part of China.
YUGOSLA, YUGOSLA2	replaced by the new SERBIA, SERBIA2, MONTENEG, MONTENE2 data sets.

Map Data Set Descriptions

Descriptive labels have been added to the map data sets in the MAPS library.

New Data Set for Military ZIP Codes

The new ZIPMIL data set in the SASHELP library contains ZIP codes for U.S. military post offices.

Changes in SAS/GRAPH Documentation

☐ Information about the DS2CSF macro has been removed. The functionality of the DS2CSF macro is available through the new GKPI procedure.

☐ Information about the META2HTM macro has been removed. To generate the Metaview applet, use the JAVAMETA device.

PART 1

SAS/GRAPH Concepts

Chapter 1 ....Introduction to SAS/GRAPH Software *3*

Chapter 2 ....Elements of a SAS/GRAPH Program *31*

Chapter 3 ....Getting Started With SAS/GRAPH *39*

Chapter 4 ....SAS/GRAPH Processing *53*

Chapter 5 ....The Graphics Output Environment *59*

Chapter 6 ....Using Graphics Devices *67*

Chapter 7 ....SAS/GRAPH Output *87*

Chapter 8 ....Exporting Your Graphs to Microsoft Office Products *113*

Chapter 9 ....Writing Your Graphs to a PDF File *123*

Chapter 10 ....Controlling The Appearance of Your Graphs *133*

Chapter 11 ....Specifying Fonts in SAS/GRAPH Programs *155*

Chapter 12 ....SAS/GRAPH Colors and Images *167*

Chapter 13 ....Managing Your Graphics With ODS *191*

Chapter 14 ....SAS/GRAPH Statements *197*

Chapter 15 ....Graphics Options and Device Parameters Dictionary *327*

CHAPTER 1

Introduction to SAS/GRAPH Software

Overview **4**
Components of SAS/GRAPH Software **4**
Device-Based Graphics and Template-Based Graphics **6**
Graph Types **7**
    Charts **7**
        Block charts **7**
        Horizontal bar charts **8**
        Vertical bar charts **8**
        Pie charts, Detailed pie charts, 3D pie charts, and Donut charts **9**
        Star charts **10**
    Bar-line Charts **10**
    Area Bar Charts **11**
    Tile Charts **12**
    Radar Charts **12**
    Two-Dimensional Plots **13**
        Two-dimensional scatter plots **13**
        Simple line plots **14**
        Regression plots **14**
        High-low plots **15**
        Bubble plots **16**
    Three-Dimensional Plots **16**
        Surface plots **16**
        Scatter plots **17**
        Contour plots **17**
    Maps **18**
        Block maps **18**
        Choropleth maps **19**
        Prism maps **19**
        Surface maps **20**
    KPI Charts **21**
    Creating Text Slide and Presentation Graphics **21**
        Text Slides **21**
        Combining Output into One Slide **22**
    Enhancing Graphics Output (graphs and text slides) **23**
        SAS/GRAPH Statements **23**
        The Annotate Facility **23**
    Creating Custom Graphics **23**
        The DATA Step Graphics Interface **23**
        Graph-N-Go **24**
About this Document **24**
    Audience **24**

Prerequisites **24**
Conventions Used in This Document **25**
    Syntax Conventions **25**
    Conventions for Examples and Output **27**
Information You Should Know **28**
    Support Personnel **28**
    Sample Programs **28**
    Map Data Sets **30**
    Annotate Macros Data Set **30**

Overview

SAS/GRAPH is the data visualization and presentation (graphics) component of the SAS System. As such, SAS/GRAPH:

- organizes the presentation of your data and visually represents the relationship between data values as two- and three-dimensional graphs, including charts, plots, and maps.
- enhances the appearance of your output by allowing you to select text fonts, colors, patterns, and line styles, and control the size and position of many graphics elements.
- creates presentation graphics. SAS/GRAPH can create text slides, display several graphs at one time, combine graphs and text in one display, and create automated presentations.
- generates a variety of graphics output that you can display on your screen or in a Web browser, store in catalogs, review, or send to a hard copy graphics output device such as a laser printer, plotter, or slide camera.
- provides utility procedures and statements to manage the output.

This chapter describes the graphs that are produced by SAS/GRAPH and explains some of the parts and features of SAS/GRAPH programs.

Components of SAS/GRAPH Software

There are several components to SAS/GRAPH software.

Device-based SAS/GRAPH procedures
: enable you to create a variety of graphs, including bar charts, pie charts, scatter plots, surface plots, contour plots, a variety of maps, and much more. The device-based SAS/GRAPH procedures include the GAREABAR, GCHART, GPLOT, GMAP, GBARLINE, GKPI, GCONTOUR, and G3D procedures, as well as others. These procedures use device drivers to generate output. SAS/GRAPH device drivers enable you to send output directly to your output device as well as create output in a variety of formats such as PNG files and interactive ActiveX controls or Java applets. This document, *SAS/GRAPH: Reference*, describes the device-based SAS/GRAPH procedures and how to use devices. See also "Device-Based Graphics and Template-Based Graphics" on page 6.

The Annotate Facility
: enables you to generate a special data set of graphics commands from which you can produce graphics output. This data set is referred to as an Annotate data set. You can use it to generate custom graphics or to enhance graphics output from

many device-based SAS/GRAPH procedures, including GCHART, GPLOT, GMAP, GBARLINE, GCONTOUR, and G3D, as well as others. For more information, see Chapter 29, "Using Annotate Data Sets," on page 641.

Network Visualization (NV) Workshop
: enables you to visualize and investigate the patterns and relationships hidden in network data (node-link data). Some common applications that use network data include supply chains, communication networks, Web sites, database schema, and software module dependencies. NV Workshop is designed for visualizing large networks. Using a combination of data tables, statistical graphs, and network graphs, NV Workshop enables you to extract information that would otherwise remain hidden. Help is available from the menu within the product. Network Visualization Workshop runs in Windows operating environments only. For additional information, see *SAS/GRAPH: Network Visualization Workshop User's Guide*.

SAS/GRAPH statistical graphics suite
: is part of ODS Statistical Graphics (referred to as ODS Graphics for short). ODS Graphics is functionality for creating statistical graphics that is available in a number of SAS software products, including SAS/STAT, SAS/ETS, SAS/QC, and SAS/GRAPH. The SAS/GRAPH statistical graphics suite provides the following features:

    SAS/GRAPH statistical graphics procedures
    : provide a simple syntax for creating graphics commonly used in exploratory data analysis and for creating customized statistical displays. These procedures include the SGPANEL, SGPLOT, and SGSCATTER procedures. In addition, the SGRENDER procedure provides a SAS procedure interface to create graphs using the Graph Template Language. These procedures are template-based procedures; they do not use devices like the device-based SAS/GRAPH procedures. For more information, see "Device-Based Graphics and Template-Based Graphics" on page 6 and *SAS/GRAPH: Statistical Graphics Procedures Guide*.

    Graph Template Language (GTL)
    : is the underlying language for the default templates that are provided by SAS for procedures that use ODS Statistical Graphics. You can use the GTL either to modify these templates or to create your own customized graphs. Templates written with the GTL are built with the TEMPLATE procedure. For more information about Graph Template Language, see *SAS/GRAPH: Graph Template Language User's Guide* and *SAS/GRAPH: Graph Template Language Reference*.

    ODS Graphics Editor
    : is an interactive editor that enables you to edit and enhance graphs that are produced by procedures that use ODS Graphics. You can use the ODS Graphics Editor to modify the existing elements of a graph such as titles and labels, or to add features such as text annotation for data points. The ODS Graphics Editor runs in Windows and UNIX operating environments only. For more information, see *SAS/GRAPH: ODS Graphics Editor User's Guide*.

    ODS Graphics Designer
    : provides a point-and-click interface for creating ODS Graphics. Using the ODS Graphics Designer does not require knowledge of ODS templates or the Graph Template Language. With the ODS Graphics Designer, you can easily create multi-cell graphs, classification panels, scatter plot matrices, and more. You can save your output as an image file or as an ODS Graphics Designer file (SGD file) that you can edit later. The ODS Graphics Designer

runs in Windows and UNIX operating environments only. For more information, see *SAS/GRAPH: ODS Graphics Designer User's Guide*. For additional information on the ODS Statistical Graphics functionality, see *SAS Output Delivery System: User's Guide* and *SAS/STAT User's Guide*.

Device-Based Graphics and Template-Based Graphics

SAS/GRAPH produces graphics using two very distinct systems. SAS/GRAPH can produce output using a device-based system or using a template-based system. The traditional system for producing graphics output that most users are familiar with is the device-based system.

device-based graphics
 are SAS/GRAPH output that is generated by a default or user-specified device (DEVICE= option). Device drivers supplied by SAS are stored in the SAS/GRAPH catalog. Examples of device drivers are GIF, PNG, ACTIVEX, SVG, and SASPRTC. Most procedures that produce device-based graphics also produce GRSEG catalog entries in addition to any image files that are produced. Common SAS/GRAPH procedures that produce device-based graphics and GRSEG catalog entries include the GCHART, GPLOT, GMAP, GBARLINE, GCONTOUR, and G3D procedures. The device-based procedures that do not produce GRSEG catalog entries are the GAREABAR, GKPI, and GTILE procedures. For device-based graphics, you can use the GOPTIONS statement to control the graphical environment.

template-based graphics
are SAS/GRAPH output that is produced from a compiled ODS template of type STATGRAPH. Templates supplied by SAS are stored in SAS/GRAPH. Device drivers and most global statements (such as SYMBOL, PATTERN, AXIS, and LEGEND) have no effect on template-based graphics. The SAS/GRAPH procedures that produce template-based graphics are the SGPLOT, SGPANEL, SGSCATTER, and SGRENDER procedures. Many SAS/STAT, SAS/ETS, and SAS/QC procedures also produce template-based graphics when you specify the ODS GRAPHICS ON statement. (Template-based graphics are frequently referred to as *ODS graphics*.) Template-based graphics are always produced as image files and never as GRSEG catalog entries. For template-based graphics, you must use the ODS GRAPHICS statement to control the graphical environment.

The *SAS/GRAPH: Reference* contains information about device-based graphics only. For information about template-based graphics, see *SAS/GRAPH: Statistical Graphics Procedures Guide* and *SAS/GRAPH: Graph Template Language Reference*.

Graph Types

SAS/GRAPH produces many kinds of charts, plots, and maps in both two- and three-dimensional versions. In addition to helping you understand the variety of graphs that are available to you, these descriptions also help you choose the correct type of graph for your data and point you to the appropriate chapter.

Charts

SAS/GRAPH uses the GCHART procedure to produce charts that graphically represent the value of a statistic for one or more variables in a SAS data set. See Chapter 36, "The GCHART Procedure," on page 989 for a complete description.

Block charts

Block charts use three-dimensional blocks to graphically represent values of statistics. Block charts are useful for emphasizing relative magnitudes and differences among data values.

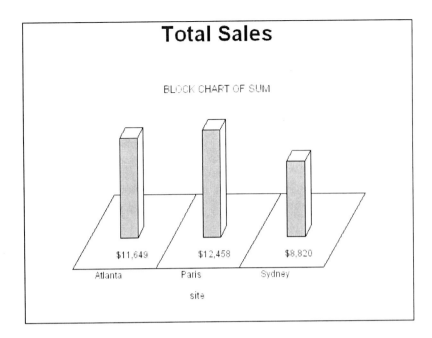

Horizontal bar charts

Horizontal bar charts use horizontal bars to represent statistics based on the values of one or more variables. Horizontal bar charts can generate a table of chart statistics and are useful for displaying exact magnitudes and emphasizing differences.

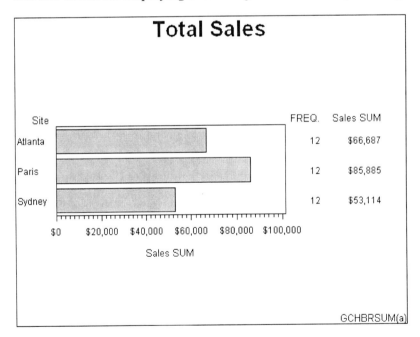

Vertical bar charts

Vertical bar charts use vertical bars to represent statistics based on the values of one or more variables. Vertical bar charts, which generate only one statistic, are useful for displaying exact magnitudes and emphasizing differences.

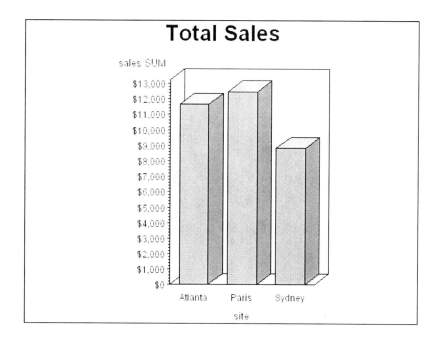

Pie charts, Detailed pie charts, 3D pie charts, and Donut charts

Pie charts, detailed pie charts, 3-D pie charts, and Donut charts use the angle of pie slices to graphically represent the value of a statistic for a data range. Pie charts are useful for examining how the values of a variable contribute to the whole and for comparing the values of several variables.

Figure 1.1 Detailed Pie Chart

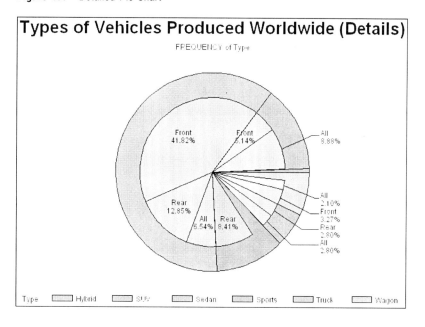

Figure 1.2 Donut Chart

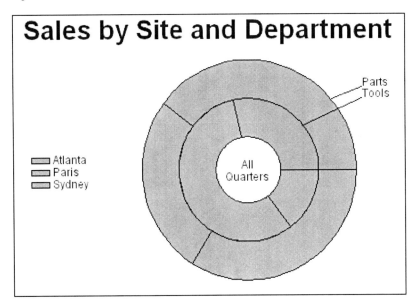

Star charts

Star charts use the length of spines to graphically represent the value of a statistic for a data range. Star charts are useful for analyzing where data are out of balance.

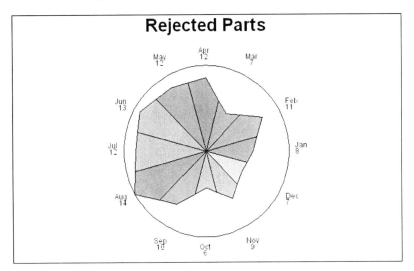

Bar-line Charts

The GBARLINE procedure produces vertical bar charts with plot overlays. These charts graphically represent the value of a statistic calculated for one or more variables in an input SAS data set. The charted variables can be either numeric or character.

See Chapter 35, "The GBARLINE Procedure," on page 947 for a complete description.

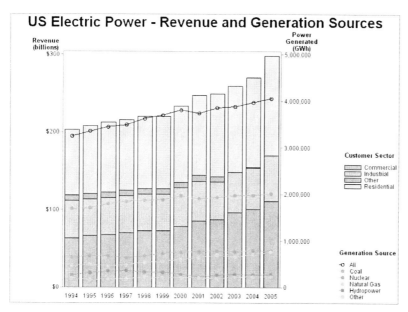

Area Bar Charts

The GAREABAR procedure produces area bar charts that show the magnitudes of *two* variables for each category of data. For example, the following area bar chart shows the sales total for each of three geographical sites. The width of each bar indicates the number of sales persons at each site. In a bar chart such as the chart shown in "Vertical bar charts" on page 8, the width is the same for each bar. In an area bar chart, the width and height of each bar is determined by the value of variables. See Chapter 34, "The GAREABAR Procedure," on page 931 for a complete description.

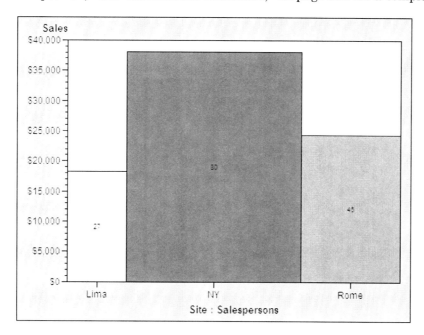

Tile Charts

The GTILE procedure produces charts that tile charts, which consist of a rectangle or square divided into tiles. The sizes of the individual tiles represent the value of the size variable. You can also specify a color variable, so that the colors of the individual tiles represent the magnitude of the color variable. Tile charts are useful for determining the relative magnitude of categories of data or the contribution of a category toward the whole.

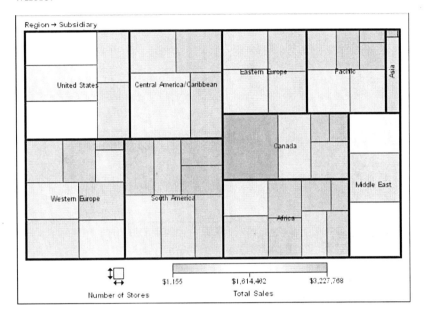

Radar Charts

The GRADAR procedure produces radar charts that show the relative frequency of data measures. On a radar chart, the chart statistics are displayed along spokes that radiate from the center of the chart. The charts are often stacked on top of one another with reference circles, thus giving them the look of a radar screen. Radar charts are frequently called star charts and are often used in quality control or market research problems.

See Chapter 47, "The GRADAR Procedure," on page 1419 for a complete description.

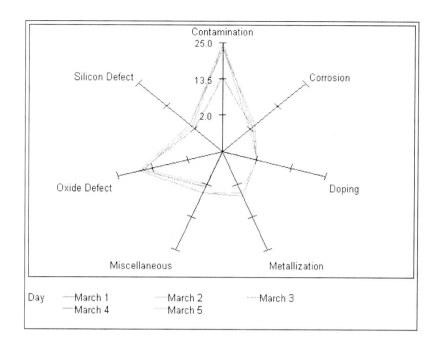

Two-Dimensional Plots

SAS/GRAPH uses the GPLOT procedure to produce two-dimensional graphs that plot one or more dependent variables against an independent variable within a set of coordinate axes. GPLOT can display the data points as individual symbols (as in a scatter plot), or use interpolation methods specified by the SYMBOL statement to join the points, request spline interpolation or regression analysis, produce various high-low plots, or generate several other types of plots.

GPLOT can also display data as bubble plots in which circles of different sizes represent the values of a third variable.

Plots are useful for demonstrating the relationship between two or more variables and frequently compare trends or data values or depict movements of data values over time.

See Chapter 45, "The GPLOT Procedure," on page 1325 for a complete description.

Two-dimensional scatter plots

Two-dimensional scatter plots show the relationship of one variable to another, often revealing concentrations or trends in the data. Typically, each variable value on the horizontal axis can have any number of corresponding values on the vertical axis.

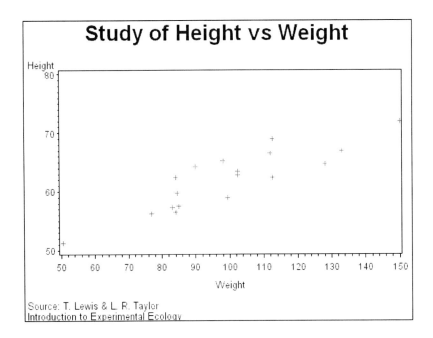

Simple line plots

Simple line plots show the relationship of one variable to another, often as movements or trends in the data over a period of time. Typically, each variable value on the horizontal axis has only one corresponding value on the vertical axis. The line connecting data points can be smoothed using a variety of interpolation methods, including the Lagrange and the cubic spline interpolation methods.

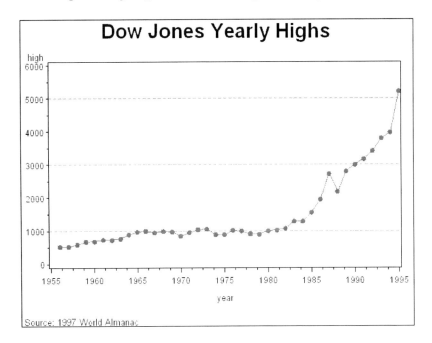

Regression plots

Regression plots specify that the plot is a regression analysis. You can specify one of three types of regression equation – linear, quadratic, or cubic, and you can choose to display confidence limits for mean predicted values or individual predicted values.

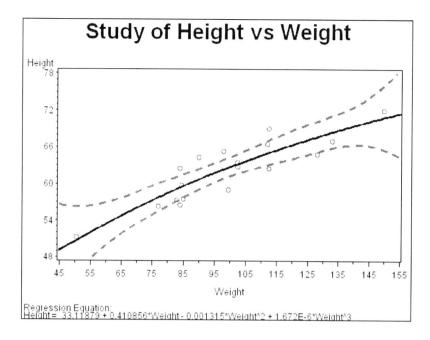

High-low plots

High-low plots show how several values of one variable relate to one value of another variable. Typically, each variable value on the horizontal axis has several corresponding values on the vertical axis. High-low plots include box, needle, and stock market plots.

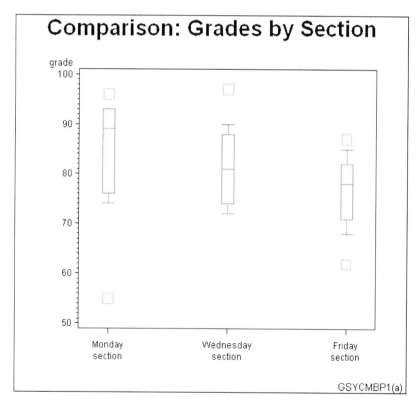

Bubble plots

Bubble plots show the relative magnitude of one variable in relation to two other variables. The values of two variables determine the position of the bubble on the plot, and the value of a third variable determines the size of the bubble.

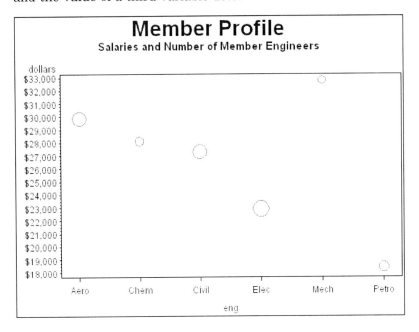

Three-Dimensional Plots

SAS/GRAPH uses the G3D procedure to produce three-dimensional surface and scatter plots that examine the relationship among three variables. Variable values are plotted on a set of three coordinate axes.

See Chapter 53, "The G3D Procedure," on page 1541 for a complete description.

Surface plots

Surface plots are three-dimensional plots that display the relationship of three variables as a continuous surface. Surface plots examine the three-dimensional shape of data.

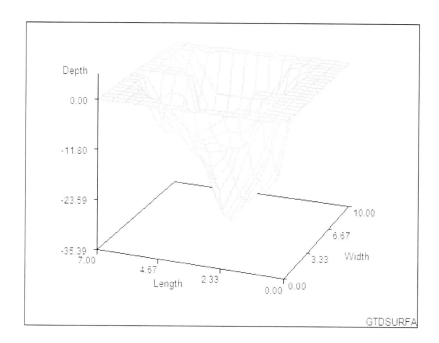

Scatter plots

Scatter plots enable you to examine three-dimensional data points instead of surfaces and to classify your data using size, color, shape, or a combination of these features.

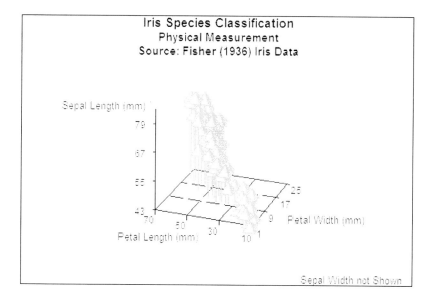

Contour plots

SAS/GRAPH uses the GCONTOUR procedure to examine three-dimensional data in two dimensions. Lines or areas in a contour plot represent levels of magnitude (z) corresponding to a position on a plane (x,y).

See Chapter 37, "The GCONTOUR Procedure," on page 1095 for a complete description.

Contour plots are two-dimensional plots that show three-dimensional relationships. These plots use contour lines or patterns to represent levels of magnitude of a contour variable plotted on the horizontal and vertical axes.

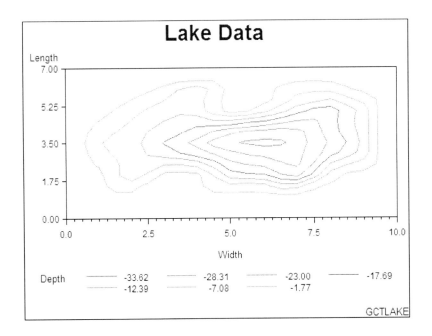

When you need to interpolate or smooth data values that are used by the G3D and GCONTOUR procedures, use the G3GRID procedure. The G3GRID procedure does not produce graphics output but processes existing data sets to create data sets that the G3D or GCONTOUR procedure can use to produce three-dimensional surface or contour plots. See Chapter 54, "The G3GRID Procedure," on page 1571 for a complete description.

Maps

SAS/GRAPH uses the GMAP procedure to produce two- and three-dimensional maps that can show an area or represent values of response variables for subareas.

SAS/GRAPH includes data sets to produce geographic maps. In addition, you can create your own map data sets.

See Chapter 43, "The GMAP Procedure," on page 1239 for a complete description.

Block maps

Block maps are three-dimensional maps that represent data values as blocks of varying height rising from the middle of the map areas.

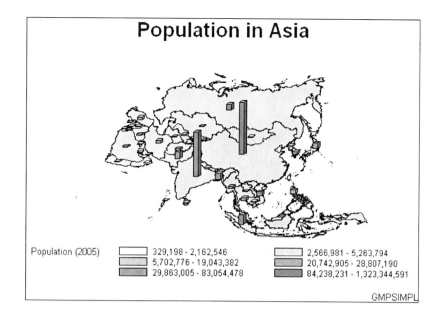

Choropleth maps

Choropleth maps are two-dimensional maps that display data values by filling map areas with combinations of patterns and color that represent the data values.

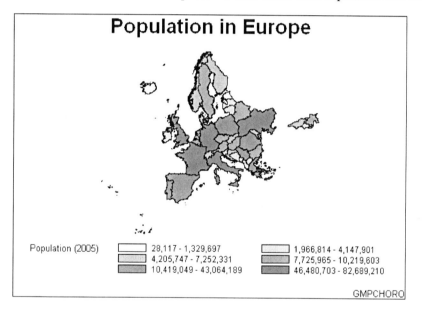

Prism maps

Prism maps are three-dimensional maps that display data by raising the map areas and filling them with combinations of patterns and colors.

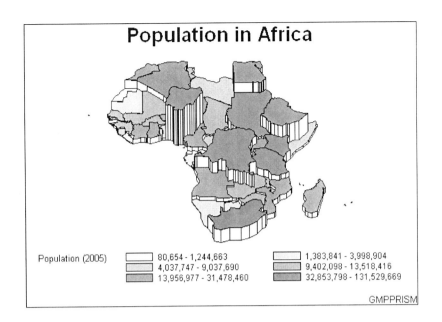

Surface maps

Surface maps are three-dimensional maps that represent data values as spikes of varying heights.

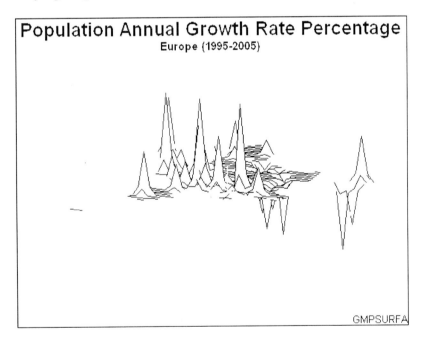

SAS/GRAPH also provides several utility procedures for handling map data.

The GPROJECT procedure lets you choose how geographic maps are projected. This is particularly important for large areas because producing a map of any large area on the Earth involves distorting some areas in the process of projecting the spherical surface of the Earth onto a flat plane. You can use the procedure to select the projection method that least distorts your map.

Map areas are constructed of joined data points. Each data point represents an observation in a SAS data set. For large maps, the amount of data can be prohibitively expensive (in terms of computing resources or time to process); the GREDUCE procedure enables you to reduce the number of points in the data set. The GREMOVE procedure enables you to remove boundary lines within a map.

KPI Charts

The GKPI procedure creates graphical key performance indicator (KPI) charts. KPIs are metrics that help a business monitor its performance and measure its progress toward specific goals. The procedure produces five KPI chart types:

- slider (vertical or horizontal)
- bullet graph (vertical or horizontal)
- radial dial
- speedometer
- traffic light (vertical or horizontal).

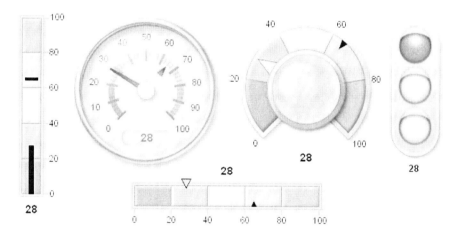

Creating Text Slide and Presentation Graphics

You can use SAS/GRAPH to create slide presentations of your graphs. With SAS/GRAPH you can

- create text slides with the GSLIDE procedure
- combine several graphs into one output with the GREPLAY procedure
- automatically or manually replay your graphs and text slides with the GREPLAY procedure.

Text Slides

Use the GSLIDE procedure to create text slides in which you can specify a variety of colors, fonts, sizes, angles, overlays, and other modifications as well as drawing lines and boxes on the output.

See Chapter 51, "The GSLIDE Procedure," on page 1517 for a complete description.

Text slides display text as graphics output. Text slides can be used as title slides for presentations, or to produce certificates, signs, or other display text.

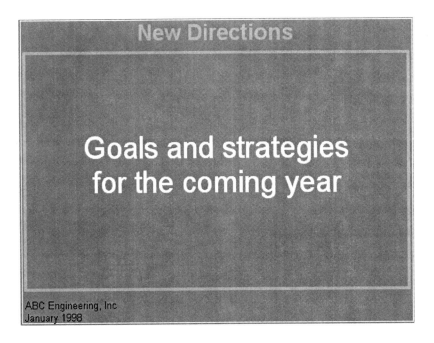

Combining Output into One Slide

Use the GREPLAY procedure to combine several graphs into a single output. You can create special effects by overlaying or rotating the graphs at any angle.

Templated graphs display two or more graphs or text slides as one output by replaying stored graphs into a template or framework. Like graphs and text slides, templated graphs can be ordered in groups and stored in catalogs for replay as part of a presentation.

Figure 1.3 Templated graphs

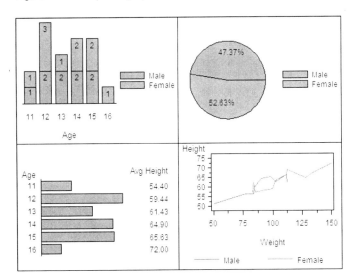

In addition, you can use the GREPLAY procedure to create an automated or user-controlled presentation of graphics output. The GREPLAY procedure enables you to name, arrange, and customize the presentation of graphs that are stored in a catalog. See Chapter 50, "The GREPLAY Procedure," on page 1473 for a complete description.

Enhancing Graphics Output (graphs and text slides)

SAS/GRAPH Statements

You can also use *global statements* and *graphics options* in SAS/GRAPH programs. With global statements, you can add titles and footnotes and control the appearance of axes, symbols, patterns, and legends. With graphics options, you can control the appearance of graphics elements by specifying default colors, fill patterns, fonts, text height, and so on.

The Annotate Facility

The Annotate facility enables you to program graphics by using certain variables in SAS data sets. It is often used to add text or special elements to the graphics output of other procedures, although it can also be used to construct custom graphics output. Text and graphics can be placed at coordinates derived from input data, as well as coordinates expressed as explicit locations on the display.

Figure 1.4 Annotated graphs

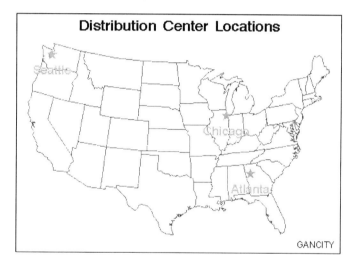

Creating Custom Graphics

The Annotate facility can also be used to generate custom graphics without using any of the SAS/GRAPH graphing procedures.

The DATA Step Graphics Interface

The DATA Step Graphics Interface provides functions and calls that produce graphics output from the DATA step, rather than from a procedure. The functions and calls are similar in form to those specified by the ISO Graphic Kernel Standard (GKS); however, the interface is not an implementation of the GKS. The form is similar enough that many GKS-compliant programs can be converted easily to run as SAS/GRAPH programs.

Graph-N-Go

To generate presentation graphs without writing any SAS/GRAPH code, you can use Graph-N-Go (not available on mainframes). You can start Graph-N-Go in several ways:

- from the menus in any SAS window, select **Solutions ▶ Reporting ▶ Graph-N-Go**
- submit either of the following from the SAS command line:

    ```
    gng
    ```

    ```
    graphngo
    ```

- use an Explorer window to directly open a GFORM entry. Double-click (or right-click and choose Open) on a GFORM entry to start a Graph-N-Go session using that entry.

Information on using the application is in Graph–N-Go help, which you can access from the application's main window in either of two ways:

- select **Help ▶ Using This Window**
- press F1 (this might not work in some operating environments).

You can also get help for the application by submitting the following command from the SAS command line:

```
help gng
```

About this Document

This document provides reference information for all facilities, procedures, statements, and options that can be used with SAS/GRAPH. This chapter describes what you need to know to use SAS/GRAPH, and what conventions are used in text and example code. To gain full benefit from using this document, you should familiarize yourself with the information presented in this chapter, and refer to it as needed.

Audience

This document is written for users who are experienced in using the SAS System. You should understand the concepts of programming in the SAS language, and you should have an idea of the tasks you want to perform with SAS/GRAPH.

Prerequisites

The following table summarizes the SAS System concepts that you need to understand in order to use SAS/GRAPH:

To learn how to	Refer to
invoke the SAS System at your site	instructions provided by the on-site SAS support personnel
use Base SAS software	*SAS Language Reference: Concepts* or *SAS Language Reference: Dictionary*
use the DATA step to create and manipulate SAS data sets	
use the SAS Text Editor to enter and edit text	

To learn how to	Refer to
allocate SAS libraries and assign librefs	documentation for using the SAS System under the operating system for the hardware at your site
create external files and assign filerefs	
manipulate SAS data sets using SAS procedures	*Base SAS Procedures Guide*

Conventions Used in This Document

This section explains the conventions this document uses for text, SAS language syntax, and file and library references. The document uses the following terms in discussing syntax:

keyword
: is a literal that is a primary part of the SAS language. (A literal must be spelled exactly as shown, although it can be entered in uppercase or lowercase letters.) Keywords in this document are procedure names, statement names, macro names, routine names, and function names.

argument
: is an element that follows a keyword. It is either literal, or it is user-supplied. It has a built-in value (for example, NODISPLAY), or it has a value assigned to it (for example, COLOR=*text-color*).

    Arguments that you must use are *required arguments*. Other arguments are *optional arguments*, or simply *options*.

value
: is an element that follows an equal sign. It assigns a value to an argument. It might be a literal, or it might be a user-supplied value.

parameter
: is a value assigned to an argument that itself takes a value, for example, the COLOR= parameter of the LABEL= option in a LEGEND statement, as shown in the following statement:

```
legend label=(color=blue);
```

Syntax Conventions

Type styles have special meanings when used in the presentation of SAS/GRAPH syntax in this document. The following list explains the style conventions for the syntax sections:

UPPERCASE
: identifies SAS keywords such as the names of statements and procedures (for example, PROC GCHART). Uppercase characters also identify arguments and values that are literals (for example, NOLEGEND and LABEL=NONE).

italic
: identifies arguments or values that you supply. Items in italic can represent user-supplied values that are either

    - nonliteral values assigned to an argument (for example, *axis-color* in COLOR=*axis-color*)
    - nonliteral arguments (for example, VBAR *chart-variable*. . . ;).

    In addition, an item in italics can be the generic name for a list of arguments or parameters from which the user can choose (for example, *appearance-options*).

The following symbols are used to indicate other syntax conventions:

< > (angle brackets)   identify optional arguments. Any argument not enclosed in angle brackets is required.

| (vertical bar)   indicates that you can choose one value from a group. Values separated by bars are mutually exclusive.

. . . (ellipsis)   indicates that the argument following the ellipsis can be repeated any number of times (*plot-request* <. . . *plot-request-n*>, for example). If the ellipsis and the following argument are enclosed in angle brackets, they are optional. In SAS/GRAPH, an ellipsis also indicates a range from which a value is selected (LINE=1 . . . 46, for example).

The following examples illustrate the syntax conventions described in this section. These examples contain selected syntax elements, not complete syntax.

PROC GANNO ANNOTATE=*Annotate-data-set*
   <DATASYS>;

- PROC GANNO is in uppercase because it is a SAS keyword, the name of a statement. The remaining elements are arguments for the statement.
- ANNOTATE= is not enclosed in angle brackets because it is a required argument. It is in uppercase to indicate that it is a literal and must be spelled as shown.
- *Annotate-data-set* is in italic because it is a value that you must supply; in this case, the value must be a data set name.
- DATASYS is enclosed in angle brackets because it is an optional argument. It is in uppercase to indicate that it is a literal and must be spelled as shown.
- The ending semicolon (;) is required because it is outside the angle brackets for the option.

SYMBOL <1 . . . 99>
   <COLOR=*symbol-color*>
   <MODE=EXCLUDE|INCLUDE>
   <*appearance-options*>;

- SYMBOL is in uppercase because it is a SAS keyword, the name of a statement. The numbers 1 . . . 99 are in angle brackets because they are optional. The ellipsis indicates that you choose one from the range of numbers 1 through 99. The remaining elements are arguments for the statement.
- COLOR= is enclosed in angle brackets because it is an optional argument.
- *Symbol-color* is in italics because it represents a value that you specify.
- MODE= is enclosed in angle brackets because it is an optional argument.
- EXCLUDE and INCLUDE are in uppercase because they are literal values and must be spelled exactly as shown. They are separated by a vertical bar (an OR bar) because you use one or the other but not both.
- *Appearance-options* is in italics because it is a generic name for a list of options that can be used in the SYMBOL statement.

HBAR *chart-variable*< . . . *chart-variable-n*>
   </ <PATTERNID=BY | GROUP | MIDPOINT | SUBGROUP>
   <*statistic-options*>>;

- *Chart-variable* is italic because it is an argument that you supply. It is required because it is not in angle brackets.
- *Chart-variable-n* is enclosed in angle brackets because additional user-supplied arguments are optional. The ellipsis before the argument indicates that it can be repeated as many times as desired.
- PATTERNID= is a literal option. The values BY, GROUP, MIDPOINT, and SUBGROUP are literal values that are mutually exclusive. You can use only one, and it must be spelled as shown.
- *Statistic-options* is in italics because it is the generic name of a list of options that affect the chart statistics.

When you are using an option, a statement, or a procedure whose syntax shows arguments or values in italics, you must supply the argument or value. When the argument or value is a font, color, or variable name, SAS/GRAPH expects valid font names, color names, and variable names. Consider the following four syntax samples:

FONT=*font*

COLOR=*color*

COLOR=*text-color*

PIE *chart-variable* < . . . *chart-variable-n*>;

- *Font* must be a valid SAS font name. (See Chapter 11, "Specifying Fonts in SAS/GRAPH Programs," on page 155 for details.)
- *Color* and *text-color* must be valid SAS/GRAPH colors. (See Chapter 12, "SAS/GRAPH Colors and Images," on page 167 for details.)
- *Chart-variable* must be a valid SAS variable name. (See *SAS Language Reference: Dictionary* for details.)

Conventions for Examples and Output

Most of the chapters in this document include examples that illustrate some of the features of a procedure or its statements. Each example contains

- a description of the highlights of the example
- the program statements that produce the output
- the actual output from the example
- an explanation of the features of the example.

The output that is shown for the examples was generated in a Windows operating environment. If you are using a different operating environment, you might need to make some minor adjustments to the example programs.

In most cases, the output was sent to the Listing destination and generated using the default style and device for that destination. Exceptions are noted in the text.

The dimensions of the graphics output area vary across devices and when using the GRAPH windows. The dimensions can affect aspects of the graphics output – for example, the appearance of axes or the position of graphics elements that use explicit coordinates in units other than percent. You might need to adjust the dimensions of your graphics output area or the size of graphics elements to correct any differences you see. Most of the images of output in this document were generated with a GOPTIONS statement that specified a size approximately equal 5.5 inches by 4.2 inches, although some images might be larger, if necessary, to accommodate the content of the graph.

```
goptions hsize=5.5inin vsize=4.2in;
```

These HSIZE= and VSIZE= settings are not shown in the example code and are not necessary for generating the output, but you might want to use similar settings if your output looks different from the output that is shown in the document.

Most examples specify these options:

RESET=ALL   sets all graphics options to default values and cancels all global statements.

BORDER   draws a border around the graphics output area.

Information You Should Know

This section outlines information you should know before you attempt to run the examples in this document.

Support Personnel

Most sites have personnel available to help users learn to run SAS System. Record the name of your on-site SAS support personnel. Also, record the names of anyone else you regularly turn to for help with running SAS/GRAPH.

Sample Programs

The documentation for each procedure, for global statements, and for features such as the Annotate facility provide examples that demonstrate these features of SAS/GRAPH. You can copy the example code from the help or the OnlineDoc and paste it into the Program Editor in your SAS session.

These same programs are included in the sample library SAS Sample Library. How you access the code in the sample library depends on how it is installed at your site.

- In most operating environments, you can access the sample code through the SAS Help and Documentation. Select **Help ▶ SAS Help and Documentation**. On the **Contents** tab, select **Learning to Use SAS ▶ Sample SAS Programs ▶ SAS/GRAPH ▶ Samples**.
- In other operating environments, the SAS Sample Library might be installed in your file system. If the SAS Sample Library has been installed at your site, ask your on-site SAS support personnel where it is located.

To access the sample programs through SAS Help and Documentation or through your file system, you must understand the naming convention used for the samples. The naming convention for SAS/GRAPH samples is G*pcxxxxx*, where *pc* is the product code and *xxxxx* is an abbreviation of the example title. The product code can be a code for a procedure, a statement, or in the case of Java and ActiveX examples, WB for "web graphs." For example, the code for the first example in the GMAP procedure chapter, Example 1 on page 1301, is stored in sample member GMPSIMPL. The sample-library member name is sometimes displayed as a footnote in the output's lower-right corner.

- In the Help system, the sample programs are organized by product. Within each product category, most of the samples are sorted by procedure. Thus, to access the code for the first example in the GMAP procedure chapter, select **Learning to Use SAS ▶ SAS/GRAPH ▶ Samples**, scroll to `GMAP Procedure`, and select `GMPSIMPL-Producing a Simple Block Map`.

□ In your file system, the files that contain the sample code have filenames that match the sample member names. For example, in a directory-based system, the code for sample member GMPSIMPL is located in a file named GMPSIMPL.SAS.

Note: For Java and ActiveX (web graph) samples, the naming convention is GWB*xxxxx*. △

Table 1.1 Product Codes for SAS/GRAPH Procedures

Procedure	Code
dsgi	DS
ganno	AN
gareabar	AB
gbarline	BL
gchart	CH
gcontour	CT
geocode	GE
gfont	FO
ginside	IN
gkpi	KP
gmap	MP
goptions	OP
gplot	PL
gproject	PJ
gradar	RR
greduce	RD
gremove	RM
greplay	RE
gslide	SL
gtile	TL
g3d	TD
g3grid	TG

Table 1.2 Product Codes for SAS/GRAPH Statements

Statement	Code
axis	AX
by	BY
footnote	FO

Statement	Code
goptions	ON
legend	LG
note	NO
pattern	PN
symbol	SY
title	TI

Map Data Sets

To run the examples that draw maps, you need to know where the map data sets are stored on your system. Depending on your installation, the map data set might automatically be assigned a libref. Ask your on-site SAS support personnel or system administrator where the map data sets are stored for your site.

Annotate Macros Data Set

To run the examples using Annotate macros, you need to know where the Annotate macro data set is stored on your system. Depending on your installation, the Annotate macro data set might automatically be assigned a fileref. Ask your on-site SAS support personnel or system administrator where the Annotate macro data set is stored for your site.

CHAPTER

2

Elements of a SAS/GRAPH Program

Overview **31**
A Typical SAS/GRAPH Program **31**
   SAS/GRAPH PROC Step **32**
      Procedure Statement **32**
      Subordinate Statement **32**
      Other Statements and Options **32**
   Global Statements **33**
   Annotate DATA Set **34**
   DSGI Functions and Routines in a DATA Step **34**
   ODS Statements **34**
      Destination Statements **34**
      ODS Statement Options **35**
   Base SAS Language Elements **35**
      FILENAME Statement **36**
      LIBNAME Statement **36**
Other Resources **36**

Overview

The elements used by SAS/GRAPH programs can include SAS/GRAPH language elements, ODS statements, and Base SAS language elements. The purpose of this chapter is to familiarize you with the overall structure of a typical SAS/GRAPH program, to define its basic parts, and to show how these parts relate to one another.

A Typical SAS/GRAPH Program

Most SAS/GRAPH programs have Base SAS statements, ODS statements, and SAS/GRAPH statements. Annotate DATA steps and DSGI functions are also used in many SAS/GRAPH programs. The sample program below identifies the basic parts of a typical SAS/GRAPH program. Each element is described in more detail in the following sections.

Display 2.1 Typical SAS/GRAPH Program

```
 1  ods html file="c:\regression.htm"  11 style=analysis;
 2  options reset=all device=activex;
 3  title "Study of Height vs Weight";
    footnote j=r h=2 "Data: SASHELP.CLASS";
 4  symbol interpol=rcclm95 value=circle;
 5  proc gplot data=sashelp.class;
 6     plot height*weight /
           10 haxis=45 to 155 by 10;
 7  run;
 8  quit;
 9  ods html close;
```

- Lines 1, 9: ODS destination statements
- Lines 1–4, 11: Global statements
- Lines 5–8: PROC step

SAS/GRAPH PROC Step

A group of SAS procedure statements is called a PROC step. The PROC step consists of all the statements, variables, and options that are contained within the (beginning) PROC and (ending) RUN statements of a procedure. These statements can identify and analyze the data in SAS data sets, generate the graphics output, control the appearance of the output, define variables, and perform other operations on your data. You can also specify global statements and options within the PROC step to customize the appearance of your graph, but it is often more efficient to specify global statements before the PROC step.

Procedure Statement

The procedure statement [5] identifies which procedure you are invoking (for example, GCHART, GMAP, or GCONTOUR) and identifies which input data set is to be used.

Subordinate Statement

Subordinate statements [6] are statements used within the procedure that perform the work of the procedure. Subordinate statements that generate graphs are called *action statements*. At least one action statement is required for a procedure to produce a graph. Examples of action statements are the HBAR statement in the GCHART procedure and the BUBBLE statement in the GPLOT procedure.

Non-action statements are those that do not generate graphs. The GRID statement in PROC G3GRID and the DELETE statement in PROC GDEVICE are examples of non-action statements.

Other Statements and Options

There are many options [10] that you can specify within the PROC step to control your graphics output. PROC step options always follow the forward slash (/) following the action statement of the procedure. These options might control such things as axis characteristics, midpoint values, statistics, catalog entry descriptions, or appearance elements of your graph. For example, the SUBGROUP= option in the BLOCK statement of the GCHART procedure tells the procedure to divide the graph's bars into segments according to the values of the SUBGROUP= variable. The HAXIS option in the PLOT statement of the GPLOT procedure, as shown in Display 2.1 on page 32, specifies where to draw the major tick mark values for the horizontal axis.

Global Statements

A *global statement* is a statement that you can specify anywhere in a SAS program. Global statements set values and attributes for all the output created from that point in the program when the statement is specified. The specifications in a global statement are not confined to the output generated by any one procedure but apply to all the output generated then point on in the program, unless they are overridden by a procedure option or another global statement. The RESET= option in the GOPTIONS statement also overrides global statements by resetting them.

Below is a list of all the SAS/GRAPH statements along with a brief description of each. See Chapter 14, "SAS/GRAPH Statements," on page 197 for a more detailed description of each of these statements.

AXIS
> modifies the appearance, position, and range of values of axes in charts and plots.

BY
> processes data and orders output according to the values of a classification (BY) variable. The BY statement in SAS/GRAPH is essentially the same as the BY statement in Base SAS, but the effect on the output is different when it is used withSAS/GRAPH procedures. When used with SAS/GRAPH procedures, the BY statement subsets the data and creates a graph for each unique value of the BY-variable.
>
> *Note:* The BY statement is an exception here because it is not a global statement. It must be specified within a DATA or PROC step. △

GOPTIONS [2]
> specifies graphics options that control the appearance of graphics elements by specifying characteristics such as default colors, fill patterns, fonts, or text height. Graphics options can also temporarily change device settings.

LEGEND
> modifies the appearance and position of legends generated by procedures that produce charts, plots, and maps.

PATTERN
> defines the characteristics of patterns used in graphs created by the GAREABAR, GBARLINE, GCHART, GCONTOUR, GMAP, and GPLOT procedures.

SYMBOL [4]
> defines the characteristics of symbols that display the data plotted by a PLOT statement used by PROC GBARLINE, PROC GCONTOUR, and PROC GPLOT as well the interpolation method for plot data. The SYMBOL statement also controls the appearance of lines in contour plots.

TITLE, NOTE, and FOOTNOTE [3]
> add text to maps, plots, charts, and text slides. They control the content, appearance, and placement of text on your graph. The FOOTNOTE statement is used to display lines of text at the bottom of the page. The TITLE statement is used to specify up to ten title lines to be printed on the title area of the output. The NOTE statement is used to add text to the procedure output area of your graph.
>
> *Note:* The NOTE statement is a local statement. It can be specified only within a PROC step, and it affects the output of that PROC step only. △

Annotate DATA Set

An Annotate DATA set is a data set containing graphics commands that can be applied to SAS/GRAPH output. See Chapter 29, "Using Annotate Data Sets," on page 641 for information on building and using Annotate data sets. The Annotate facility can be used to create a completely new graph or to annotate existing PROC output. See Chapter 30, "Annotate Dictionary," on page 667 for a complete description of all Annotate functions and variables. Below is an example of how the Annotate facility can be used to add text labels and symbols to a graph that was created using the GMAP procedure.

Display 2.2 Using Annotate with GMAP Procedure Output

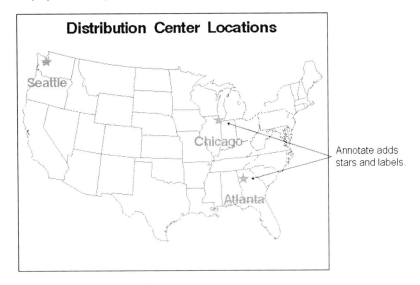

DSGI Functions and Routines in a DATA Step

The DATA Step Graphics Interface (DSGI) enables you to create graphics output within the DATA step or from within an SCL application. Through DSGI, you can call the graphics routines used by SAS/GRAPH to generate a custom graph, or to rescale and replay existing graphs into viewports. DSGI GASK routines can be used to query current system or graphics area settings. For more information on DSGI functions and routines, see Chapter 31, "The DATA Step Graphics Interface," on page 769.

ODS Statements

Destination Statements

Like Base SAS, SAS/GRAPH uses ODS destination statements ([1], [9]) to control where the output goes and how it looks. While ODS statements are not required in every SAS/GRAPH program, they are necessary if you want to generate graphs for destinations other than the default listing destination. Some other destinations include HTML, RTF, and PDF. For more information about ODS destinations, see "Understanding ODS Destinations" in *SAS Output Delivery System: User's Guide*.

As shown in Display 2.1 on page 32, the ODS destination statement is used at the beginning and end of the program to open and close the destination, respectively. If you do chose to use a destination other than the default and need to use the ODS destination statement, you should always open the destination before calling the procedure. To conserve system resources, you should also use the ODS destination statement to close the LISTING destination if you do not need LISTING output.

ODS Statement Options

You can use the STYLE= option [11] on the ODS destination statement to change the style that is applied to your output. For more information about the STYLE= option, see Chapter 10, "Controlling The Appearance of Your Graphs," on page 133.

Base SAS Language Elements

The following Base SAS language statements are also part of SAS/GRAPH:

FORMAT statement
: assigns a format to a variable. SAS/GRAPH procedures use formatted values to determine such aspects of the graph as midpoints, axis labels, tick-mark values, and legend entries.

FILENAME
: associates a SAS fileref with an external text file or output device. See "FILENAME Statement" on page 36 for a more detailed description of this statement.

RUN statement [7]
: executes the statements in the PROC step.

LABEL statement
: assigns a descriptive text string (a "label") to a variable. The label appears in place of the variable name on the axis and legend.

LIBNAME
: associates a libref with a SAS library. See "LIBNAME Statement" on page 36 for a more detailed description of this statement.

ODS statements
: control the output of SAS/GRAPH procedures, where the output is sent (destination), the appearance of the output (STYLE=), and the output file type (DEVICE=). See Chapter 3, "Getting Started With SAS/GRAPH," on page 39 for information on using ODS with SAS/GRAPH procedures.

OPTIONS statement
: changes the value of one or more SAS system options.

QUIT statement [8]
: executes any statements that have not executed and ends the procedure. It also ends a procedure that is using RUN-GROUP processing.

WHERE statement
: specifies observations from SAS data sets that meet a particular condition. You can use a WHERE statement to easily subset your data.

For a complete description of these statements, see "Statements" in *SAS Language Reference: Dictionary*.

FILENAME Statement

The FILENAME statement associates a SAS fileref with an external text file or output device. With SAS/GRAPH software, you can use a FILENAME statement to to the following tasks:

- point to a text file that you want to use for data input or output.
- assign the destination of a graphics stream file (GSF). This destination can be either a single, specific file or an aggregate file storage location, such as directory or PDS. See "Exporting Your Output" on page 111 for information on creating graphics stream files.

You can also use the FILENAME statement to route input to and from other devices. For details, see the SAS documentation for your operating environment.

A FILENAME statement that points to an external file has this general form:

FILENAME *fileref 'external-file'*;

fileref
    is any SAS name.

external-file
    is the physical name of the external file or aggregate file storage location you want to reference. For details on specifying the physical names of external files, see the SAS documentation for your operating environment.

LIBNAME Statement

The LIBNAME statement associates a libref with a SAS library. A SAS library can be either temporary or permanent. Typically, SAS libraries used with SAS/GRAPH software contain the following items:

- SAS files for data input and output.
- SAS catalogs that contain SAS/GIS maps, fonts, GRSEG, CMAP, TEMPLATE, or device entries.
- SAS catalogs that contain graphics output. These catalogs are often stored in permanent libraries. See "Controlling Where Your Output is Stored" on page 97 for information on storing graphics output in a permanent catalog.

The LIBNAME statement has this general form:

LIBNAME *libref 'SAS-library'*;

libref
    is any SAS name.

SAS-library
    is the physical name for the SAS library on your host system. For details on specifying *SAS-library*, see the SAS documentation for your operating environment.

The libref WORK is reserved; it always points to an area where temporary data sets and catalogs are kept. The contents of WORK are deleted when you exit a SAS session.

Other Resources

- For more information on using and managing SAS/GRAPH programs to create graphics output, see Chapter 3, "Getting Started With SAS/GRAPH," on page 39.

□ For more information on bringing SAS/GRAPH output to the Web, see Chapter 16, "Introducing SAS/GRAPH Output for the Web," on page 439.

□ For information on using and managing SAS/GRAPH output, see Chapter 7, "SAS/GRAPH Output," on page 87.

CHAPTER

3

Getting Started With SAS/GRAPH

Introduction **39**
Introduction to ODS Destinations and Styles **40**
  Opening And Closing Destinations **40**
  The LISTING Destination **41**
  Introduction to Styles **41**
  Specifying a Style **42**
Generating Output With SAS/GRAPH Procedures **43**
  Sending Output to the GRAPH Window (LISTING Destination) **43**
  Sending Output to a File **44**
  Sending Output to a Web Page **45**
  Sending Output to an RTF File (Microsoft Word Document) **46**
  Sending Output to a PDF File **47**
Controlling the Graphics Output Format With the DEVICE= Option **48**
  Overview of Devices and Destinations **48**
  Specifying the DEVICE= Graphics Option **49**
Summary of Default Destinations, Styles, and Devices **49**
Sending Output To Multiple Open Destinations **51**
  Closing Destinations To Save System Resources **51**
  Specifying Devices And Styles With Multiple Open Destinations **51**
Related Topics **52**

Introduction

Like other SAS procedures, the output from SAS/GRAPH procedures is controlled by ODS (Output Delivery System). ODS controls where your output is sent, which could be to a file, to the GRAPH window, directly to a printer, and so on. By default, ODS also applies a style to your output. Styles set the overall appearance of your output; that is, the colors and fonts that are used.

SAS/GRAPH uses device drivers to generate graphics output. SAS/GRAPH device drivers control the format of your graphics. For example, they determine whether SAS/GRAPH produces a PNG file, an SVG file, or an ActiveX control.

Note: This document deals only with device-based graphics. See "Device-Based Graphics and Template-Based Graphics" on page 6. △

Each ODS destination is associated with a default style and a default graphics device to optimize your output for that destination. However, using ODS statements and SAS/GRAPH statements and options, you can customize all of the aspects of your output, including where your output is sent, its appearance, and the format of your graphics.

ODS destination The ODS destination controls where your output is sent, such as to a file or directly, to a printer, and so on. The ODS destination is specified by the ODS destination statement.

ODS style The ODS style controls the appearance of your output, including colors and fonts. The ODS STYLE= attribute in the ODS destination is specified by the ODS style statement.

SAS/GRAPH device The SAS/GRAPH device controls the format of your graphics output such as PNG, GIF, SVG , and so on. The SAS/GRAPH DEVICE= option is specified in the GOPTIONS SAS/GRAPH DEVICE= statement

Note: The LISTING destination is unique. For the LISTING destination, the device controls where your output is sent. △

The following sections discuss these concepts of SAS output and describe how you can use SAS/GRAPH and ODS statements and options to create the graphic output you want.

For complete information on ODS, see also *SAS Output Delivery System: User's Guide*.

Introduction to ODS Destinations and Styles

ODS destinations determine where your SAS/GRAPH output is sent. For example, the LISTING destination sends output to the GRAPH window (by default), and the HTML destination sends output to an HTML file. By default, ODS styles determine the overall appearance of your output.

Opening And Closing Destinations

A *destination* is a designation that ODS uses to determine where to send your output. Valid destinations include LISTING (the GRAPH window, by default), HTML, RTF, and PDF, but other destinations are also available.

To generate output from SAS, a valid ODS destination must be open. By default, the LISTING destination is open, but you can open other destinations as needed by specifying an ODS destination statement. Depending on the options available for the destination, you can specify options such as the filename or the path to an output directory. With the exception of the LISTING destination, you must also close the destination before output is generated.

```
ods destination <options>;   /* opens the destination */
    /* procedure statements and other program elements here */
ods destination close;   /* closes the destination */
```

For example, to send output to the HTML destination, you would specify

```
ods html;
    /* procedure statements and other program elements here */
ods html close;
```

For more information on ODS destinations, see "Managing ODS Destinations" on page 191 and "ODS Destination Statement Options" on page 192.

The LISTING Destination

The LISTING destination is open by default. If you are sending output to other destinations and are not interested in the output that is sent to the LISTING destination, you should close it to conserve resources. The usual practice is to close LISTING at the beginning of your program and to reopen it at the end. This practice ensures that you always have one open destination. See "Closing Destinations To Save System Resources" on page 51 for more information.

The LISTING destination is somewhat different from other ODS destinations. For the LISTING destination, if you do not specify a device, then your output is sent to the GRAPH window. However, if you specify a device, then where your output is sent is determined by the device. For example, the PNG device sends output to a PNG file instead of the GRAPH window. Your company might have device drivers specific to your site that send output directly to a certain printer. Where your output is sent is controlled by the device entry in the SASHELP.DEVICES catalog. See "Controlling the Graphics Output Format With the DEVICE= Option" on page 48 and Chapter 6, "Using Graphics Devices," on page 67 for more information about devices.

The LISTING destination is the only destination that does not have to be closed before output can be generated.

Introduction to Styles

By default, ODS applies a *style* to all output. A style is a template, or set of instructions, that determines the colors, font face, font sizes, and other presentation aspects of your output. SAS ships many predefined styles in the STYLES item store, such as Analysis, Statistical, and Journal. Examples of some of these predefined styles are shown in Table 3.1 on page 42. Many additional styles (see "Viewing the List of Styles Provided by SAS" on page 141) are available in the STYLES item store in SASHELP.TMPLMST.

Each destination has a default style associated with it. For example, the default style for the PDF destination is Printer, and the default style for the HTML destination is Default. See "ODS Destinations and Default Styles" on page 135 and "Recommended Styles" on page 136 for more information.

Table 3.1 Examples of Styles Available in SASHELP.TMPLMST

Display 3.1 Style=Statistical

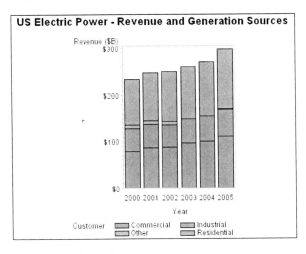

Display 3.2 Style=Analysis

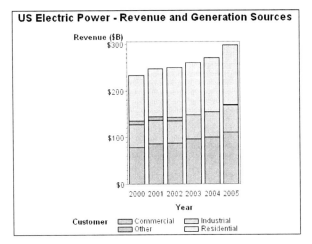

Display 3.3 Style=Ocean

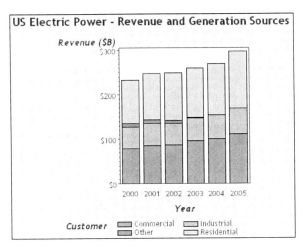

Display 3.4 Style=Harvest

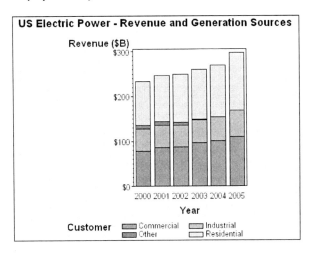

Display 3.5 Style=Gears

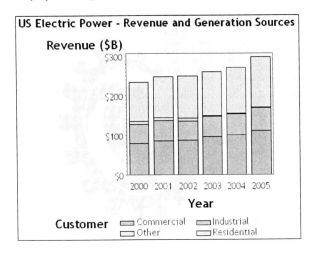

Display 3.6 Style=Banker

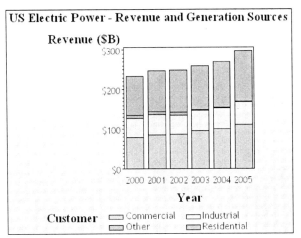

Specifying a Style

To change the style that is applied to your output, specify the STYLE= option on your ODS destination statement. For example, if you want to change the overall look of your graph for the HTML destination to the Analysis style, you would specify `style=analysis` in the ODS HTML destination statement as follows:

```
ods html style=analysis;
```

See "About Style Templates" on page 135 and "Specifying a Style" on page 139 for more information.

Note: You can turn off the use of styles by default by specifying the NOGSTYLE option. See "Changing the Appearance of Output to Match That of Earlier SAS Releases" on page 154 and the GSTYLE system option in *SAS Language Reference: Dictionary* for more information. △

Generating Output With SAS/GRAPH Procedures

ODS provides many destinations to which you can send output. Some of the most often used destinations are LISTING, HTML (a Web page), RTF (an Microsoft Word document), and PDF. As described in "Introduction to Styles" on page 41, each destination is associated with a default style. The following topics each show the default output for each of the destinations listed above.

Each destination is also associated with a default device driver for generating graphics output. Device drivers determine the form that your graphics output takes. For example, the PNG device driver generates PNG image files, and the JAVA device driver generates Java applets that can be run from within HTML pages.

Each destination is associated with a default style and a default device, so you do not need to specify either one to get professional-quality output. You can even send output to several destinations at the same time without specifying either a device or a style.

Sending Output to the GRAPH Window (LISTING Destination)

When working in an interactive environment such as Windows, the LISTING destination is the GRAPH window. By default, the LISTING destination is open, so to send output to it, you simply submit your SAS/GRAPH program. The following example is a simple GCHART program that produces the output shown in Display 3.7 on page 44.

```
goptions reset=all border;
title "US Electric Power - Revenue and Generation Sources";

proc gchart data=sashelp.electric (where=(year >= 2000)) ;
   vbar year / discrete sumvar=Revenue subgroup=Customer;
run;
quit;
```

Display 3.7 LISTING Destination Output Using the Listing Style (Shown in the GRAPH Window)

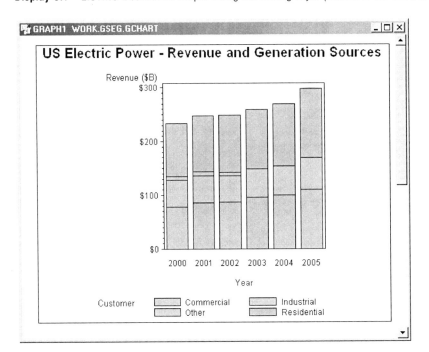

The default style applied to output sent to the LISTING destination is the Listing style. When you send output to the LISTING destination, SAS/GRAPH uses a default device driver that generates output for the GRAPH window. This device driver does not write an image file to disk.* For the LISTING destination, the default device driver varies by operating environment. In a Display Manager Session (DMS), the default device driver on Windows systems is WIN. On UNIX systems, the default device driver is XCOLOR, and on z/OS systems, the default device driver is IBMPCGX.

Sending Output to a File

To send output to disk file, send your output to the ODS LISTING destination, but specify a graphics output device using the DEVICE= graphics option. You can use a FILENAME statement and the GSFNAME= graphics option to specify a name and location for the graphics output file. If you do not specify a name with the GSFNAME= graphics option, the default name for the procedure or the name specified with the NAME= option is used as the filename.

To create a GIF file with the graph shown in Display 3.7 on page 44, in the procedure code, add a FILENAME statement to create a file reference to the desired output file. Then, add the DEVICE=GIF and GSFNAME=*FileRef* graphics options to the GOPTIONS statement, where *FileRef* is the file reference that you created in the FILENAME statement.

```
filename gout "./revgensrcs.gif";
goptions reset=all device=gif gsfname=gout border;
title  "US Electric Power - Revenue and Generation Sources";
```

---

* SAS/GRAPH procedures create GRSEG catalog entries when you send output to the LISTING destination, but the GRSEG file format is an internal file format specific to SAS/GRAPH. It cannot be used as if it was an image file such as a PNG, GIF, or JPEG file.

```
proc gchart data=sashelp.electric (where=(year >= 2000)) ;
   vbar year / discrete sumvar=Revenue subgroup=Customer;
run;
quit;
```

By default, the Listing style is applied to the graph as shown in Display 3.7 on page 44. In the FILENAME statement, the current directory is the default SAS output directory.

For more information on sending graphics output to a file, see "Controlling Where Your Output is Stored" on page 97.

Sending Output to a Web Page

Tosend output to a Web page, send your output to the HTML destination by specifying the ODS HTML statement. This statement opens the HTML destination so that it can receive output. You must also close the HTML destination before output can be generated.

To create a Web page with the graph shown in Display 3.7 on page 44, add the ODS HTML statements around the procedure code.

```
ods listing close;
ods html;
goptions reset=all border;
title  "US Electric Power - Revenue and Generation Sources";

proc gchart data=sashelp.electric (where=(year >= 2000)) ;
   vbar year / discrete sumvar=Revenue subgroup=Customer;
run;
quit;
ods html close;
ods listing;
```

Display 3.8 HTML Destination Output Using the Default Style (Styles.Default)

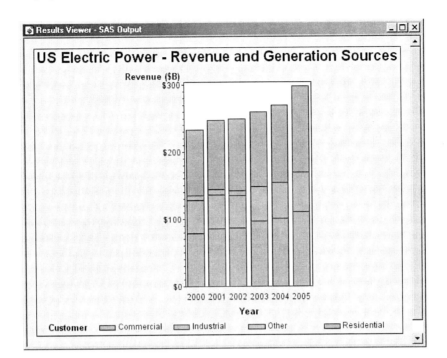

By default, SAS/GRAPH creates a PNG file that contains the graph and an HTML page that references the PNG file. You can use the BODY= and PATH= options in the ODS HTML statement to specify a specific filename and location for the HTML and PNG files. SAS/GRAPH displays the HTML page in the Results Viewer. You can also view the graph outside of your SAS session by displaying the HTML page in your browser. The default device driver is PNG, and the default style is Default (STYLES.DEFAULT).

Sending Output to an RTF File (Microsoft Word Document)

To send output to an RTF file, send your output to the RTF destination by specifying the ODS RTF statement. This statement opens the RTF destination so that it can receive output. You must also close the RTF destination before output can be generated.

To create an RTF document that contains the graph shown in Display 3.7 on page 44, add the ODS RTF statements around the procedure code.

```
ods listing close;
ods rtf;
goptions reset=all border;
title "US Electric Power - Revenue and Generation Sources";

proc gchart data=sashelp.electric (where=(year >= 2000)) ;
   vbar year / discrete sumvar=Revenue subgroup=Customer;
run;
quit;
ods rtf close;
ods listing;
```

Display 3.9 RTF Output Using the Rtf Style

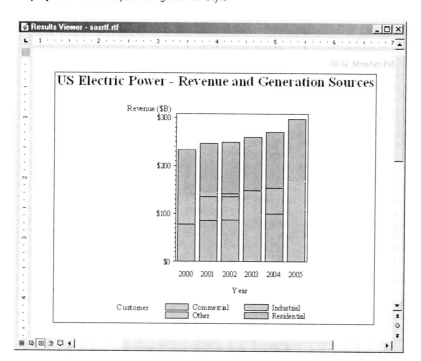

By default, SAS/GRAPH creates an RTF file with the graph embedded in it and displays this RTF file in the Results Viewer. When you send output to the RTF destination, SAS/GRAPH does not write a separate image file to disk. The default device driver is the SASEMF driver, and the default style is Rtf.

Sending Output to a PDF File

To send output to a PDF file, send your output to the PDF destination by specifying the ODS PDF statement. This statement opens the PDF destination so that it can receive output. You must also close the PDF destination before output can be generated.

To create a PDF document that contains the graph shown in Display 3.7 on page 44, add the ODS PDF statements around the procedure code.

```
ods listing close;
ods pdf;
goptions reset=all border;
title "US Electric Power - Revenue and Generation Sources";

proc gchart data=sashelp.electric (where=(year >= 2000)) ;
   vbar year / discrete sumvar=Revenue subgroup=Customer;
run;
quit;
ods pdf close;
ods listing;
```

Display 3.10 PDF Output Using the Printer Style

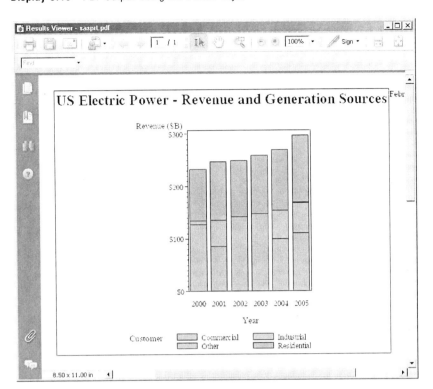

By default, SAS/GRAPH creates a PDF file and displays this PDF file in the Results Viewer. When you send output to the PDF destination, SAS/GRAPH does not write a separate image file to disk. The default device driver is the SASPRTC device driver, and the default style applied to output sent to the PDF destination is Printer.

Controlling the Graphics Output Format With the DEVICE= Option

Overview of Devices and Destinations

SAS/GRAPH procedures use device drivers to generate graphics output. Device drivers determine the format of your graphics output. For example, the GIF device driver generates GIF image files. The ACTIVEX device driver generates ActiveX controls that can be run within HTML pages or RTF documents. The SASPRTC device generates images for the current printer as determined by the PRINTERPATH= system option (or the SYSPRINT= system option on Windows).

Every ODS destination has a default device driver associated with it. For example, the default device driver for the HTML destination is PNG. By default, when you send output to the HTML destination, your graphics output is rendered as a PNG file. (An HTML file is also generated. This HTML file contains any non-graphical output generated by your application plus an <IMG> tag that inserts the PNG output that was generated.)

Each destination supports several devices. For example, the HTML destination supports the SVG, PNG, GIF, JAVA, and ACTIVEX devices, in addition to several others. "Viewing The List Of All Available Devices" on page 70 describes how to display the entire list of devices that are available. Table 3.2 on page 50 lists the default and supported devices for the LISTING, HTML, RTF, and PDF destinations.

Specifying the DEVICE= Graphics Option

You can change the device, and therefore the format of your graphics output, by changing the device driver that SAS uses. You can specify a device with either the OPTIONS statement or the GOPTIONS statement. For example, to use the GOPTIONS statement to change the device, submit this code:

```
goptions device=device-entry;
```

Devices that you might specify include PNG, GIF, JPEG, SVG, ACTIVEX, ACTXIMG, JAVA, JAVAIMG, and many others. For all open destinations, SAS/GRAPH attempts to use the device that you specify. If the device that you specify is not valid for an open destination, SAS/GRAPH switches to the default device for that destination.

For details, see "GOPTIONS Statement" on page 220. "Summary of Default Destinations, Styles, and Devices" on page 49 describes the supported devices for the LISTING, HTML, RTF, and PDF destinations. See also Chapter 6, "Using Graphics Devices," on page 67.

Summary of Default Destinations, Styles, and Devices

Each destination has a default device and default style that are used if you do not specify otherwise. Also, each destination has a set of recommended devices. Table 3.2 on page 50 summarizes this information for the LISTING, HTML, RTF, and PDF destinations.

You can use any style with any destination. If you specify a device with the GOPTIONS DEVICE= option, you should specify a device that is compatible with all of the destinations that you have open.

Table 3.2 Default Devices and Styles for Commonly Used ODS Destinations

ODS Destination	Default Device	Default Style	Default Output	Recommended Devices
LISTING	WIN (Windows) XCOLOR (UNIX) IBMPCGX (z/OS)	Listing	Graphics output is displayed in the GRAPH window[1]	All devices[2] except JAVA and ACTIVEX
HTML	PNG	Default (Styles.Default)	HTML and PNG file	PNG GIF JPEG JAVA JAVAIMG ACTIVEX ACTXIMG SVG JAVAMETA GIFANIM
RTF	SASEMF	Rtf	RTF file (with embedded metafile)	SASEMF PNG JPEG JAVAIMG ACTIVEX ACTXIMG
PDF	SASPRTC	Printer	PDF file	SASPRTC (color) SASPRTG (gray scale) SASPRTM (monochrome)
PRINTER	SASPRTC	Printer	Controlled by the PRINTERPATH= system option (and by the SYSPRINT= system option on Windows)[3]	SASPRTC (color) SASPRTG (gray scale) SASPRTM (monochrome)

1 The default devices for the LISTING destination do not write image files to disk.
2 JAVAMETA is supported for the LISTING destination, but its output requires processing with the Metaview applet.
3 In Windows, if the PRINTERPATH= option is not specified, then SAS uses the setting of the SYSPRINT= system option. If neither the SYSPRINT= nor the PRINTERPATH= option has been set, then SAS uses the default Windows printer.

Note: SASHELP.DEVICES also has high resolution versions of the PNG and JPEG devices, PNG300 and JPEG300. These devices are not appropriate choices for the HTML destination. Web browsers cannot display images in high resolution, so high resolution images appear very large. △

Sending Output To Multiple Open Destinations

When you are sending output to more than one destination at the same time, you should remember two points:

- You should close any open destinations whose output you are not interested in. Doing so saves system resources.
- If you specify a device that is not supported for an open destination, SAS/GRAPH switches to the default device for that destination and prints a warning to the SAS log.

Closing Destinations To Save System Resources

SAS/GRAPH creates output for every open destination. The LISTING destination is open by default, and you can open as many additional destinations as needed. For example, you can open the HTML and PDF destinations, and generate output for all three destinations by submitting your SAS code only once. However, SAS/GRAPH goes through the process of generating GRSEG catalog entries and graphics output files for each open destination. This process uses system resources. Each open destination increases system resources required for by your application. If you are not interested in the output of a destination, it is recommended that you close that destination.

Specifying Devices And Styles With Multiple Open Destinations

Unless you specify a different device or different style, SAS/GRAPH uses the default device and default style for each open destination. For example, suppose your application specifies the following:

```
ods listing close;
ods html;
ods rtf;
    /* procedure statements and other program elements here */
ods html close;
ods rtf close;
ods listing;
```

SAS/GRAPH uses the PNG device and the Default style to generate output for the HTML destination, and it uses the SASEMF device and the Rtf style to generate output for the RTF destination.

If you specify a different device with the DEVICE= option in the GOPTIONS statement, SAS/GRAPH attempts to use that device to generate output for every open destination. If you want to use a different style for all output, you need to specify that style on each ODS destination statement. For example, to use the ACTIVEX device and the ANALYSIS style for all output sent to both the HTML and RTF destinations, you would specify the GOPTIONS statement and the STYLE= option as follows:

```
goptions device=activex;
ods listing close;
ods html style=analysis;
ods rtf style=analysis;
    /* procedure statements and other program elements here */
ods html close;
ods rtf close;
ods listing;
```

If you specify a device that is not supported for an open destination, SAS/GRAPH switches to the default device for that destination and prints a warning to the SAS log.

Related Topics

Additional information is available on all of the SAS/GRAPH output concepts that are described in this topic. For more information on generating output with the SAS/GRAPH procedures, see the following topics:

- Chapter 7, "SAS/GRAPH Output," on page 87
- Chapter 16, "Introducing SAS/GRAPH Output for the Web," on page 439

For more information on the ODS, ODS destinations, and ODS styles, see the following topics:

- Chapter 13, "Managing Your Graphics With ODS," on page 191
- Chapter 10, "Controlling The Appearance of Your Graphs," on page 133

For more information on using the graphics devices, see Chapter 6, "Using Graphics Devices," on page 67.

CHAPTER
4

SAS/GRAPH Processing

Running SAS/GRAPH Programs **53**
SAS Data Sets **54**
Specifying an Input Data Set **54**
    Using a Library Reference **54**
    Using a File Specification **55**
    Input Data Set Requirements **55**
    Automatic Data Set Locking **56**
Using Engines with SAS/GRAPH Software **56**
RUN-Group Processing **56**
    RUN-group Processing with global and local statements **56**
    RUN-group Processing with BY statements **57**
    RUN-group Processing with the WHERE Statement **57**

Running SAS/GRAPH Programs

Here are the environments and modes in which you can run a SAS/GRAPH program:
- The *SAS windowing environment* provides a text editor for submitting programs, windows for the SAS log and SAS output, and many other facilities. For more information on the SAS windowing environment see "Introduction to the SAS Windowing Environment" in *SAS Language Reference: Concepts.*
- *Interactive line mode* enables you to submit programs one line at a time in response to prompts from the SAS/GRAPH system. In interactive line mode, the SAS/GRAPH program can display graphics output on your monitor as well as store the output in a file.
- *Noninteractive mode* enables you to issue a SAS command that executes a SAS/GRAPH program that is stored in an external file. This mode is valid only in your current terminal session. In this mode, the SAS/GRAPH program can display graphics output on your monitor as well as store the output in a file.
- *Batch mode* enables you to execute a SAS program (stored in a file) in a separate terminal session. In batch mode, the graphics output is not displayed on your monitor. In this case, your program must send the graphics output to a printer or plotter, permanent catalog, or an external file.

Note: Certain fonts called device-resident fonts are specific to the device being used and therefore are not portable between devices when running in batch mode. See "Overview" on page 1175 for more information on using fonts in batch mode. △

Regardless of how you run your programs, SAS/GRAPH software applies ODS styles by default to your graphics output. For more information on ODS styles see Chapter 10, "Controlling The Appearance of Your Graphs," on page 133.

See Chapter 7, "SAS/GRAPH Output," on page 87 for more information about SAS/GRAPH output.

SAS Data Sets

Many SAS/GRAPH procedures use and create SAS data sets. SAS data sets are files stored in SAS libraries and can be either temporary or permanent.

When you create a SAS data set, it is stored automatically in the WORK library. Unless you specify a different library, the WORK library serves as a temporary holding place for all the data sets you access and create for the duration of a SAS session. By default, the WORK library and all the data sets stored in it will be removed after the SAS session ends.

You can also create permanent SAS libraries that can be saved in a specified location on your computer. Permanent libraries are not deleted when the SAS session terminates and are available for processing in subsequent SAS sessions.

For more information on SAS data sets and other data processing details, see *SAS Language Reference: Concepts*.

For a complete discussion of SAS data set options and SAS system options, see *SAS Language Reference: Dictionary*.

Specifying an Input Data Set

You can specify an input data set by using one of the following methods:

- a library reference
- a file specification

When using either of these methods, you usually specify the DATA= option in the procedure statement, as shown in this example:

```
proc gplot data=stocks;
```

If you omit the DATA= option, then the procedure uses the SAS data set that was most recently used or created in the current SAS session.

If you do not specify a SAS data set and no data set has been created in the current SAS session, an error occurs and the procedure stops.

Most of the procedures that read data sets or create output data sets accept data set options. SAS data set options appear in parentheses after the DATA= option specification, as shown in this example:

```
proc gplot data=stocks(where=(year=1997));
```

Using a Library Reference

A SAS library is a storage location for SAS data sets in your operating environment. Data sets stored in a SAS Library are created and referenced using either a one- or two-level name. SAS data sets stored in the temporary WORK library are usually specified using a one-level name. Procedures assume that SAS data sets that are specified with a one-level name are to be read from or written to the WORK library. Since temporary SAS data sets are typically stored by default in the WORK data library, you can specify them using a one-level name and SAS knows where to find them. For example, this statement specifies the data set stocks that resides in the WORK library:

```
proc gplot data=stocks;
```

To specify a permanent data set you typically use a two-level name. A permanent library reference is specified in the form *libref.SAS-data-set-name* in which libref identifies a storage location on your host system. A LIBNAME statement associates a libref with the storage location. See also "LIBNAME Statement" in *SAS Language Reference: Dictionary*. For example, these statements specify a permanent data set:

```
libname   reflib 'my-SAS-library';
     proc gplot data=reflib.stocks;
run;
```

You can use a one-level name for permanent SAS data sets if you specify a USER data library. In this case, the procedure assumes that data sets with one-level names are in the User library instead of in the WORK data library. You can assign a User library with a LIBNAME statement or the USER= SAS system option. For example, these statements use a single-level name to specify a permanent data set that is stored in the library identified as the User library:

```
options user='my-SAS-library';
proc gplot data=stocks;
```

For more information on SAS Libraries see "SAS Libraries" in *SAS Language Reference: Concepts*.

Using a File Specification

To use a file specification for specifying a data set, enclose the file specification in single quotation marks. The specification can be a filename, or a path and filename. The specification must follow the file naming conventions of your operating environment.

For example, the following code creates a file named *mydata* in the default storage location, which is the location where the SAS session was started:

```
data 'mydata';
```

The quotes are required for a file specification; if omitted, SAS treats the specification as a library reference. In the above example, if the quotes are omitted, SAS creates the data set in the temporary WORK catalog and identifies it by the name WORK.MYDATA.

To create the file in a location other than the default location, the quoted file specification must include the full path to the desired location.

You cannot use quoted file specifications for the following items:

- SAS catalog names
- MDDB and FDB references
- the _LAST_= system option

Input Data Set Requirements

SAS/GRAPH procedures often have certain requirements for the input data sets they use. Some procedures might require the input data set to be sorted in a certain way while others might require the data set to contain certain variables or types of information. If necessary, you can use DATA steps and Base SAS procedures in your program to manipulate the data appropriately. For more information on the requirements of any given procedure, see the "Concepts" section which is included at the beginning of each procedure overview.

Automatic Data Set Locking

All SAS/GRAPH procedures that produce graphics output automatically lock the input data sets during processing. By locking a data set, SAS/GRAPH software prevents another user from updating the data at the same time you are using it to produce a graph. If data in a data set changes while you are using it to draw a graph, unpredictable results can occur in the graph or your program can end with errors.

Using Engines with SAS/GRAPH Software

In SAS, procedures use *engines* to access data. Characteristics of these engines vary; generally, they enable SAS procedures to access a data library in a particular way. Engines can specify the expected format for the SAS data file, the type of read or write activity that can occur in SAS data files, and so on. In most cases, you use the default engine for the current SAS version and do not need to specify an engine.

For more information about SAS engines, see "Library Engines" in *SAS Language Reference: Concepts*.

RUN-Group Processing

You can use RUN-group processing with the GAREABAR, GBARLINE, GCHART, GKPI, GMAP, GPLOT, GRADAR, GREPLAY, GSLIDE, and GTILE procedures to produce multiple graphs without restarting the procedure every time.

To use RUN-group processing, you start the procedure and then submit multiple RUN-groups. A *RUN-group* is a group of statements that contains at least one action statement and ends with a RUN statement. The procedure can contain other SAS statements such as AXIS, BY, GOPTIONS, LEGEND, TITLE, or WHERE. As long as you do not terminate the procedure, it remains active and you do not need to resubmit the PROC statement.

To end RUN-group processing and terminate the procedure, submit a QUIT or RUN CANCEL statement, or start a new procedure. If you do not submit a QUIT or RUN CANCEL statement, SAS/GRAPH does not terminate RUN-group processing until it reaches another step boundary.

Note: When using SAS/GRAPH with the ODS statement, it is best to use a QUIT statement after each procedure that uses RUN-group processing, rather than relying on a new procedure to end the processing. Running too many procedures without an intervening QUIT statement can use up too much memory. Also, note that failing to submit a QUIT statement before submitting an ODS CLOSE statement results in the process memory not being freed at all. △

RUN-group Processing with global and local statements

Global statements and NOTE statements that are submitted in a RUN-group affect all subsequent RUN-groups until you cancel the statements or exit the procedure. For example, each of these two RUN-groups produces a plot and both plots display the title defined in the first RUN-group:

```
/* first run group*/
proc gplot data=sales;
```

```
   title1 "Sales Summary";
   plot sales*model_a;
run;

      /* second run group */
   plot sales*model_b;
run;
quit;
```

RUN-group Processing with BY statements

BY statements persist in exactly the same way as global and local statements. Therefore, if you submit a BY statement within a RUN-group, the BY-group processing produces a separate graph for each value of the BY variable for the RUN-group in which you submit it and for all subsequent RUN-groups until you cancel the BY statement or exit the procedure. Thus, as you submit subsequent action statements, you continue to get multiple graphs (one for each value of the BY variable). For more information, see "BY Statement" on page 216.

RUN-group Processing with the WHERE Statement

The WHERE statement enables you to graph only a subset of the data in the input data set. If you submit a WHERE statement with a RUN-group, the WHERE definition remains in effect for all subsequent RUN-groups until you exit the procedure or reset the WHERE definition.

Using a WHERE statement with RUN-group processing follows most of the same rules as using the WHERE statement outside of RUN-group processing with these exceptions:

- With the GMAP procedure, the WHERE variable must be in the input data set.
- With a procedure that is using an Annotate data set, the following requirements must be met:
  - The ANNOTATE= option must be included in the action statement.
  - The WHERE variable must occur in both the input data set and the Annotate data set.

CHAPTER 5

The Graphics Output Environment

Overview **59**
The Graphics Output and Device Display Areas **59**
Controlling Dimensions **60**
Controlling Display Area Size and Image Resolution **61**
Units **62**
    Cells **62**
    Other Units **64**
Maintaining the Quality of Your Image Across Devices **65**
    Maintaining Proportions **65**
    Getting the Colors You Want **65**
    Previewing Your Output **65**
How Graphic Elements are Placed in the Graphics Output Area **65**
How Errors in Sizing Are Handled **66**

Overview

The result of most SAS/GRAPH procedures is the graphic display of data in the form of graphics output. *Graphics output* consists of commands that tell a graphics device how to draw graphic elements. A *graphics element* is a visual element of graphics output—for example, a plot line, a bar, a footnote, the outline of a map area, or a border.

To generate graphics output, your program uses a device driver that directs the graphics output to a display device (a monitor or terminal), a hard-copy device, or a file. Even though all graphics devices do not understand the same commands, SAS/GRAPH can produce graphics output on many types of graphics devices.

Your program controls this process as well as the environment in which the graphics appear. This section describes this graphics environment and how you can modify it and make your programs work for different output devices.

The Graphics Output and Device Display Areas

When SAS/GRAPH produces graphics output, it draws the graphic elements inside an area called the *graphics output area*. The graphics output area is contained within the *device display area*. Characteristics of both the graphics output area and the device display area are determined by the values of specific device parameters. In many cases the dimensions of the graphics output area equals those of the device display area. This is particularly true for display devices such as monitors and terminals. Hard-copy devices, such as a printed output, create a margin since the dimensions of the graphics output area are smaller than those of the device's display area.

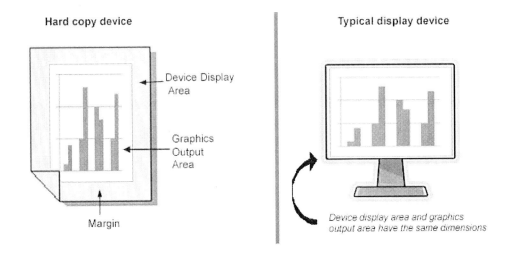

You can modify some of the characteristics of the graphics output area and the device display area by using graphics options to change the values of the device parameter.

This section describes how you can change the dimensions of the output and display areas, how these changes in dimension affect the output, and the types of units you can specify for your output. For a description of the graphics options and device parameters referred to in this section, see Chapter 15, "Graphics Options and Device Parameters Dictionary," on page 327.

Controlling Dimensions

The outer dimensions of the device's display area are controlled by the values of the XMAX and YMAX device parameters. XMAX sets the maximum horizontal dimension; YMAX sets the maximum vertical dimension.

The outer dimensions of the graphics output area are controlled by the values of the HSIZE and VSIZE device parameters.

Since the dimensions of the device display area are typically the same as the dimensions of the graphics output area, the default value of HSIZE and VSIZE is 0. However, for hard-copy devices, because the XMAX, YMAX values represent the outer boundaries of the output medium (such as a sheet of paper), these devices might need a margin. Therefore, HSIZE, VSIZE, HORIGIN, and VORIGIN are assigned default values and the default graphics output area is somewhat smaller than the device's display area. Figure 5.1 on page 61 illustrates such a device.

Note: The default unit of measurement for the XMAX and YMAX options is inches. △

Figure 5.1 Default Dimensions of the PSCOLOR Device

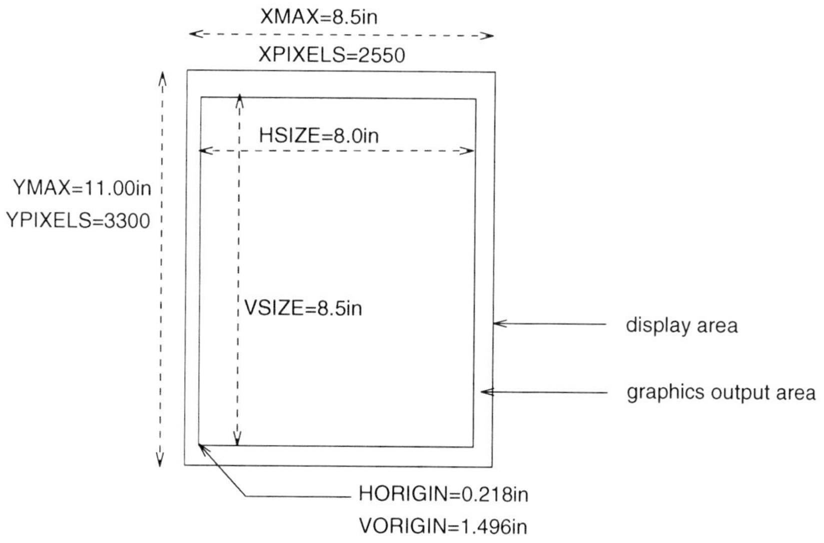

```
                    XMAX=8.5in
              <------------------->
                   XPIXELS=2550
          ┌───────────────────────────┐
          │   ┌───────────────────┐   │
          │   │     HSIZE=8.0in   │   │
          │   │<----------------->│   │
          │   │                   │   │
YMAX=11.00in  │                   │   │
YPIXELS=3300  │                   │   │
          │   │  VSIZE=8.5in      │   │ ←──── display area
          │   │                   │   │
          │   │                   │   │ ←──── graphics output area
          │   │                   │   │
          │   │                   │   │
          │   └───────────────────┘   │
          └───────────────────────────┘
               HORIGIN=0.218in
               VORIGIN=1.496in
```

For further discussion of how the default values for HSIZE and HORIGIN are determined using the value of the LEFTMARGIN option, see "HSIZE" on page 384 and "HORIGIN" on page 382.

Note that HORIGIN and VORIGIN define the left margin and bottom margin, respectively. The right margin and top margin are calculated by the device driver as follows:

right-margin = XMAX − (HSIZE + HORIGIN)

top-margin = YMAX − (VSIZE + VORIGIN)

You cannot specify values for *right-margin* and *top-margin*.

You can change the dimensions of the graphics output area for a SAS session or for a single graph with the HSIZE= and VSIZE= graphics options. Changing the size of the graphics output area does not change the dimensions of the device's display area or affect the resolution. The values of HSIZE= and VSIZE= cannot exceed the maximum dimensions for the device as specified by XMAX and YMAX. Furthermore, you cannot specify values for graphics options HSIZE= and VSIZE= that exceed the HSIZE and VSIZE values for that device.

Controlling Display Area Size and Image Resolution

The resolution of an image is the number of pixels per inch. Resolution is determined by the values of the device parameters XMAX, YMAX, XPIXELS, and YPIXELS, and is calculated by dividing the number of pixels by the corresponding outer dimension. For example:

x-resolution = XPIXELS / XMAX

Therefore, the X resolution of the PSCOLOR device illustrated in Figure 5.1 on page 61 is 300dpi (dots per inch).

Ordinarily, you do not want to change the image resolution because changing it might distort your image. However, you might want to change the size of the display area. To do so without changing the resolution, use the GOPTIONS statement to change either the values of XPIXELS= and YPIXELS=, or the values of XMAX= and YMAX=. SAS/GRAPH automatically calculates the correct value for the unspecified parameters so that the device retains the default resolution.

For information on controlling the resolution of your image see "Using the XPIXELS=, XMAX=, YPIXELS=, and YMAX= Graphics Options to Set the Resolution for Device-Based Graphics" on page 96.

Units

Cells

Within the graphics output area, SAS/GRAPH defines an invisible grid of rows and columns. This grid consists of *character cells* as shown in Figure 5.2 on page 62.

The size and shape of these cells affect the size and appearance of your graph since each graphic element is drawn using units of cells. The size and shape of the cells are determined by both the size of the graphics output area and by the number of rows and columns that SAS/GRAPH has defined in the grid. You can control the number of rows by specifying the LROWS device parameter (for a landscape orientation) or the PROWS device parameter (for a portrait orientation). Similarly, the number of columns is controlled by the LCOLS (landscape) or PCOLS (portrait) device parameter.

It is not recommended that you change the number of rows and columns in the grid from the default for your device. If you must do so, you can specify the HPOS= and VPOS= graphics options. HPOS= overrides the value of LCOLS or PCOLS and sets the number of columns in the graphics output area. VPOS= overrides the value of LROWS or PROWS and sets the number of rows in the graphics output area.

Figure 5.2 on page 62 illustrates how device parameter settings for the size of the output area relate to the parameter settings for the number of character cells in the output area.

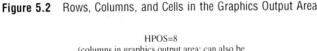

Figure 5.2 Rows, Columns, and Cells in the Graphics Output Area

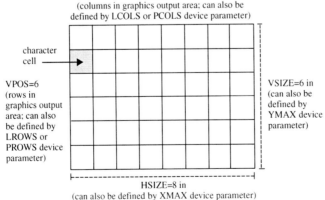

Changing only the outer dimensions of the graphics output area (HSIZE= and VSIZE=) retains the cell size but causes SAS/GRAPH to automatically recalculate the number of rows and columns, as illustrated in Figure 5.3 on page 63.

Figure 5.3 Changing HSIZE= and VSIZE= Changes Dimensions and Recalculates the Number of Rows and Columns

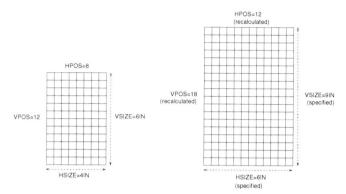

Changing only the number of rows and columns (HPOS and VPOS) changes the size of the cells without altering the overall size of the output. Figure 5.4 on page 63 shows how increasing the number of rows and columns reduces the size of the individual cells.

Figure 5.4 Changing HPOS= and VPOS= Changes Cell Size

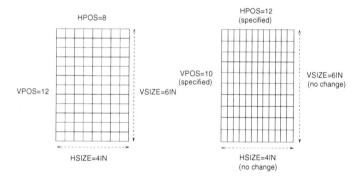

If you use units of CELLS to control the size of the text in your graph while also changing the number of rows and columns, then the size of the text changes. If the cells are large (that is, HPOS= and VPOS= have small values), the text might not fit. If the cells are too small, the text might be too small to read. In this case, you can adjust the size of the text with the HEIGHT= statement option or the HTEXT= graphics option.

To change all the attributes of the graphics output area, specify values for all four options, as shown in Figure 5.5 on page 64.

Figure 5.5 Changing HSIZE=, VSIZE=, HPOS=, and VPOS= Changes Dimensions and the Number and Size of Cells

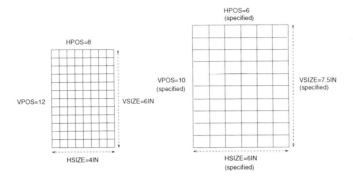

Table 5.1 on page 64 summarizes the interaction of the HSIZE=, VSIZE=, HPOS=, and VPOS= graphics options.

Table 5.1 Interaction of Graphics Options Affecting Cells

Options Specified	Options Not Specified	Result
HSIZE= and VSIZE=	HPOS= and VPOS= (or specify HPOS=0 and VPOS=0)	changes the external dimensions of the graphics output area and recalculates the number of rows and columns in order to retain cell size and proportions.
HPOS= and VPOS=	HSIZE= and VSIZE=	keeps the external dimensions but changes the cell size according to the number of rows and columns.
HSIZE=, HPOS=, VSIZE=, and VPOS=		changes the dimensions of the graphics output area, the number of rows and columns, and recalculates the cell size.

Other Units

By default, most graphic elements are drawn using units of CELLS to determine their size. For example, the default character height for the TITLE1 definition is two cells; for all other text the default height is one cell.

Changing the cell size to control the size of one element, such as text, can distort other parts of your graph. Instead, you might want to change the type of units that SAS/GRAPH uses to control the size of the graphic elements. In addition to CELLS you can use the following units:

- inches (IN)
- centimeters (CM)
- points (PT)
- percent (PCT)

The percent unit specification is often a good choice because it changes the height of the graphic elements in proportion to the size of the graphics output area.

You can specify the unit for individual graphic elements, or you can use the GUNIT= graphics option to set the units for most graphic element heights.

Maintaining the Quality of Your Image Across Devices

When you want to write a program that produces the same graphics output on two different devices, you can use features in SAS/GRAPH to simplify the process.

Maintaining Proportions

You can use percent of the graphics output area (PCT) as the unit of measure when specifying text size to make sure that text is proportional across devices. For example, a one-inch-high title might be appropriate on a standard piece of paper, but a title of this size uses almost all of the display area of a slide. To make units of percentage the default for size specifications, use the GUNIT= graphics option:

```
goptions gunit=pct;
```

You can also specify PCT anywhere you specify a size:

```
axis1 label=(height=3 pct 'Year');
```

See "GUNIT" on page 378 for a complete description of the GUNIT= graphics option.

Getting the Colors You Want

Since ODS styles are designed to provide optimal results for a variety of devices, you use the STYLE= option in the ODS statement to chose a style best suited for your device. For example, you might want to chose the ODS style *Journal* since it works well with black and white devices. You can also set a different style for each ODS output destination. For information on ODS styles and destinations see "Specifying Devices And Styles With Multiple Open Destinations" on page 51. You can compare colors and patterns for different devices and choose the device that has the fewest colors. A slide camera, for example, offers over 16 million colors from which to chose, but some graphics monitors display significantly fewer colors.

Previewing Your Output

You can preview the appearance of the output on a different device with the TARGETDEVICE= graphics option. For example, to see how the output looks on a color PostScript printer, specify as follows:

```
goptions targetdevice=pscolor;
```

How Graphic Elements are Placed in the Graphics Output Area

By default, SAS/GRAPH software positions certain graphics elements in predefined locations in the graphics output area. Figure 5.6 on page 66 shows the graphics output area and the areas within it that are used by the following graphic elements:

- Titles are placed in the title area at the top of the graphics output area.
- Footnotes are placed in the footnote area at the bottom of the graphics output area.
- The graph itself uses the *procedure output area*, which is the area left after the titles and footnotes have been drawn.

□ Legends use the procedure output area and can affect the amount of space available for the graph. By default, space is reserved for the legend below the axis area of a graph and above the footnote area. However, you can position the legend in the part of the procedure output area that is reserved for the graph. For details, see "LEGEND Statement" on page 225.

Note: Titles and footnotes can be positioned elsewhere on the graph as well, with different effects on space allocation. See "TITLE, FOOTNOTE, and NOTE Statements" on page 279 for details. For destinations other than the listing destination, some graphics elements, such as the title and footnote, can appear in the graphics output instead of the procedure output area. △

Figure 5.6 Default Locations for Graphic Elements in the Graphics Output Area

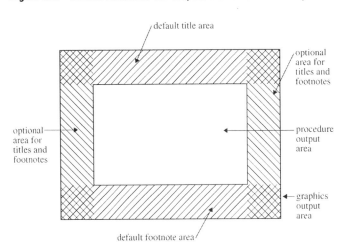

Note: If the titles, footnotes, and legend are very large, they can make the procedure output area too small for the graph. You can control the size of title and footnote text and of most legend elements with statement options. For details, see Chapter 14, "SAS/GRAPH Statements," on page 197 for a description of the appropriate statement. In addition, the section "GOPTIONS Statement" on page 220 lists the graphics options that control the size of various graphic elements. △

How Errors in Sizing Are Handled

Sometimes SAS/GRAPH cannot fit one or more graphic elements on the graph. This can happen if an element is too big for the available space (for example, the title is too long), or if you have too many elements to fit in a given space (for example, a bar chart has too many bars). In these cases, SAS/GRAPH does one of the following:

□ resizes the graphics element and issues a warning explaining what it did

□ issues an error message and does not attempt to produce the graph

For example, it adjusts the size of titles to make them fit but it does not drop bars in order to produce a readable bar chart. If you get unexpected results or no graph, check the SAS log for notes, warnings, and errors.

CHAPTER

6

Using Graphics Devices

Overview **67**
What Is a SAS/GRAPH Device? **68**
Commonly Used Devices **68**
Default Devices For ODS Destinations **69**
Viewing The List Of All Available Devices **70**
Deciding Which Device To Use **71**
Overriding the Default Device **72**
Device Categories And Modifying Default Output Attributes **72**
Using Universal Printer Shortcut Devices **75**
Using Scalable Vector Graphics Devices **77**
    What Is an SVG Document? **77**
    Why Create SVG Documents? **78**
    The SVG Devices and the Output That They Create **79**
    Example: Placing Images Behind SVG Documents **79**
    Example: Generating A Single SVG Document With Multiple Pages and Page Controls **81**
    Implementing Drill-Down Functionality With the SVG Devices **83**
    Web Server Content Type for SVG Documents **83**
    Browsers That Support SVG Documents **83**
    Controlling Graph Resolution With The SVG Devices **84**
    Controlling Graph Size With the SVG Devices **84**
    SAS System Options and SVG Output **84**
Viewing and Modifying Device Entries **85**
    Viewing the Contents of a Device Entry **85**
    Modifying Device Entry Parameters **85**
Creating a Custom Device **86**
Related Topics **86**

Overview

SAS/GRAPH procedures that produce graphics output require a device to create the output. The following topics discuss the role of devices in generating SAS/GRAPH output, provide directions for selecting and specifying them, and explain how you can change the settings of device parameters.

Note: SAS/GRAPH produces graphics using two very distinct systems. SAS/GRAPH can produce output using a device-based system or using a template-based system. Template-based graphics (ODS graphics) do not use SAS/GRAPH devices. See "Device-Based Graphics and Template-Based Graphics" on page 6. △

What Is a SAS/GRAPH Device?

A SAS/GRAPH device generates graphical output in a specified format. It might send output to a file on disk, such as a PNG file or a GIF file, or it might send output directly to a hardware device, such as a Postscript printer or a display. A SAS/GRAPH device consists of two parts: a device entry and device driver.

Device entry
: A device entry is a SAS catalog entry of type DEV. Every device that is shipped with SAS/GRAPH has a device entry in the SASHELP.DEVICES catalog. Device entries contain parameters that control the following:
  - the appearance of the output when styles are not in effect, such as dimensions and orientation, cell size, colors, and default SAS/GRAPH or device-resident fonts
  - where output is sent (when you send output to the LISTING destination and use a SAS/GRAPH device)
  - communications between the operating environment and the device
  - host commands that are issued before and after its driver produces output
  - the device driver that is used to generate graphics output

  See also "Viewing and Modifying Device Entries" on page 85.

Device driver
: A device driver is the executable module that produces device-specific commands that a device can understand. Every device entry specifies the name of the executable module (device driver) that is to be used to generate output. The device driver uses the parameters specified in the device entry or the current style to tell it exactly how to do so.

When you specify a device, you are specifying the name of a device entry. SAS/GRAPH uses that device entry to determine which device driver to use in order to generate final output. However, most users do not ever need to deal directly with device drivers, so for simplicity, this document simply refers to "devices".

Commonly Used Devices

The following table lists some of the more commonly used SAS/GRAPH devices and describes the output they produce.

Table 6.1 SAS/GRAPH Devices and the Output They Generate

Device	External Files
ACTIVEX	This device is used with the ODS HTML and ODS RTF destinations. It generates an HTML or RTF file that contains XML code that is consumed by the ActiveX control. When the HTML or RTF file is viewed in a browser, the SAS/GRAPH output is displayed as an interactive ActiveX control.
ACTXIMG	A PNG file that contains a static image of the graph that is generated with the ACTIVEX device.
BMP	A BMP file that contains the graph

Device	External Files
CGM	A CGM file that contains the graph.
CGMOF97L	A CGM file suitable for inserting into Microsoft Word or PowerPoint presentations.
EMF	An EMF file that contains the graph.
GIF	A GIF file that contains the graph.
JAVA	This device is used with the ODS HTML destination. It generates a JavaScript that ODS includes in the HTML file. When the HTML file is viewed in a browser, the SAS/GRAPH output is displayed as an interactive Java applet.
IBMPCGX	Display device. This device is available on z/OS hosts only.
JAVAIMG	A PNG file that contains a static graph that is generated with the JAVA device.
JPEG	A JPG file that contains the graph.
PCL5	A PCL file that contains the graph.
PDF	A PDF file that contains one or more graphs and tables.
PNG	A PNG file that contains the graph.
PSCOLOR	A PostScript file that contains one or more graphs.
PSL	A PostScript file that contains the graph in gray scale.
SASEMF	An EMF file that contains the graph. This device is the default device for the ODS RTF destination.
SVG	An SVG file that contains the graph.
TIFFP	A TIFF file that contains the graph in color.
WIN	Display device. This device is available on Windows hosts only.
XCOLOR	Display device. This device is available on UNIX hosts only.

Note: Chapter 16, "Introducing SAS/GRAPH Output for the Web," on page 439 describes any requirements or limitations associated with using the ActiveX, Java, and SVG devices. △

Default Devices For ODS Destinations

Each ODS destination has a default device. Table 6.2 on page 70 summarizes this information for the most commonly used destinations. These default devices are used to generate output for each open destination unless you override the default device as described in "Overriding the Default Device" on page 72.

Table 6.2 Default Devices and Styles for Commonly Used ODS Destinations

ODS Destination	Default Device	Default Style	Default Output	Recommended Devices
LISTING	WIN (Windows) XCOLOR (UNIX) IBMPCGX (z/OS)	Listing	Graphics output is displayed in the GRAPH window[1]	All devices[2] except JAVA and ACTIVEX
HTML	PNG	Default (Styles.Default)	HTML and PNG file	PNG GIF JPEG JAVA JAVAIMG ACTIVEX ACTXIMG SVG JAVAMETA GIFANIM
RTF	SASEMF	Rtf	RTF file (with embedded metafile)	SASEMF PNG JPEG JAVAIMG ACTIVEX ACTXIMG
PDF	SASPRTC	Printer	PDF file	SASPRTC (color) SASPRTG (gray scale) SASPRTM (monochrome)
PRINTER	SASPRTC	Printer	Controlled by the PRINTERPATH= system option (and by the SYSPRINT= system option on Windows)[3]	SASPRTC (color) SASPRTG (gray scale) SASPRTM (monochrome)

1. The default devices for the LISTING destination do not write image files to disk.
2. JAVAMETA is supported for the LISTING destination, but its output requires processing with the Metaview applet.
3. In Windows, if the PRINTERPATH= option is not specified, then SAS uses the setting of the SYSPRINT= system option. If neither the SYSPRINT= nor the PRINTERPATH= option has been set, then SAS uses the default Windows printer.

Viewing The List Of All Available Devices

You can view the complete list of devices that are available in any of the following ways:

- Use the SAS Explorer window to display the contents of the default device catalog, SASHELP.DEVICES, or any other device catalog.
- Use the GDEVICE procedure to open the GDEVICE DIRECTORY window, which lists all of the devices in the current catalog. By default, the current catalog is SASHELP.DEVICES. To specify a catalog, include the CATALOG= option, as shown in the following statement:

  ```
  proc gdevice catalog=sashelp.devices;
  run;
  ```

  If you do not specify a catalog, and you have defined a libref named GDEVICE0, then the GDEVICE procedure looks in the GDEVICE0 catalog first. See "Using the GDEVICE Windows" on page 1136 for details.

- Use GDEVICE procedure statements to write the list of device drivers to the Output window. For example:

  ```
  proc gdevice catalog=sashelp.devices nofs browse;
      list;
  run;
  quit;
  ```

  The NOFS option in the PROC GDEVICE statement causes the procedure not to use the GDEVICE window.

  If you want to write the list of devices to an external file you can do either of the following actions:

  - save the contents of the Output window.
  - use the PRINTTO procedure to redirect the GDEVICE procedure output to an external file. See *Base SAS Procedures Guide* for a description of the PRINTTO procedure.

Deciding Which Device To Use

The default device for each ODS destination generates optimal results for that destination. It is recommended that you use the default device whenever possible. If you do not specify a device, then SAS/GRAPH automatically uses the default device listed in "Default Devices For ODS Destinations" on page 69 for each open destination.

Note: If you are working with multiple open destinations, see "Specifying Devices And Styles With Multiple Open Destinations" on page 51. △

If you need to specify a different device, you should specify one of the recommended devices in the table in "Default Devices For ODS Destinations" on page 69. If you specify a device that cannot be used with an open destination, SAS/GRAPH switches to a device that produces similar results as the device that you specified.

The SAS/GRAPH device that you specify should be appropriate for your specific output device. For example, if you are using a color PostScript printer and you select a device for a black and white PostScript printer, your graph will not print in color.

When you are sending output to the HTML destination, there are several devices that you can specify. See "Selecting a Type of Web Presentation" on page 447 for information and recommendations on which device to use.

Note: High resolution devices such as PNG300 and JPEG300 are not appropriate choices for the HTML destination. Web browsers cannot display images in high resolution, so high resolution images appear very large. These drivers are appropriate for high resolution output that can be inserted into other software applications. △

Overriding the Default Device

You can override the default device in a SAS session in the following ways:

- Specify the name of a device entry with the DEVICE= option in a GOPTIONS statement. For example:

    ```
    goptions device=gif;
    ```

    For details, see "GOPTIONS Statement" on page 220.

- Specify the name of a device entry with the DEVICE= option in an OPTIONS statement. For details, see "DEVICE= System Option" in *SAS Language Reference: Dictionary*.

- Enter **OPTIONS** on the SAS command line, or select **Tools ▶ Options ▶ System** to open the SAS System Options window. Expand **Graphics**, and select **Driver settings**. Right-click on **Device**, select **Modify value**, and specify the name of the graphics device that you want to use.

- Enter the device name in the DEVICE prompt window. The DEVICE prompt window opens automatically if you submit a SAS/GRAPH program that produces graphics output, no device has been specified, and you are running outside of the SAS windowing system environment.

If you specify a device in more than one way, the last specification that SAS/GRAPH encounters is the one that it uses. The device specification stays in effect until you specify another device, submit the graphics option RESET=GOPTIONS or RESET=ALL, or end your SAS session.

If you use the same device for most or all of your SAS/GRAPH programs, you can put the GOPTIONS DEVICE= statement in an AUTOEXEC file. See the SAS companion for your operating environment for details on setting up an AUTOEXEC file.

You can also specify a device for previewing or printing your output with the TARGETDEVICE= graphics option. For details, see "Printing Your Graph" on page 110.

If you submit a SAS procedure without specifying a device and your display device does not support the GRAPH window or you are running outside the SAS windowing system, then SAS/GRAPH prompts you for a device.

Device Categories And Modifying Default Output Attributes

There are four general categories of devices that are distributed with SAS/GRAPH. The type of device determines how you control certain aspects of your output.

Note: Chapter 10, "Controlling The Appearance of Your Graphs," on page 133 describes the recommended methods for controlling the attributes of your SAS/GRAPH output. Modifying device parameters should be attempted only in unusual circumstances when modifying parameters and options in the GOPTIONS statement is not sufficient. If you need to modify a device entry, consider contacting SAS Technical Support for assistance first. △

Native SAS/GRAPH devices
  produce output in the native language of the device. For example, TIFFP, PS300, SASEMF, JPEG, CGMC, and GIF are native SAS/GRAPH devices. With native SAS/GRAPH devices, you can specify default attributes for your output by customizing the device entry (the DEV catalog entry). For example, by editing the DEV catalog entry for the device, you can change the default size and resolution of your output and the default colors and fonts that are used when styles are turned

off. Native SAS/GRAPH devices do not set and or use the SYSPRINT= or PRINTERPATH= system options.

Java and ActiveX devices
: produce output using different technologies than the native SAS/GRAPH devices. These devices are the JAVA, JAVAIMG, ACTIVEX, and ACTXIMG devices. These devices do not use information specified in the device entry.

Universal Printer shortcut devices
: use the Universal Printing system to generate output. Universal Printing is a printing system that provides printing capabilities to SAS applications and procedures on all the operating environments that are supported by SAS. It is part of Base SAS. For information on universal printing, see "Printing With SAS" in *SAS Language Reference: Concepts*.

    Universal Printer shortcut devices can generate output in the following formats: PDF, PostScript, PCL, PNG, GIF, and SVG. For example, PNG and SVG are Universal Printer shortcut devices. Any device whose name begins with the letter U, such as UGIF or UPSL, is also a Universal Printer shortcut device. The description of a Universal Printer shortcut device generally says "Universal Printer" when you view the contents of the SASHELP.DEVICES catalog (see "Viewing The List Of All Available Devices" on page 70). The list of all Universal Printer shortcut devices is shown in Table 6.4 on page 76.

    Universal Printer shortcut devices are designed to emulate a native SAS/GRAPH device, which means that these devices behave as much as possible like native SAS/GRAPH devices. For example, these devices set the value of PRINTERPATH= so that you need only specify the device name with the GOPTIONS statement. However, for these devices there are some attributes of your output, such as default resolution, that cannot be changed by modifying the DEV catalog entry. See "Using Universal Printer Shortcut Devices" on page 75 for more information.

Interface devices
: are devices that, in some operating environments, use the facilities of the operating environment, and, in other operating environments, use Universal Printing to generate output. There are three subcategories of interface devices: printer, display, and metafile.

    The printer interface devices are the SASPRTC, SASPRTG, and SASPRTM devices (and the WINPRT* devices on Windows systems). In Windows operating environments, if the PRINTERPATH= system option has not been set, these devices use the setting of the SYSPRINT= system option to determine the default output device and the Windows Print Manager to control the generation of output. In Windows operating environments, the Universal Printing System is used if the PRINTERPATH= system option is specified or if the UPRINT system option has been specified at invocation. Otherwise, they use the setting of the PRINTERPATH= system option to determine the default output device and the Universal Printing system to control the generation of output.

Table 6.3 Device Categories, GOPTIONS, and DEV Entries

Device Category		Examples	Honor GOPTIONS specifications?	Honor Device (DEV) entry specifications?
Native SAS/GRAPH devices		GIF TIFFP JPEG CGM BMP[1] SASBMP[2] EMF, WMF[1] SASEMF, SASWMF[2] JAVAMETA ZPNG IBMPCGX	Yes[4]	Yes
Java and ActiveX devices		JAVA ACTIVEX JAVAIMG ACTXIMG	Yes, except as noted in the documentation for specific graphics options. Also, resolution is controlled by the operating environment.	no
Shortcut devices		PNG UGIF SVG	Yes[3], except for resolution[5]	Yes, except for size, resolution, and fonts
Interface devices	Printer[6]	SASPRTC SASPRTG SASPRTM	Yes, except for resolution[7]	Yes, except for size, resolution, and fonts
	Display	WIN XCOLOR	Yes, except for resolution[9]	Yes, except for size[8], resolution[9], and fonts
	Metafile	BMP[1] EMF, WMF[1]	Yes	Yes, except for resolution[9] and fonts

1 On Windows, BMP, EMF, and WMF are interface metafile devices. In all other operating environments, BMP, EMF, and WMF are native SAS/GRAPH devices.
2 In operating environments other than Windows, SASEMF, SASWMF, and SASBMP are copies of EMF, WMF, and BMP, respectively.
3 With SVG devices, the XMAX= and YMAX= graphics options set the size of the page, and the HSIZE= and VSIZE= graphics options set the size of the SVG output. With other devices,

all four options set the size of the graphics output, and if all four are specified, the smaller specifications are used.

4 Some native devices have a set resolution, and others have a fixed set of supported resolutions that you can specify.

5 Shortcut devices use Universal Printers. Universal Printers have a fixed set of supported resolutions that can be selected through the Print Setup dialog box or through the PRINTDEF procedure.

6 The WINPRT* devices are identical to the SASPRT* devices. They differ in name only.

7 The interface printer devices use a mix of host printing facilities and Universal Printing, depending on the operating environment. On Windows systems, use the Windows Print Manager to change the default resolution and size. On other systems, resolution and size are set through the Print Setup dialog box or through the PRINTDEF procedure.

8 The device is queried. The size is constrained by the window.

9 Display resolution is set in the display properties for the operating environment. The device is queried, and the resolution is set according to the value returned.

See also "Viewing and Modifying Device Entries" on page 85.

Using Universal Printer Shortcut Devices

Universal Printer shortcut devices enable you to generate SAS/GRAPH output using the Universal Printing system without specifying ODS statements or an OPTIONS PRINTERPATH= statement. The shortcut devices were created primarily for use with the LISTING and HTML destinations. They perform two functions:

- set the PRINTERPATH= system option. These options determine which Universal Printer is used to generate your final output. See "Using Universal Printer Shortcut Devices" on page 75.

- convert SAS/GRAPH GRSEG output into instructions understood by Universal Printers.

Using a Universal Printer shortcut device requires that there is a Universal Printer with the same name in the SAS registry. Universal printers have already been defined for all of the shortcut devices that are shipped with SAS. However, if you create your own device by copying one of the Universal Printer shortcut device entries, then you must make sure that you define a Universal Printer with the same name as your new device entry. For information on creating a new SAS/GRAPH device, see "Creating a Custom Device" on page 86. For information on defining a new Universal Printer, see "Define a New Printer" in the Printing With SAS section of *SAS Language Reference: Concepts*.

An example of the differences in specifying a shortcut device and in specifying a Universal Printer directly (without going through the shortcut device) is shown in Table 6.4 on page 76.

Table 6.4 Differences In Using Shortcut Devices And Universal Printers

Using a shortcut device	Using a Universal Printer directly

`goptions device=PNG;`
`/* procedure step */`
	The following two sets of code are equivalent.
	
	`ods printer printer=PNG;` `options printerpath=PNG;`
	`  /* procedure step */` `ods printer;`
	`ods printer close;` `  /* procedure step */`
	                                                 `ods printer close;`

The device is set to PNG by the GOPTIONS statement. The default output filename is controlled by the procedure, for example, sasgraph.png.	The device is set to SASPRTC because SASPRTC is the default device for the PRINTER destination. The Universal Printer is set to PNG by the PRINTER= or PRINTERPATH= option. The default output filename is controlled by ODS, for example, sasprt.png.

Table 6.5 on page 76 lists all of the Universal Printer shortcut devices that are provided by SAS.

Table 6.5 Universal Printer Shortcut Devices

Name	Description
PCL5	PCL5 Universal Printer
PCL5C	PCL5c Universal Printer
PCL5E	PCL5e Universal Printer
PDF	PDF Version 1.3 — color
PDFA	Archive PDF - ISO-19005-1/b
PDFC	PDF Version 1.3 — color
PNG	PNG Universal Printer
PNG300	PNG Universal Printer-300 dpi
PNGT	PNG Universal Printer with Transparency
PSCOLOR	PostScript Level 1 (Color)
PSL	PostScript Level 1 (Gray Scale)
PSLEPSF	PostScript EPS (Gray Scale)
PSLEPSFC	PostScript EPS (Color)
SVG	SVG Universal Printer
SVGT	SVG Transparency Universal Printer
SVGVIEW	SVG Printer w/ Control Buttons
SVGZ	SVG Compressed Universal Printer
UEPS	PostScript EPS (Gray Scale)
UEPSC	PostScript EPS (Color)
UGIF	GIF Universal Printer
UPCL5	PCL5c Universal Printer
UPCL5C	PCL5c Universal Printer

Name	Description
UPCL5E	PCL5e Universal Printer
UPDF	PDF Version 1.3 – color
UPNG	PNG Universal Printer
UPNGT	PNG Universal Printer with Transparency
UPSL	PostScript Level 1 (Gray Scale)
UPSLC	PostScript Level 1 (Color)

Using Scalable Vector Graphics Devices

Scalable Vector Graphics (SVG) is an XML language for describing two-dimensional vector graphics. SAS creates SVG documents based on the World Wide Web Consortium (W3C) recommendation for SVG documents. SAS SVG files are created using the UNICODE standard encoding.

Note: Animation is not supported in SAS 9.2. △

SAS can create SVG documents by using either the SVG Universal Printers or SAS/GRAPH SVG devices. There are four SVG devices: SVG, SVGT, SVGVIEW, and SVGZ. These devices are Universal Printer shortcut devices and are, therefore, intended mainly for use with the LISTING and HTML destinations. With the PRINTER destination, it is recommended that you use the SVG Universal Printers directly (see Table 6.4 on page 76).

The information provided here is limited to creating SVG documents using SAS/GRAPH devices in the LISTING and HTML destinations. For information about creating SVG documents in the PRINTER destination using the SVG Universal Printers, see "Creating Scalable Vector Graphics Using Universal Printing" in *SAS Language Reference: Concepts*.

For detailed information about the SVG standard, see the W3C documentation at http://www.w3.org/TR/SVG.

What Is an SVG Document?

An SVG document produced by SAS/GRAPH is an XML file that contains an `<svg>` element.

SVG document fragment
: any number of SVG graphic or container elements enclosed an `<svg>` element. Typical SVG graphics elements include circle, line, text, image, and many others. These elements draw the graphics that comprise the SVG document.

SVG document
: an SVG document fragment that can stand by itself. The SVG devices produce stand-alone SVG documents. When you send output to the HTML destination, the SVG document is embedded in an HTML document using the `<embed>` tag.

For example, the following code produces an SVG file named **europepop.svg** and an HTML file named **europe.htm**:

```
goptions reset=all device=svg;
ods listing close;
ods html file="europe.htm";
```

```
title "Population in Europe";
proc gmap map=maps.europe(where=(id ne 405 and id ne 845))
             data=sashelp.demographics(where=(cont=93)) all;
    id id;
    choro pop / name="europePop";
run;
quit;
ods html close;
ods listing;
```

You can view the SVG coding by opening the SVG document, **europepop.svg**, in a text editor. When you view the SVG document in an SVG-enabled browser (see "Browsers That Support SVG Documents" on page 83), the browser renders the image.

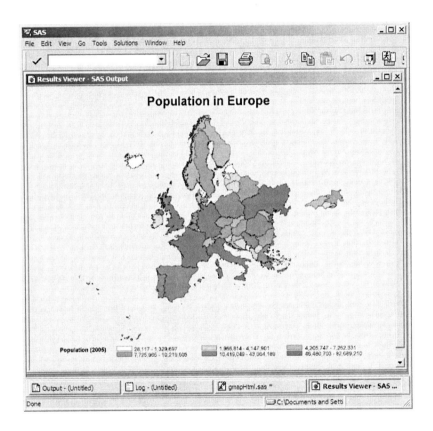

Why Create SVG Documents?

Because SVG graphics are vector graphics, they can be resized without losing quality. A single SVG document can be scaled to any size or transformed to any resolution without compromising the clarity of the document. Bitmap images such as PNG and GIF lose quality any time they are resized.

Also, if you need to display the same graphic at multiple sizes or resolutions, you would need multiple bitmap images, but only one SVG document. SVG documents display clearly at any size in any viewer or browser that supports SVG. The user can zoom in to view details in a complicated SVG graphic.

An SVG document might also be smaller in file size than the same graphic created by a bitmap (image) device such as GIF or PNG.

SVG documents are ideal for producing documents to display on a computer monitor, PDA, or cell phone; or documents to be printed.

The SVG Devices and the Output That They Create

There are four SVG devices:

SVG
> produces SVG 1.1 documents. When used in the HTML destination, if your procedure produces multiple graphs, the SVG device produces separate SVG documents for each graph. When used in the LISTING destination, the SVG device produces one SVG file, and the pages are in a continuous layout.

SVGT
> produces SVG 1.1 documents that are transparent (no background). These documents are useful when you want to overlay several graphs on top of each other and you want all of the graphs to be visible. The SVGT device is intended for use when a procedure produces multiple graphs and is best used in conjunction with the ODS PRINTER destination. See "Creating Overlaid Transparent SVG Documents" in *SAS Language Reference: Concepts* for more information.

SVGZ
> produces compressed SVG 1.1 documents, which are useful when file size is an issue. However, some browsers do not support compressed SVG documents, and you cannot view these files in a text editor. (See also "Browsers That Support SVG Documents" on page 83.)

SVGVIEW
> produces SVG1.1 documents with navigational controls when the SVG file contains multiple pages. This device is primarily for use in the LISTING destination with procedures that produce multiple graphs. The navigational controls enable you to page through the graphs. See "Example: Generating A Single SVG Document With Multiple Pages and Page Controls" on page 81. When used in the HTML destination, the SVGVIEW device produces separate SVG documents for each graph, just like the SVG device.

Example: Placing Images Behind SVG Documents

You can use the IBACK= graphics option in the GOPTIONS statement to specify the graphics file that you want to be placed behind the SVG graphic. SAS/GRAPH creates a PNG file from the image file that you specify. This PNG file is used as the background image and is referenced in the SVG with an `<image>` tag. The `<image>` tag specifies a relative (not absolute) pathname to the PNG file. If the SVG file is moved, the PNG file must also be moved to the same location. If many images are referenced in an SVG file, it is recommended that you create a new directory and store your SVG file and any images it references in the directory. Then, the entire directory can be moved as a package.

```
/* Reset existing options, specify the SVG device, and   */
/* set the size of the SVG document.  Specify the        */
/* background image with the IBACK= option. Replace      */
/* external-image-file with the name of an image that    */
/* resides on your system.                               */

goptions reset=all device=svg hsize=4.8in vsize=3.2in
```

```
                         imagestyle=fit iback="external-image-file";

/* Close the LISTING destination to conserve resources. */
/* Open the HTML destination and specify */
/* the name of the HTML output file.                    */

ods listing close;
ods html file="carType.htm";

/* Specify the title for the graphic file and */
/* define response axis characteristics.       */
title h=2 "Types of Vehicles Produced Worldwide";
axis1 label=none major=none minor=none;

/* Generate the bar chart.  The NAME= option */
/* specifies the name of the SVG file.                  */
proc gchart data=sashelp.cars;
   vbar type / raxis=axis1 outside=freq
               noframe name="carType";
run;
quit;

/* Close the HTML destination and    */
/* reopen the LISTING destination. */
ods html close;
ods listing;
```

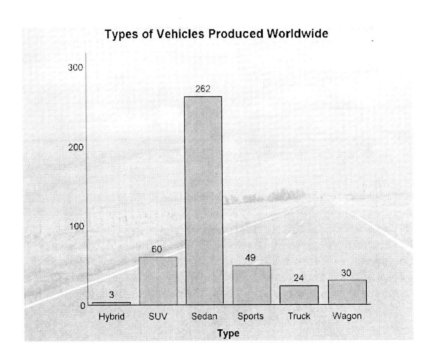

For additional information, see "Displaying an Image in a Graph Background" on page 182.

Example: Generating A Single SVG Document With Multiple Pages and Page Controls

The SVGVIEW device is designed to be used when in the LISTING destination. It is useful when a single procedure produces multiple graphs, such as with BY-group processing. When used in the LISTING destination, the SVGVIEW device creates a single SVG document with multiple pages. Each graph produced by the procedure is on a different page. The SVG document, by default, has control buttons that enable you to navigate forward and backward through the graphs as well as display an index page that shows a thumbnail image of each page in the document.

For example, the following display shows the initial graph that is produced by the program in Example Code 6.1 on page 82. The program produces six graphs. You can page through them clicking on using the **Prev** and **Next** buttons.

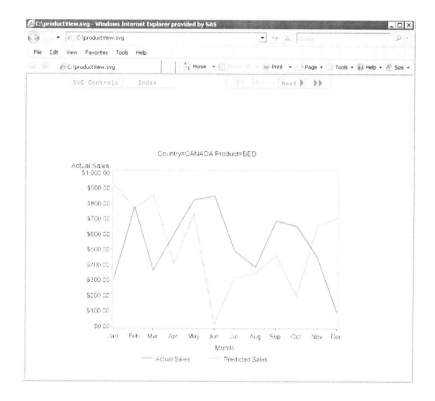

The **Index** button displays a page of thumbnail images. There is one thumbnail for each page in the SVG document.

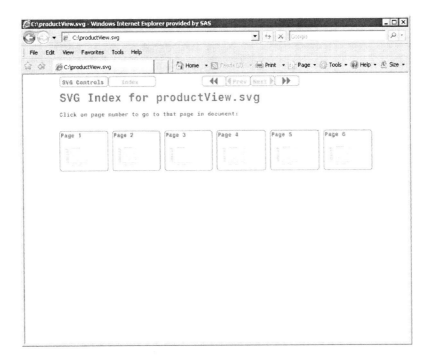

The program that generates this SVG document is as follows:

Example Code 6.1 Program Code: Using SVGVIEW Device With BY-Group Processing

```
/* Subset the data set SASHELP.PRDSALE. */
/* Output the subset to WORK.PRODSUB.   */

data prodsub;
  set sashelp.prdsale;
  where year=1994 and
     (country = "U.S.A." or country = "CANADA")
      and region="EAST" and division="CONSUMER" and
     (product in ("SOFA", "TABLE", "BED"));
run;

/* Sort WORK.PRODSUB. */

proc sort data=prodsub;
   by country product;
run;

/* Define a fileref for the SVG document. */
/* Use the GSFNAME= option to send the    */
/* output of the LISTING destination to   */
/* that fileref.                          */

filename mysvg "productView.svg";
goptions reset=all device=svgview
         gsfmode=replace gsfname=mysvg;

/* Join the data points and change the */
/* line style for the predicted sales  */
/* to a dashed line.                   */
```

```
symbol1 interpol=join line=1 color=_style_;
symbol2 interpol=join line=2 color=_style_;
legend1 label=none;

/* Generate a graph for each unique    */
/* combination of country and product. */

proc gplot data=work.prodsub;
   by country product;
   plot actual*month predict*month /
        overlay legend=legend1;
run;
quit;
```

When used in the HTML destination, the SVGVIEW device produces separate SVG documents for each graph, just like the SVG device.

For additional information, see "Multi-Page SVG Documents in a Single File" and "Multi-Page SVG Documents in a Single File" in *SAS Language Reference: Concepts*.

Implementing Drill-Down Functionality With the SVG Devices

You can implement drill-down links in SVG documents that are generated in the HTML and LISTING destinations. In both cases, you use the HTML= option or the HTML_LEGEND= option (or both options) to specify variables in your input data that define the drill-down URLs. See "Adding Links with the HTML= and HTML_LEGEND= Options" on page 601 for information on implementing drill-down links, including defining link variables.

Implementing drill-down links in SVG documents that are generated in the LISTING destination has an additional requirement: you must specify the IMAGEMAP= option in the PROC statement. This option makes the image map generated by the procedure available to the SVG device. For example:

```
proc gchart data=sashelp.prdsale imagemap=myimgmap;
```

Web Server Content Type for SVG Documents

If the mime content type setting for your Web server does not have the correct setting for SVG documents, your Web browser might render SVG documents as text files or SVG documents might be unreadable.

To ensure that SVG documents are rendered correctly, you can configure your Web server to use this mime content type:

```
image/svg+xml
```

Browsers That Support SVG Documents

In order to view SVG documents, you need a viewer or browser that supports Scalable Vector Graphics. Some browsers, such as Mozilla Firefox, have built-in support for SVG documents. Other browsers, such as Microsoft Internet Explorer, require an SVG plug-in to view SVG documents. One such plug-in is available from Adobe Systems, Inc.

The following table lists some browsers and viewers that support SVG documents. See "Browser Support for Viewing SVG Documents" in *SAS Language Reference: Concepts* for additional information.

Table 6.6 SVG Browser Support

Browser or Viewer	Company
Adobe SVG Viewer 3[1]	Adobe Systems, Inc.
Batik SVG Toolkit	Apache Software Foundation
eSVG Viewer and IDE	eSVG Viewer for PC, PDA, Mobile
GPAC Project	GPAC
Mozilla Firefox[2]	Mozilla Foundation
Opera	Opera Software
TinyLine	TinyLine

1 Adobe SVG Viewer 3 works in Internet Explorer 7. Check **www.adobe.com** for information on support by Adobe Systems, Inc. for the SVG viewer.
2 Mozilla Firefox does not support compressed SVG documents or font embedding. To avoid font mapping problems, specify the NOFONTEMBEDDING system option. Zooming and panning features are not currently implemented. Also, if you select **View ▶ Page Style ▶ No Style**, all graphs appear as a black rectangle.

Controlling Graph Resolution With The SVG Devices

The default resolution for the SVG devices is 96 dpi. Because the SVG devices are Universal Printer shortcut devices, you cannot change the resolution using options in the GOPTIONS statement. To change the resolution for these devices, you must use either the Print Setup dialog box or the PRTDEF procedure to change the resolution for the Universal Printer. Universal Printers have a fixed set of supported resolutions.

To use the Print Setup dialog box, select **File ▶ Print Setup**, and select the printer for which you want to change the resolution. Select `Properties` and click on the `Advanced` tab. Select the resolution that you want to use from the list.

For information on using the PRTDEF procedure, see "The PRTDEF Procedure" in *Base SAS Procedures Guide*.

Controlling Graph Size With the SVG Devices

The default graph size for the SVG output is 600 x 800 pixels. You can change the size of your graph with the HSIZE= and VSIZE= graphics options. You can change the paper size by specifying the XMAX= and YMAX= or the XPIXELS= and YPIXELS= graphics options. Specifying a value for the XMAX=, YMAX=, XPIXELS=, or YPIXELS= graphics options changes the setting of the PAPERSIZE= system option. See Chapter 15, "Graphics Options and Device Parameters Dictionary," on page 327 and "SAS System Options and SVG Output" on page 84.

SAS System Options and SVG Output

Because the SVG devices are Universal Printer shortcut devices, there are several SAS system options that affect the way the SVG devices generate output. These options include SVGHEIGHT=, SVGWIDTH=, SVGVIEWBOX=, SVGCONTROLBUTTONS, and PAPERSIZE=, among others. These options and their interactions are described in several topics in *SAS Language Reference: Concepts* under "Creating Scalable Vector Graphics Using Universal Printing" . Before reviewing the topics that deal with the various system options, you should review the topic "SVG Terminology" .

Topics dealing primarily with SAS system options are as follows:
- "SAS System Options That Effect Stand-alone SVG Documents"
- "Scaling an SVG Document to the Viewport"
- "Setting the ViewBox"
- "Preserving the Aspect Ratio"
- "Interaction between SAS SVG System Options and the SVG Tag Attributes"

Viewing and Modifying Device Entries

As described in "What Is a SAS/GRAPH Device?" on page 68, device entries contain parameters that control much of the default behavior and default output attributes of a device. However, even though a device entry exists for every device, the information in it is not always used. See "Device Categories And Modifying Default Output Attributes" on page 72 for more information.

Viewing the Contents of a Device Entry

SAS/GRAPH provides device entries for your operating environment in the SASHELP.DEVICES catalog. If your site has created custom device entries, they might also be stored in SASHELP.DEVICES, although custom devices are typically stored in the catalog GDEVICE0.DEVICES. For more information about custom device entries, see "Device Catalogs" on page 1126 or ask your on-site SAS support personnel.

Use any of the following methods to view the contents of a device entry:

- Use the SAS Explorer window to display the contents of the DEVICES catalog in the SASHELP library. Double-click a device entry to display the contents of the device entry in the Output window.
- Run the GDEVICE procedure in program mode. For example, the following statements list in the Output window the contents of the PSCOLOR device entry:

  ```
  proc gdevice catalog=sashelp.devices nofs browse;
     list pscolor;
  run;
  quit;
  ```

- Run the GDEVICE procedure in windowing mode. The following statements open the GDEVICE directory window that lists the available devices:

  ```
  proc gdevice catalog=sashelp.devices;
  run;
  ```

  From the GDEVICE Directory window, select the device name to open the GDEVICE Detail window. From there you can move to the other GDEVICE windows for the entry, either by selecting windows from the **Tools** menu or entering commands on the command line. For details, see "Using the GDEVICE Windows" on page 1136.

Modifying Device Entry Parameters

Use the GDEVICE procedure to modify the properties of an existing device entry. See Chapter 38, "The GDEVICE Procedure," on page 1125.

The modifications made to a device entry are in effect for all SAS sessions.

The new values that you specify for device parameters must be within the device's capabilities. For example, devices are limited in the size of the output they can display.

Some output devices cannot display color. If you try to increase the size of the display past the device's capability or if you specify colors for a device that cannot display them, you will get unpredictable results. You cannot force a device to act as a device with different capabilities by choosing a different device driver

Note: The device driver that is associated with a device entry is shown in the `Module` field in the device entry. It is recommended that you do not change the device driver associated with a device entry. Please contact SAS Technical Support before changing the device driver associated with a device entry. △

Note: If you run SAS/GRAPH software in a multi-user environment, you should not change the device entries in the SASHELP.DEVICES catalog unless you are the system administrator or other on-site SAS support personnel. △

If you need to change a device entry in SASHELP.DEVICES, copy it into a personal catalog named DEVICES, and then modify the copy. To use the new device, assign the libref GDEVICE0 to the library that contains the modified copy. See "Creating or Modifying Device Entries" on page 1142 for details.

Creating a Custom Device

You can use the GDEVICE procedure to create a custom device. For each new device, you need to create a new device entry. Device entries that you create or modify are typically stored in the catalog GDEVICE*n*.DEVICES.

If you want to create a custom device, it is recommended that you copy an existing device and modify it as needed. If you cannot find a device that is suitable for your purposes, contact SAS Technical Support.

See "Modifying Device Entry Parameters" on page 85 and Chapter 38, "The GDEVICE Procedure," on page 1125 for more information.

Related Topics

Other tasks related to devices are discussed in the following topics:

"Devices" on page xvii
  describes changes in device support for the current release.

Chapter 7, "SAS/GRAPH Output," on page 87
  provides general information about graphics output formats and the SAS/GRAPH output process, setting the size and resolution of your graphics output, previewing on one device how output will look on another device, sending output directly to a printer or other hardcopy device, and replaying output.

Chapter 16, "Introducing SAS/GRAPH Output for the Web," on page 439
  describes the options available for creating a Web presentation. Several devices can be used to create a Web presentation, including JAVA and ACTIVEX, which create interactive presentations.

Chapter 8, "Exporting Your Graphs to Microsoft Office Products," on page 113
  describes how to choose a device for output that you want to use in Microsoft Office products.

Chapter 38, "The GDEVICE Procedure," on page 1125
  describes how to create and modify devices.

- creating files in graphics formats that can be viewed with a Web browser with other applications (see "Graphics Output Files" on page 92).

CHAPTER 7

SAS/GRAPH Output

About SAS/GRAPH Output 88
    SAS/GRAPH Output Terminology 88
    Supported Graphics Formats 88
    Output Types 89
    About GRSEGs 89
    What You Can Do With SAS/GRAPH Output 90
Specifying the Graphics Output File Type for Your Graph 91
    About the Output Delivery System (ODS) 91
    About the Graphics Output Devices 91
        The Output that Each Device Generates 91
        Graphics Output Files 92
        About File Extensions 93
The SAS/GRAPH Output Process 93
    All Devices Except JAVA, JAVAIMG, ACTIVEX, and ACTXIMG 93
    JAVA or ACTIVEX Device 93
    JAVAIMG or ACTXIMG Device 94
Setting the Size of Your Graph 94
    Using the HSIZE= and VSIZE= Graphics Options to Set the Size of Your Graphics Area 94
    Using the XPIXELS= and YPIXELS= Graphics Options to Set the Size of Your Graph 95
Setting the Resolution of Your Graph 95
    Using the XPIXELS=, XMAX=, YPIXELS=, and YMAX= Graphics Options to Set the Resolution for Device-Based Graphics 96
    Using a Device Variant to Set the Size or Resolution of Your Graph 97
Controlling Where Your Output is Stored 97
    Specifying the Name and Location of Your ODS Output 97
    Specifying the Name and Location of Your Graphics Output File 98
    About Filename Indexing 99
    Specifying the Catalog Name and Entry Name for Your GRSEGs 100
        Using the Default Catalog and Entry Name 100
        Specifying a Name for Your GRSEG with the NAME= Option 101
        Specifying the Catalog and GRSEG Name with the GOUT= and NAME= Options 101
        Where GRSEGs are Stored When Multiple ODS Destinations are Used 102
    Summary of How Output Filenames and GRSEG Names are Handled 102
Replacing an Existing Graphics Output File Using the GSFMODE= Graphics Option 104
Storing Multiple Graphs in a Single Graphics Output File 104
    Using Graphics Options to Store Multiple Graphs in One Graphics Output File 105
    Using the GREPLAY Procedure to Store Multiple Graphs in One Graphics Output File 105
Replaying Your SAS/GRAPH Output 106
    Replaying Your Output Using the GREPLAY Procedure 106
    Replaying Output Using the DOCUMENT Procedure 107
        Creating Your ODS Document 107

>   *Replaying Your ODS Document* **108**
> *Previewing Output* **109**
> *Printing Your Graph* **110**
>   *Sending Your Graph Directly to a Printer* **110**
>   *Saving and Printing Your Graph* **110**
> *Exporting Your Output* **111**

About SAS/GRAPH Output

The result of most SAS/GRAPH procedures is the graphic display of data in the form of *graphics output*, which is distinct from *SAS output*. Whereas SAS output consists of text, graphics output consists of commands that tell a graphics device how to draw graphic elements. A *graphics element* is a visual element of graphics output—for example, a plot line, a bar, a footnote, the outline of a map area, or a border.

This chapter discusses how to display, print, store, and export SAS/GRAPH output after you have created it.

SAS/GRAPH Output Terminology

The following terms are used when describing SAS/GRAPH output:

Graphics output file
: A file that contains bitmapped or vector graphic information. See "Supported Graphics Formats" on page 88.

Image file
: A file that contains bitmapped graphic information. Examples include GIF, PNG, and JPEG files. Image files are a subset of graphics output files.

Document file
: A file output by the Output Delivery System (ODS) that contains an image or is used to view an image. Examples include HTML, PDF, RTF, SVG, and PostScript files.

Supported Graphics Formats

You can export your SAS/GRAPH output in many different graphics file formats. SAS/GRAPH supports the following image file formats:

BMP	Windows Bitmap
GIF	Graphics Interchange Format
JPEG	Joint Photographic Experts Group
PNG	Portable Network Graphics
TIFF	Tagged Image Format File

SAS/GRAPH supports the following vector file formats:

CGM	Computer Graphics Metafile
EMF	Microsoft Enhanced Metafile
EPS	Encapsulated PostScript
PCL	Printer Control Language
PDF	Portable Document Format

PS   PostScript

SVG   Scalable Vector Graphics

The vector-based formats
- are usually smaller than image files
- can be edited with third–party software (except for EPS)
- support system fonts
- support font embedding with the PDF, SVG, 1, and PostScript devices
- provide a clear image on high-resolution devices.

The type of graphics file format that you choose depends on how you are going to use the output. For example, you are planning to import the graph into other software applications, such as Microsoft Excel, Word or Power Point, you might prefer to create a CGM file. The vector-based files are usually smaller than image files, they support TrueType fonts, and except for EPS, they can be edited with third-party software. In addition, they use device-resident fonts and provide a clear image on high-resolution devices.

If you want to display the graph on a Web page, or import it into software that cannot accept vector graphics. You must create an image file such as PNG or GIF.

Most software applications that process graphics input can accept one or more of these file formats. Check the documentation for the hardware or software product to which you want to send the graph to determine what file formats it can use.

For a complete list of graphics file formats that are available with SAS/GRAPH in your operating environment, refer to the Device Help for SAS/GRAPH in the SAS Help facility.

Output Types

The SAS graphics procedures can generate the following types of output:
- a GRSEG (except for procedures GKPI, GTILE, and GAREABAR)
- a graphics output file that contains the graph (BMP, JPG, GIF, PNG, and so on)
- an HTML file that contains XML code that is consumed by the ActiveX control or Java applet

In addition, the SAS Output Delivery System (ODS) creates document files, which include the following types of output:
- an HTML file that displays a graph
- an RTF file that contains a graph
- a PCL file that contains a graph
- a PDF file that contains a graph
- a PostScript file that contains a graph
- an SVG file that contains one or more graphs

About GRSEGs

A GRSEG is a SAS catalog entry that contains graphics commands in a generic, device-independent format. There are few cases in which you would be concerned with the GRSEGs. One case for using the GRSEGs is when combining multiple graphs into a single graphics output file using the GREPLAY procedure (see "Using the GREPLAY Procedure to Store Multiple Graphs in One Graphics Output File" on page 105). Beyond this case, there are few reasons to use the GRSEGs. If you plan to use the GRSEGs, you must understand when they are generated and where they are stored.

GRSEGs are supported by the SAS/GRAPH procedures that use the graphics output devices with some exceptions. The procedures that are supported by only the JAVA, JAVAIMG, ACTIVEX, and ACTXIMG devices, such as GKPI, GTILE, and GAREABAR, do not support GRSEGs.

A procedure that generates a GRSEG produces output in two steps:

1 It creates a GRSEG in a SAS catalog.
2 It uses a graphics output device to translate the commands from the GRSEG to commands that a particular graphics device understands. This is called device-dependent output.

This method enables you to produce graphics output on several types of graphics output devices.

A GRSEG is stored in a catalog in the SAS temporary directory. The graphics instructions that are contained in the GRSEG are understood only by the SAS/GRAPH software. You cannot use third-party graphics applications to view the graphic in a GRSEG. The SAS/GRAPH software provides devices that enable you to output a GRSEG to standard graphics formats such as GIF, PNG, and PDF, which you can view using third-party applications.

SAS/GRAPH software always assigns a name and a description to each GRSEG so that you can identify it. By default, the names and descriptions are determined by the procedure. For example, a GRSEG produced by the GCHART procedure is assigned the name GCHART and a description such as PIE CHART OF MONTH.

By default, SAS/GRAPH appends each new GRSEG to the catalog. If you create more than one graph with a procedure during a SAS session and the GRSEGs are stored in the same catalog, SAS/GRAPH software appends a number to the end of the name of subsequent GRSEGs. This number makes the names unique within the catalog. For example, if you create three graphs with the GCHART procedure during the same SAS session, the GRSEGs are named GCHART, GCHART1, and GCHART2. SAS/GRAPH software uses this naming convention whether GRSEGs are being stored in a temporary or a permanent catalog.

You can supply a name and description when you create the graph by using the NAME= and DESCRIPTION= options. If you create more than one graph of the same name, the SAS/GRAPH software increments the specified name just as it does the default names.

What You Can Do With SAS/GRAPH Output

By default, SAS/GRAPH procedures that produce graphics output display the output on your computer screen using either the GRAPH window or the direct-display method. Using the SAS ODS and the graphics options, you can direct graphics output to a variety of other destinations. Specifically, you can do the following with your graphics output:

- send it directly to a graphics hard-copy device, such as a printer. For details, see "Printing Your Graph" on page 110.
- save it in a temporary or permanent SAS catalog for later replay. See "Replaying Your SAS/GRAPH Output" on page 106.
- export it to a graphics output file using different graphics file formats. For example, you can save SAS/GRAPH output in formats such as CGM or PostScript for use with other software applications. For details, see "Exporting Your Output" on page 111.

Regardless of the destination of a graph, a GRSEG is created for those SAS/GRAPH procedures that support GRSEGs. The GRSEG is stored in the WORK.GSEG catalog unless you specify a different catalog with the GOUT= procedure option. To generate only GRSEGs and suppress all other forms of graphics output, use the NODISPLAY graphics option. See "DISPLAY" on page 353.

After your graphics output is saved in a catalog, you can do the following with your graphics:

- transport them in catalogs from one operating environment to another. For details, see Appendix 4, "Transporting and Converting Graphics Output," on page 1659.
- convert them for use with a different version of SAS by converting the catalog containing the graphics output. For details, see "Converting Catalogs to a Different Version of SAS" on page 1662.
- export them to graphics output files using different graphics file formats. For details, see "Exporting Your Output" on page 111.

Specifying the Graphics Output File Type for Your Graph

About the Output Delivery System (ODS)

The SAS ODS sends your graph output to a default destination or a destination that you specify, such as your monitor, a printer, or a graphics output file. Each destination has a default style and graphics output device associated with it. You can use the STYLE= ODS option to specify a different style, and you can use the DEVICE= graphics option to specify a different device that is supported by the ODS destination that you are using.

See Chapter 16, "Introducing SAS/GRAPH Output for the Web," on page 439 for more information on using the ODS destinations, styles, and supported devices.

About the Graphics Output Devices

The Output that Each Device Generates

By default, the SAS/GRAPH ODS outputs to the LISTING destination, which displays your graph on your monitor and creates a GRSEG in the catalog. You can specify a graphics output device other than your monitor for the ODS LISTING destination, or you can specify a different ODS destination and device. For information on the ODS destinations and the devices that each supports, see Chapter 16, "Introducing SAS/GRAPH Output for the Web," on page 439.

The following table lists the common graphics output devices, and the default output that each generates.

Table 7.1 SAS/GRAPH Devices and the Output They Generate

Device	External Files
ACTIVEX	This device is used with the ODS HTML and ODS RTF destinations. It generates an HTML or RTF file that contains XML code that is consumed by the ActiveX control. When the HTML or RTF file is viewed in a browser, the SAS/GRAPH output is displayed as an interactive ActiveX control.
ACTXIMG	A PNG file that contains a static image of the graph that is generated with the ACTIVEX device.

Device	External Files
BMP	A BMP file that contains the graph
CGM	A CGM file that contains the graph.
CGMOF97L	A CGM file suitable for inserting into Microsoft Word or PowerPoint presentations.
EMF	An EMF file that contains the graph.
GIF	A GIF file that contains the graph.
JAVA	This device is used with the ODS HTML destination. It generates a JavaScript that ODS includes in the HTML file. When the HTML file is viewed in a browser, the SAS/GRAPH output is displayed as an interactive Java applet.
IBMPCGX	Display device. This device is available on z/OS hosts only.
JAVAIMG	A PNG file that contains a static graph that is generated with the JAVA device.
JPEG	A JPG file that contains the graph.
PCL5	A PCL file that contains the graph.
PDF	A PDF file that contains one or more graphs and tables.
PNG	A PNG file that contains the graph.
PSCOLOR	A PostScript file that contains one or more graphs.
PSL	A PostScript file that contains the graph in gray scale.
SASEMF	An EMF file that contains the graph. This device is the default device for the ODS RTF destination.
SVG	An SVG file that contains the graph.
TIFFP	A TIFF file that contains the graph in color.
WIN	Display device. This device is available on Windows hosts only.
XCOLOR	Display device. This device is available on UNIX hosts only.

Graphics Output Files

When you export SAS/GRAPH output, you run the output through a device that creates a graphics output file. A graphics output file is a file that contains vector or bitmap graphics commands. Typically, you select a device that produces the type of graphics file format that you want, such as PNG, CGM, PS or EPS, GIF, or TIFF. You can select a device that sends the output directly to a printer or other hard-copy device without creating a graphics output file. You can specify the exact name and location of each file or assign a default location to which all files are sent.

You can also use the ODS to generate SAS/GRAPH output as HTML that you can view with a Web browser. Details are discussed in Chapter 16, "Introducing SAS/GRAPH Output for the Web," on page 439.

Once you have created a graphics output file, you can do the following:

- print the file using host commands
- view the file with an appropriate viewer or browser
- edit the file with the appropriate editing software
- import the file into other software applications

Note: A graphics output file is different from a SAS/GRAPH GRSEG. A graphics output file is a file that is independent of SAS, and a GRSEG is a type of SAS catalog file. Consequently, you use host commands to manipulate a graphics output file independent of the SAS System, whereas you must use the SAS System to manipulate SAS GRSEGs. The GREPLAY procedure can be used to replay graph entries stored in catalogs and display them in the GRAPH window. △

About File Extensions

When you send SAS/GRAPH output to an aggregate file storage location, SAS/GRAPH generates the name of the graphics output file. This is done by taking the GRSEG name and adding the appropriate file extension. Most devices provide a default extension. If a device does not generate an extension, then SAS/GRAPH uses the default extension .gsf. To specify a different extension from the one SAS/GRAPH provides, use the EXTENSION= graphics option. (For details, see "EXTENSION" on page 357).

The SAS/GRAPH Output Process

All Devices Except JAVA, JAVAIMG, ACTIVEX, and ACTXIMG

The following diagram illustrates the output process for all of the SAS/GRAPH graphics output devices except JAVA, JAVAIMG, ACTIVEX, and ACTXIMG.

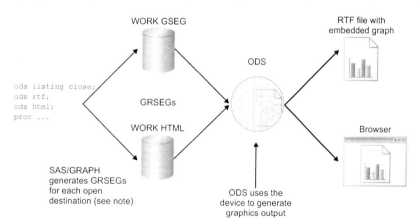

Note: The image size, color, and font information is obtained from the device entry and incorporated into the GRSEG. △

JAVA or ACTIVEX Device

The following diagram illustrates the output process for the JAVA and ACTIVEX graphics output devices.

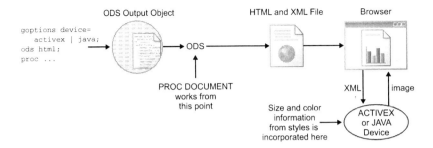

JAVAIMG or ACTXIMG Device

The following diagram illustrates the output process for the JAVAIMG and ACTXIMG graphics output devices.

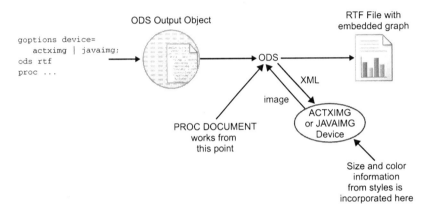

Setting the Size of Your Graph

You can use graphics options to control the size of your graph. Each device uses a default size for the graphics that they generate. You can use the HSIZE= and VSIZE= graphics options to override the default size of your graphics area, or the XPIXELS= and YPIXELS= graphics options to override the default size of your graph.

Using the HSIZE= and VSIZE= Graphics Options to Set the Size of Your Graphics Area

You can use the HSIZE= and VSIZE= graphics options to change the default size of the graphics area for the device that you are using. The HSIZE= option sets the horizontal dimension, while the VSIZE= option sets the vertical dimension. You can specify the dimension in inches (in), centimeters (cm), or points (pt). The default unit is inches (in). Here is an example that creates a 20 centimeter wide by 10 centimeter high GIF image of a graph.

```
option gstyle;
ods listing style=statistical;
goptions reset=all device=gif hsize=20cm vsize=10cm;

proc gchart data=sashelp.cars;
```

```
   vbar Make;
      where MPG_Highway >= 37;
   run;
quit;
```

Using the XPIXELS= and YPIXELS= Graphics Options to Set the Size of Your Graph

For devices other than the default display devices and the Universal Printing devices, you can use the XPIXELS= and YPIXELS= graphics options to change the default size of the display area for your graph without having to modify the device.

Note: The XPIXELS= and YPIXELS= graphics options are not supported by the default display devices. They are also not supported by Universal Printer devices (including the shortcut devices). The options are partially supported by the ACTIVEX and JAVA devices. △

Setting only the XPIXELS= and YPIXELS= options affects the size of the graph, but does not affect the resolution. Here is an example that creates a 600 pixel wide by 800 pixel high GIF image of a graph.

```
option gstyle;
ods listing style=statistical;
goptions reset=all device=gif xpixels=600 ypixels=800;

proc gchart data=sashelp.cars;
   vbar Make;
      where MPG_Highway >= 37;
   run;
quit;
```

Notice that XMAX= and YMAX= are not set. In this example, the SAS/GRAPH software recomputes the XMAX= and YMAX= values to retain the original resolution for the new graph size.

Setting the Resolution of Your Graph

To set the resolution of your template-based graphics:
- use the IMAGE_DPI= option in your ODS statement to specify the resolution in DPI.

    See *SAS/GRAPH: Graph Template Language Reference* and *SAS/GRAPH: Statistical Graphics Procedures Guide*.

To set the resolution of your device-based graphics, use one of the following methods:
- For devices other than the default display devices and the Universal Printer devices (including the shortcut devices), use the XPIXELS=, XMAX=, YPIXELS=, and YMAX= graphics options to set the resolution for graphics formats that support variable resolution.
- Use a device variant to set the resolution of your graph to a specific resolution.

Using the XPIXELS=, XMAX=, YPIXELS=, and YMAX= Graphics Options to Set the Resolution for Device-Based Graphics

For devices other than the default display devices and the Universal Printer devices, you can use the XPIXELS=, XMAX=, YPIXELS=, and YMAX= graphics options to set the resolution of your graph.

Note: The XPIXELS=, YPIXELS=, XMAX=, and YMAX= graphics options are not supported by the default display devices and the Universal Printer devices, including the shortcut devices. These graphics options are partially supported by the ACTIVEX and JAVA devices. △

Note: The resolution of GIF and BMP images is fixed and cannot be changed using this method. △

The XPIXELS= and YPIXELS= graphics options set the number of pixels for the X and Y axes respectively. The XMAX= and YMAX= graphics options set the maximum boundaries of the output on the X and Y axes respectively. The SAS/GRAPH software computes the resolution as follows:

X-resolution = XPIXELS / XMAX
Y-resolution = YPIXELS / XMAX

Table 7.2 on page 96 summarizes the affect of the XPIXELS=, XMAX=, YPIXELS=, and YMAX= graphics options have on the image resolution.

Table 7.2 Interactions of Graphics Options That Affect Resolution

Options Specified	Options Not Specified	SAS/GRAPH Action
XPIXELS= and YPIXELS=	XMAX= and YMAX=	Changes the dimensions and recalculates the value of XMAX= and YMAX= in order to retain the resolution.
XMAX= and YMAX=	XPIXELS= and YPIXELS=	Changes the dimensions and recalculates the value of XPIXELS= and YPIXELS= in order to retain the resolution.
XMAX= and XPIXELS=		Changes the horizontal dimension and recalculates the resolution.
YMAX= and YPIXELS=		Changes the vertical dimension and recalculates the resolution.

For example, for the graphics option settings XPIXELS=800 and XMAX=8in, the resulting X resolution is 100 DPI.

You can set the X resolution, the Y resolution, or both. Here is an example that sets the resolution of a 1000-pixel-wide-by-1200-pixel-high TIFF image of a graph to 200 DPI.

```
option gstyle;
ods listing style=seaside;
goptions reset=all device=tiffp xpixels=1000 xmax=5in ypixels=1200 ymax=6in;

proc gchart data=sashelp.cars;
   vbar Make;
```

```
        where MPG_Highway >= 37;
    run;
quit;
```

Using a Device Variant to Set the Size or Resolution of Your Graph

Some of the graphics output devices have variants that produce graphics of a specific size or resolution for a given format. Table 7.3 on page 97 lists the GIF device variants that produce images of a specific size.

Table 7.3 GIF Device Variants that Produce Images of a Specific Size

Device Variant	Default Image Size
GIF160	160 x 120
GIF260	260 x 195
GIF373	373 x 280
GIF570	570 x 430
GIF733	733 x 550

The PNG300 and JPEG300 device variants produce 300 DPI images in the PNG and JPEG format respectively.

Note: The PNG300 and JPEG300 devices are not appropriate for use with the ODS HTML destination. These devices are used when a high-resolution graph (300 DPI) in the PNG or JPEG format is required for printing purposes. Because most browsers do not use the resolution value stored in the PNG or JPEG file, images produced by the PNG300 and JPEG300 devices appear very large when they are viewed in the browser. △

See "Overview" on page 67.

Controlling Where Your Output is Stored

Specifying the Name and Location of Your ODS Output

By default, ODS output is stored in the default SAS output directory. You can use the FILE= option in your ODS statement to specify where your ODS output files are stored. For the HTML destination, you can also use the PATH=, GPATH=, and the BODY= options to specify a different location for the HTML output file and the graphics output files. Here is an example that uses the FILE= ODS option with the PDF destination to send the PDF output to file mygraph.pdf in the default SAS directory.

```
goptions reset=all;
ods listing close;
ods pdf style=money file="mygraph.pdf";
proc gchart data=sashelp.prdsale;
    vbar Product / sumvar=actual;
        title1 "First Quarter Sales in Canada";
        where Quarter=1 and Country="CANADA";
```

```
        run;
     quit;
     ods pdf close;
     ods listing;
```

Here is an example that uses the PATH=, GPATH=, and the BODY= ODS options with the HTML destination to send the HTML output to file mygraph.html in the current directory, and the graphics output file to the **images** subdirectory.

```
     goptions reset=all;
     ods listing close;
     ods html style=banker path="./" gpath="images" body="mygraph.html";
     proc gchart data=sashelp.prdsale;
        vbar Product / sumvar=actual;
           title1 "First Quarter Sales in Canada";
           where Quarter=1 and Country="CANADA";
        run;
     quit;
     ods html close;
     ods listing;
```

For more information on the PATH=, GPATH=, and BODY= options, see *SAS Output Delivery System: User's Guide*.

Specifying the Name and Location of Your Graphics Output File

When you use the ODS LISTING destination, you can use the GSFNAME= graphics option to send your output to a graphics output file that you specify. The GSFNAME= option requires a FILENAME statement that creates a file reference that points to a file or an aggregate file storage location. The syntax of the FILENAME statement is as follows:

```
FILENAME RefName "DirectoryOrFile"
```

If the file reference points to an aggregate file storage location, the graphics output files are named according to the NAME= option, if specified, or the default naming convention. If the file reference points to a file, the file specified in the FILENAME statement is used, even if the NAME= option is specified. See "Summary of How Output Filenames and GRSEG Names are Handled" on page 102.

Here is an example that shows how to send the output of the GCHART procedure to file mychart.png in the MyGraphs directory.

```
     filename graphout "MyGraphs";
     goptions reset=all device=png gsfname=graphout;
     proc gchart data=sashelp.cars;
        pie Make / name="MYCHART";
           where MSRP <= 15000;
        run;
     quit
```

If a MYCHART GRSEG entry does not already exist in the temporary catalog, the device sends the output to file mychart.png in the Mygraphs directory. If a MYCHART GRSEG entry already exists, the device uses an incremented name such as MYCHART1. In the previous example, you can replace the aggregate file location with a filename in the FILENAME statement and omit the NAME= option and get the same result.

If you specify the filename in the FILENAME statement, you must include the proper file extension. See "About File Extensions" on page 93.

You can also store your output in a graphics output file on a remote host using FTP. Here is an example that uses FTP to store multiple PNG graphs in directory **/public/sas/graphs** on the remote UNIX host unixhost73.

```
filename grafout ftp "/public/sas/graphs" dir host="unixhost73" fileext
   user="anonymous";
ods listing style=banker;
goptions reset=all device=png gsfname=grafout;

/* Create our data set by sorting sashelp.cars by type */
proc sort data=sashelp.cars out=work.cars;
   by type;
run;

/* Generate the graphs */
proc gchart data=work.cars;
   vbar Make;
      title1 "30 MPG or Better";
      where MPG_Highway >= 30;
      by type;
   run;
quit;
```

This example creates four PNG files in directory **/public/sas/graphs** on host unixhost73. Since the GCHART procedure uses BY-group processing, the FILENAME statement includes the DIR option, which defines an aggregate file storage location. If you need to create only one graph, remove the DIR option and specify the absolute path to your graphics output file in your FILENAME statement.

About Filename Indexing

When duplicate names occur in graphics output filenames, SAS/GRAPH procedures use indexing systems to determine unique names for new graphics output files. (Numbers are added to the end of the filename to create new filenames.) Two indexing systems are used: ODS Statistical Graphics indexing and catalog-based indexing. ODS Statistical Graphics indexing is used in all ODS Statistical Graphics output and by the procedures listed in Table 7.4 on page 100. All of the other procedures use catalog-based indexing.

Table 7.4 Filename Indexing Systems Used by SAS/GRAPH Procedures

Procedure Type	Indexing System	How To Control Graphics Filenames	Procedure Name
Device-based	Catalog-based	NAME= option in the procedure action statement	All procedures not listed below.
	ODS Statistical Graphics		GAREABAR
			GKPI
			GTILE
Template-based	ODS Statistical Graphics	IMAGENAME= option in the ODS GRAPHICS statement	SGDESIGN
			SGPLOT
			SGPANEL
			SGSCATTER
			SGRENDER

Note: See "Device-Based Graphics and Template-Based Graphics" on page 6 for a description of the procedure types. △

Because two independent indexing systems are used by the SAS/GRAPH procedures, it is possible that graphics output files can be overwritten if you specify the same graphics filename both for procedures that use catalog-based indexing and for procedures that use ODS Statistical Graphics indexing. To avoid this problem, make sure that you specify different names for the procedures that use ODS Statistical Graphics indexing and the procedures that use catalog-based indexing. For example, if your application uses both the GMAP procedure and the GAREABAR procedure, and you are using the NAME= option to specify output filenames, make sure you specify different filenames for each procedure.

Specifying the Catalog Name and Entry Name for Your GRSEGs

Using the Default Catalog and Entry Name

If you omit the NAME= and GOUT= options, the SAS/GRAPH software uses the default naming convention to name the GRSEG entry and stores the entry in the default WORK.GSEG catalog. The GRSEG naming convention uses up to eight characters of the default name for the procedure as the base name for the GRSEG. If the name generated by the procedure duplicates an existing GRSEG, the name is incremented such as GCHART, GCHART1, GCHART2, and so on. For details, see the description of the NAME= option for a specific procedure.

If you specify a filename for the graphics output file and omit the NAME= option, the graphics output filename is the name specified in the FILENAME statement, and the GRSEG entry name is the default procedure name. When you specify the filename, make sure that you include the appropriate file extension, such as .cgm, .gif, or .ps.

If you specify an aggregate file storage location instead of a specific filename and you omit the NAME= option, the name of both the GRSEG entry and the graphics output file is the default procedure name, and SAS/GRAPH supplies the appropriate file extension.

See "Summary of How Output Filenames and GRSEG Names are Handled" on page 102 for examples.

Specifying a Name for Your GRSEG with the NAME= Option

You can use the NAME= option to change the name of your output. Here is an example that shows how to change the name of the GCHART procedure output to MYCHART.

```
filename outfile "./";
goptions reset=all device=png gsfname=outfile;
proc gchart data=sashelp.cars;
   pie Make / name="MYCHART";
      where MSRP <= 15000;
   run;
quit;
```

This example creates the file mychart.png in the SAS default output directory, and it creates the GRSEG Mychart in the SAS temporary catalog.

See "Summary of How Output Filenames and GRSEG Names are Handled" on page 102 for additional information on output naming.

Specifying the Catalog and GRSEG Name with the GOUT= and NAME= Options

By default, GRSEGs are stored in the WORK.GSEG temporary catalog under the default name of the procedure that was used to generate the graph. The GRSEG name can be specified using the NAME= option, and the output catalog can be changed using the GOUT= procedure option. GRSEG names are limited to eight characters. If the NAME= option is set to a name that is more than eight characters in length, the GRSEG name is truncated to eight characters.

The name of the library and catalog in which the GRSEG is stored can be changed with the GOUT= procedure option. The GOUT= procedure option is assigned the catalog name in the format *libref.catalog* for the desired catalog. The name can be a one-level or a two-level name. If a one-level name is used, the GRSEG is stored in the temporary WORK library under the specified catalog name. A two-level name can be used to specify a permanent catalog.

Here is an example that shows how to store a GRSEG generated by the GCHART procedure under entry MYCHART in the MYGRAPHS.CARS catalog.

```
LIBNAME Mygraphs "Mygraphs";
ods listing style=banker;

proc gchart data=sashelp.cars gout=Mygraphs.cars;
   vbar Make / name="Mychart";
      where MPG_Highway >= 37;
   run;
quit;
```

Table 7.5 on page 101 summarizes the location of the GRSEG based on the NAME= and GOUT= procedure using the GCHART procedure as an example.

Table 7.5 How NAME= and GOUT= Affect the GRSEG Location

NAME=	GOUT=	GRSEG Location
Not specified	Not specified	Gchart in WORK.GSEG
Not specified	CARS	Gchart in WORK.CARS
Not specified	MYGRAPHS.CARS	Gchart in MYGRAPHS.CARS
MYCHART	Not specified	Mychart in WORK.GSEG

NAME=	GOUT=	GRSEG Location
MYCHART	CARS	Mychart in WORK.CARS
MYCHART	MYGRAPHS.CARS	Mychart in MYGRAPHS.CARS

Where GRSEGs are Stored When Multiple ODS Destinations are Used

When you send output to multiple ODS destinations, a catalog is created for the GRSEGs for each of the destinations. If the GOUT= procedure option is not specified, by default, the GRSEGs for the first destination that was opened are sent to the WORK.GSEG catalog. The GRSEGs for the subsequently opened ODS destinations are sent to a catalog that is named after the destination itself. For example, if you open the ODS LISTING, HTML, and RTF destinations, in that order, the GRSEGs are stored in the catalogs that are shown in the following table.

Catalog Name	Content
WORK.GSEG	The GSEGs for ODS LISTING
WORK.HTML	The GSEGs for ODS HTML
WORK.RTF	The GSEGs for ODS RTF

In the default case, the GRSEGs for the first destination that is opened are stored in the WORK.GSEG catalog, regardless of the destination.

If you use the GOUT= procedure option to specify a catalog name, the GRSEGs for the first destination that you opened are sent to the catalog that is specified by the GOUT= procedure option. The GRSEGs for the subsequently opened ODS destinations are sent to a catalog that is named after the destination itself. For example, if you open the ODS HTML, LISTING, and RTF destinations, and you use the GOUT=MyGraphs.Sales procedure option, the GRSEGs are stored in the catalogs that are shown in the following table.

Catalog Name	Content
MYGRAPHS.SALES	The GRSEGs for ODS HTML
MYGRAPHS.LISTING	The GRSEGs for ODS LISTING
MYGRAPHS.RTF	The GRSEGs for ODS RTF

The GRSEGs for the first destination are stored in the catalog that is specified by the GOUT= procedure option.

Summary of How Output Filenames and GRSEG Names are Handled

Table 7.6 on page 103 summarizes how SAS/GRAPH generates names for catalog entries and graphics output files, depending on 1) whether the NAME= option is used, and 2) the file reference specification in the FILENAME statement. This illustration assumes that the GCHART procedure is used with the DEVICE=GIF graphics option. It describes the case where a GRSEG and output file of the same name do not already exist, and the case where they do already exist.

Table 7.6 How SAS/GRAPH Generates Initial GRSEG Names and Filenames

NAME=	Condition	Result
NAME="FRED"	GSFNAME= points to a file named "MYGRAPH.GIF" and the catalog is empty.	GRSEG name: FRED external filename: MYGRAPH.GIF
NAME="FRED"	GSFNAME= points to an aggregate file storage location and the catalog is empty.	GRSEG name: FRED external filename: FRED.GIF
NAME="WEATHEROBS"	GSFNAME= points to an aggregate file storage location and the catalog is empty.	GRSEG name:WEATHERO external filename: WEATHEROBS.GIF
NAME= (not specified)	GSFNAME= points to a file named "MYGRAPH.GIF" and the catalog is empty.	GRSEG name: GCHART external filename: MYGRAPH.GIF
NAME= (not specified)	GSFNAME= points to an aggregate file storage location and the catalog is empty.	GRSEG name: GCHART external filename: GCHART.GIF

Note: When the file reference points to an aggregate file storage location, the name of the GRSEG *always* determines the name of the graphics output file. It does not matter whether the GRSEG name is the default name or a name assigned by the NAME= option. △

CAUTION:
If the graph created by your program already exists in the catalog, a new GRSEG with an incremented name is created. A new graphics output file might be created, which leaves your old graphics output file in place. △

Although GRSEG names cannot be more than eight characters in length, the NAME= option supports long names. When the NAME= option is assigned a name of more than eight characters and the file reference points to an aggregate file location, the GRSEG name is the NAME= value truncated to eight characters, and the graphics output filename is the complete NAME= value. This is demonstrated by the NAME="WEATHEROBS" example in Table 7.6 on page 103.

When a GRSEG of the same name already exists in the catalog, the SAS/GRAPH software combines the NAME= option value with a number to create an incremented name of no more than eight characters. If the GSFNAME= graphics option is used and the file reference points to an aggregate file location, the new graphics output filename is also incremented, but the filename is the full value of the NAME= option with a number appended. The same number is used for the GRSEG name and the graphics output filename.

If the GSFNAME= graphics option points to a file, the graphics output filename remains the same and the original file is replaced with the new graph by default.

Table 7.7 on page 104 demonstrates how the SAS/GRAPH software increments the GRSEG name and the graphics output filenames when a GRSEG and graphics output file of the same name already exist.

Table 7.7 How SAS/GRAPH Increments GRSEG Names and Filenames

NAME=	Condition	Result
NAME="FRED"	GSFNAME= points to a file named "MYGRAPH.GIF" and GRSEG FRED already exists.	GRSEG name: FRED1 external filename: MYGRAPH.GIF
NAME="FRED"	GSFNAME= points to an aggregate file storage location and GRSEG FRED already exists.	GRSEG name: FRED1 external filename: FRED1.GIF
NAME="WEATHEROBS"	GSFNAME= points to an aggregate file storage location and GRSEG WEATHERO already exists.	GRSEG name:WEATHER1 external filename: WEATHEROBS1.GIF
NAME= (not specified)	GSFNAME= points to a file named "MYGRAPH.GIF" and GRSEG GCHART already exists.	GRSEG name: GCHART1 external filename: MYGRAPH.GIF
NAME= (not specified)	GSFNAME= points to an aggregate file storage location and GRSEGs GCHART and GCHART1 already exist.	GRSEG name: GCHART2 external filename: GCHART2.GIF

You cannot replace individual GRSEGs in a catalog. To replace a GRSEG, you must delete the GRSEG, and then re-create it. Therefore, even though the contents of the graphics output file are replaced, the GRSEG is not. Each time you submit the program, a new GRSEG is created, and the GRSEG name is incremented.

Replacing an Existing Graphics Output File Using the GSFMODE= Graphics Option

You can use the GSFMODE= graphics option to replace an existing graphics output file with a new graph. To replace an existing graphics output file, the GSFMODE= option must be set to REPLACE, which is the default value for this option. When you run a SAS program that creates a graphics output file and the graphics option GSFMODE=REPLACE is used, the existing graphics output file is replaced with the new graph. However, a unique GRSEG is still generated each time you run the procedure.

See "Introduction" on page 327.

Storing Multiple Graphs in a Single Graphics Output File

If you want to store multiple graphs in a single graphics output file, you can use either the GSFMODE=APPEND and GSFNAME= graphics options, or the GREPLAY procedure.

Using Graphics Options to Store Multiple Graphs in One Graphics Output File

You can use the GSFMODE=APPEND and the GSFNAME= graphics options to store multiple graphs in one graphics output file. When the GSFMODE= graphics option is set to APPEND and the GSFNAME= option points to a file, if the graphics output file specified by the GSFNAME= option already exists, the SAS/GRAPH software appends the new graph to the graphics output file. Otherwise, it creates the graphics output file and stores the graph in it.

Note: Although a file can contain multiple graphs, some viewers can view only one graph. This can make it appear that a file containing multiple graphs contains only one graph. △

A common application of the GSFMODE=APPEND option is in the production of animated GIFs. See "Developing Web Presentations with the GIFANIM Device" on page 519.

Using the GREPLAY Procedure to Store Multiple Graphs in One Graphics Output File

You can use the GOUT= procedure option with the GREPLAY procedure to store multiple graphs in one graphics output file. This involves the following steps:

1 Create a file reference for your output file. For example:

```
filename myfile "MyOutputFile.ps";
```

2 Run the procedure to generate your charts and store them in a catalog.

3 Add the GSFNAME=*FileRefName* to your GOPTIONS statement.

4 Run the GREPLAY procedure as follows:

```
proc greplay
   igout=<CatalogName>
      replay _all_;
   run;
quit;
```

Replace <CatalogName> with the name of the catalog in which your graphs are stored. The REPLAY _ALL_ action statement replays all of the entries in the catalog.

Here is an example that replays five graphs to one PostScript file for printing.

```
/* Specify graphics output file name */
filename psout "multicharts.ps";

/* Specify style and graphics options */
ods listing style=banker;
goptions reset=all device=pscolor gsfname=psout nodisplay;

/* Generate the graphs */
proc gchart data=sashelp.cars gout=Work.Mygraphs;
   vbar Make;
      title1 "30 MPG or better";
      where MPG_Highway > 30;
   run;
```

```
      vbar Make;
         title1 "Between 25 MPG and 29 MPG";
         where MPG_Highway >= 25 AND MPG_Highway <= 29;
   run;

   vbar Make;
      title1 "Between 20 MPG and 24 MPG";
      where MPG_Highway >= 20 AND MPG_Highway <= 24;
   run;

   vbar Make;
      title1 "Between 15 MPG and 19 MPG";
      where MPG_Highway >= 15 AND MPG_Highway <= 19;
   run;

   vbar Make;
      title1 "Less than 15 MPG";
      where MPG_Highway < 15;
   run;
quit;

/* Enable display, and then replay all of the graphs to psout */
goptions display;
proc greplay
   igout=Work.Mygraphs nofs;
      replay _all_;
   run;
quit;
```

Replaying Your SAS/GRAPH Output

You can use the GREPLAY procedure or the ODS DOCUMENT destination and the DOCUMENT procedure to replay your SAS/GRAPH output.

Replaying Your Output Using the GREPLAY Procedure

For the SAS/GRAPH procedures that support GRSEGs, you can use the GREPLAY procedure to replay your graph GRSEGs without having to rerun your DATA step and procedures. You can replay all of your graphs or only the ones you select. When you replay your graphs, use the same device that you used when you generated the original graphs. If you use a different device, your replayed graphs might be distorted.

You can replay your graphs to the GRAPH window for viewing or to a graphics output file. Here is an example that replays all of the graphs in the WORK.GSEG catalog to the GRAPH window for viewing:

```
ods listing;
goptions reset=all;
proc greplay igout=work.gseg nofs;
   replay _all_;
run;
quit;
```

You can also use the GREPLAY procedure to replay multiple graphs to a single file for the graphic and document formats that support multiple images per file. See "Using the GREPLAY Procedure to Store Multiple Graphs in One Graphics Output File" on page 105 and Chapter 21, "Generating Web Animation with GIFANIM," on page 519.

For information on the GREPLAY procedure, see Chapter 50, "The GREPLAY Procedure," on page 1473.

Replaying Output Using the DOCUMENT Procedure

For all of the SAS/GRAPH procedures, you can use the DOCUMENT procedure to replay output that you created. Use the ODS DOCUMENT destination, without having to rerun your DATA step and procedures. The ODS DOCUMENT destination creates ODS output objects for your output. You can replay the output objects at any time to your monitor or to a different device.

Creating Your ODS Document

To create an ODS document for your output, do the following in your SAS program:
1 Open ODS DOCUMENT and specify the name of the output catalog with write permissions.
2 Close ODS LISTING.
3 Open the ODS destinations that you want to send your output to.
4 Specify the device that you want to use using the DEVICE= graphics option.
5 Generate your chart.
6 Close the ODS destinations that you opened in step 3.
7 Close ODS DOCUMENT.
8 Open ODS LISTING.

Here is an example that shows how to create an ODS document containing three pie charts and how to store it in catalog Mygraphs.Mydocs. The pie charts are generated with the JAVA device.

```
/* Create the Mygraphs catalog */
LIBNAME Mygraphs "./";

/* Open the DOCUMENT destination. Specify catalog */
/* Mygraphs.Mydocs for the output and give it write permission */
ods document name=Mygraphs.Mydocs(write);

/* Close the LISTING destination */
ods listing close;

/* Open the HTML destination, and specify the JAVA device. */
ods html style=seaside;
goptions reset=all device=java;

/* Generate the charts */
proc gchart data=sashelp.cars gout=Mygraphs.Mydocs;
   pie Make / other=2;
      title1 "30 MPG or Better";
      where MPG_Highway >= 30;
   run;
   pie Make / other=3;
```

```
        title1 "Between 20 MPG and 29 MPG";
        where MPG_Highway < 30 and MPG_Highway >=20;
    run;
    pie Make / other=3;
        title1 "19 MPG or less";
        where MPG_Highway < 20;
    run;
quit;

/* Close the HTML and DOCUMENT destinations */
ods html close;
ods document close;

/* Reopen the LISTING destination */
ods listing;
```

Replaying Your ODS Document

After you create your ODS document, use the DOCUMENT procedure to replay it. You can replay all of the graphs in your document or only those that you select. To see a list of the graphs in an ODS document, use a LIST statement with the DOCUMENT procedure. Here is an example that shows how to list the graphs in Mygraphs.Mydocs.

```
proc document name=Mygraphs.Mydocs;
    list / levels=all;
run;
quit;
```

A list of the graphs in the document is displayed in the Output window as shown in the following example:

```
Listing of: \Mygraphs.Mydocs\
Order by: Insertion
Number of levels: All

   Obs    Path                                             Type
   -------------------------------------------------------------
     1  \Gchart#1                                          Dir
     2  \Gchart#1\Gchart#1                                 Graph
     3  \Gchart#1\Gchart#2                                 Graph
     4  \Gchart#1\Gchart#3                                 Graph
```

In this example, the graphs are listed in the order in which they were inserted into the catalog. To replay individual graphs, you must know the path to the graphs, which is shown in the Path column.

To replay the output:

1 Close the ODS LISTING destination.

2 Open the ODS destinations that you want to send the output to.

3 Use the DEVICE= graphics option to specify the graphics output device that you want to use to generate the graphs.

4 Run the DOCUMENT procedure with one or more REPLAY statements to replay your graphs. Specify the path to each graph, and use the DEST= option to specify the output destination.

   Note: If you want to display all of the graphs, do not specify a path. △

5 Close the ODS destinations that you opened in step 2.

6 Open the ODS LISTING destination.

Here is an example that shows how to play the first and the third graphs in the Mygraphs.Mydocs catalog to the ODS RTF destination using the ACTIVEX device.

```
goptions reset=all device=activex;
ods listing close;
ods rtf style=money;
proc document name=Mygraphs.Mydocs;
   replay \Gchart#1\Gchart#1 / levels=all dest=rtf;
   replay \Gchart#1\Gchart#3 / levels=all dest=rtf;
   run;
quit;
ods rtf close;
ods listing;
```

To replay all of the graphs in the catalog, use one REPLAY statement that does not specify a path. For example:

```
proc document name=Mygraphs.Mydocs;
   replay / levels=all dest=rtf;
   run;
```

For more information on using the ODS DOCUMENT destination and the DOCUMENT procedure, see *SAS Output Delivery System: User's Guide*.

Previewing Output

If you want to preview how a graph is going to appear on another device before you send it to that device, use the TARGETDEVICE= graphics option. For example, to preview output on your display as it would appear on a color PostScript printer, include TARGETDEVICE= in a GOPTIONS statement and specify the device for the printer:

```
goptions targetdevice=pscolor;
```

How output is displayed on your screen depends on the following:

- the orientation of the target device. As a result, the graph might not cover the entire display area of the preview device.
- the values of either the LCOLS and LROWS pair or the PROWS and PCOLS pair, depending on the orientation of the target device.
- the default color list of the target device.
- the values of the HSIZE and VSIZE device parameters for the target device. The HSIZE and VSIZE values are scaled to fit the display device, but they retain the target device aspect ratio.
- the value of the CBACK device parameter for the target device.

All other device parameter values, including the destination of the output, come from the current device entry. Therefore, the output displayed by TARGETDEVICE= might not be an exact replication of the actual output, but it is as close as possible.

See "TARGETDEVICE" on page 424 for a complete description of TARGETDEVICE=.

Printing Your Graph

You can print your SAS/GRAPH output on hard-copy devices such as a printer. Regardless of the destination, you can create a hard copy of your graph in one of the following ways:

- Print the SAS/GRAPH program output directly to a hard-copy device.
- Print the SAS/GRAPH program output by creating a graphics output file, HTML file, or PDF file, and then printing the file using host commands or host application commands.
- Print the displayed graph directly from the GRAPH or Results Viewer window or the Graphics Editor window.
- Print the displayed graph directly from a browser that supports the SVG format.

Sending Your Graph Directly to a Printer

You can send graphics output directly to a hard-copy device by sending the graphics commands directly to the device or to a device port. On most systems you can use any of the following methods to print directly to a device:

- Use the ODS PRINTER destination to send your output directly to the default printer. Use the PRINTER= option if you want to direct your output to a printer other than the default printer or if a default printer is not defined.

    See the *SAS Output Delivery System: User's Guide* for information on the ODS PRINTER statement.

    See the *SAS Language Reference: Concepts* for information on how to define a default printer for the Universal Printer.

- Use a FILENAME statement, a GOPTIONS statement, and a SAS/GRAPH device. The FILENAME statement defines a file reference that points to the print commands to send your output to any available hard-copy device. The GOPTIONS statement references the file reference, assigns the device, and specifies any additional parameters.
- Use the GDEVICE procedure to modify a SAS/GRAPH device entry to spool output directly to a printer. See Chapter 38, "The GDEVICE Procedure," on page 1125 for information on adding host commands to a device entry.
- Use the Universal Printing interface.

For detailed instructions on each of these methods, refer to the SAS Help facility for SAS/GRAPH.

Saving and Printing Your Graph

You can save your graph to a graphics output file, and then print the file using host commands. You can perform these two steps separately or combine them by incorporating the host printing commands into your program or graphics output device. In any case, you must choose a graphics file format that is compatible with your printer. For example, if you are using a PostScript printer, be sure to create a PostScript file using the appropriate device for the printer.

You can use any of the following methods to create and print a graphics output file:

- Use FILENAME and GOPTIONS statements to create the graphics output file, and then use a host command to spool the file to a spooler for the device.

- Use an ODS PRINTER statement to produce a Postscript, PDF, PCL, SVG, PNG, or GIF file. Then use a host command or a host application command to send the file to the printer.
- Use the GDEVICE procedure to modify a SAS/GRAPH device to save the output to a graphics output file and spool the output directly to a printer. See Chapter 38, "The GDEVICE Procedure," on page 1125 for information on modifying device entries.
- Use the Universal Printing interface.

    Note: On Windows platforms, the ODS PRINTER destination uses the Universal Printing interface in addition to the Windows system printers. △

For detailed instructions on each of these methods, refer to the SAS Help facility for SAS/GRAPH.

Exporting Your Output

You can export your SAS/GRAPH output to other formats or to other software applications such as Microsoft Office. See the following topics for more information.

- "Replaying Output Using the DOCUMENT Procedure" on page 107
- Chapter 9, "Writing Your Graphs to a PDF File," on page 123
- Chapter 8, "Exporting Your Graphs to Microsoft Office Products," on page 113

CHAPTER

8

Exporting Your Graphs to Microsoft Office Products

What to Consider When Choosing an Output Format **113**
  Graphics Formats Versus Document Formats **113**
  Image Resolution and Size **114**
  Color Depth **114**
  Fonts **115**
  Multiple-Image Graphics Files **115**
  Ability to Edit: Vector Versus Raster Formats **115**
Comparison of the Graphics Output **116**
  Working Around the EMF and CGM Transparency Limitation **119**
  About the Default CGM Filter for Microsoft Office **120**
Enhancing Your Graphs **120**
Importing Your Graphs into Microsoft Office **120**
  Importing Graphs into Microsoft Word **120**
  Importing Graphs into Microsoft Excel **121**
  Importing Graphs into Microsoft PowerPoint **122**

What to Consider When Choosing an Output Format

When choosing a format for your SAS/GRAPH output to use with Microsoft products, you must consider the following:

- whether you need output in a graphics format or a document format
- the resolution and size of your graphs
- the color depth required for your graphs
- the fonts you want to use
- whether you need multiple graphs per page
- whether you need to edit your graphs using Microsoft products or using other third-party software

Graphics Formats Versus Document Formats

The SAS/GRAPH software supports output in both graphics format and document format. The graphics format includes graphics information and some text, such as titles, footnotes, and legends. The graphics format includes:

EMF

WMF

CGM

PNG

JPEG

TIFF

GIF

BMP

The document format can include both text and graphics in a single document. These documents store graphics in one of the following ways:

- in the format of the document
- in a graphics format embedded in the document
- in an external file that the document links to

To include images in a document, the images must be compatible with the document. Here is a summary of the compatibility between the SAS/GRAPH document and graphics formats:

Document Format	Compatible Graphics Formats
HTML	PNG, GIF, JPEG, SVG, and ActiveX
RTF	EMF, PNG, JPEG, and ActiveX

Image Resolution and Size

Each of the SAS/GRAPH graphics output devices has a default size and resolution setting for the graphics they generate. For information on the default settings for each device, see "Overview" on page 67. If you are using a raster format for your graphs, resizing the graph after it is imported into a Microsoft application might degrade the quality of the graph. To preserve the qualify of your raster image, when you create your graph in SAS, set the size to the size you need in the Microsoft application so that it does not have to be resized after it is imported. See "Setting the Size of Your Graph" on page 94. You can also change to one of the vector formats, which can be resized with no loss of quality.

If you need a high-resolution image, many of the graphics output devices enable you to use the graphics options to change their default resolution. Some of the devices have device variants that you can use to generate high-resolution images. See "Setting the Resolution of Your Graph" on page 95

Color Depth

Another consideration when choosing a graphics format is color depth, which is the number of bits that are used to represent each color in an image. Color depth can affect the smoothness, clarity, and color trueness of the elements in a rasterized image. A greater color depth means that more distinct colors are available to represent elements such as gradient shading and antialiasing in text.

Most of the graphics file formats support Truecolor, which provides a 24-bit color depth. The GIF format provides only an 8-bit color depth, which can represent up to 256 distinct colors in a single image. For many graphics, 8-bit color depth is sufficient. However, if your output includes background images, color gradients, or other color-intensive elements, consider using a format that supports Truecolor. The formats that support Truecolor include the following:

BMP

CGM

EMF

EPS

PNG

SVG

WMF

See "Overview" on page 67 for information on the color depth supported by each of the graphics output devices.

Fonts

Microsoft Office products use fonts that are native to the Windows operating system, which include TrueType and OpenType fonts. The SAS/GRAPH graphics output devices might support the fonts that you are using in your Microsoft applications. See "Introduction" on page 1643 for information on the fonts that the SAS/GRAPH graphics output devices use.

Multiple-Image Graphics Files

If you need to store more than one graph in a file, you can use one of the following methods:

- Use the GREPLAY procedure to replay multiple graphs to a file of the same format that was used to generate the original graphs.
- Use the ODS DOCUMENT destination and the DOCUMENT procedure to replay multiple graphs to a file of any supported format
- Use the ODS PRINTER destination with a Universal Printer device that supports multiple-page documents.
- Use the GIFANIM procedure to insert multiple graphs into an animated GIF.

See "Using the GREPLAY Procedure to Store Multiple Graphs in One Graphics Output File" on page 105 and "Exporting Your Output" on page 111 for information on replaying your graphs. See "Developing Web Presentations with the GIFANIM Device" on page 519 for information on using the GIFANIM device.

Ability to Edit: Vector Versus Raster Formats

If you need the ability to edit your graphs using Microsoft or other third-party software, choose a graphics format that enables you to perform the type of editing that you need to do. For vector formats, such as WMF, EMF, SVG, and CGM, you can edit individual text and graphic elements using graphics editing software. Although EPS contains vector graphs, Microsoft products cannot edit an EPS image. For raster images, some programs such as Microsoft Paint enable you to edit the image. However, in Microsoft Office products, editing is limited to changing only the global attributes of the image, such the size, contrast, brightness, and so on.

Comparison of the Graphics Output

The SAS/GRAPH software can generate the following types of graphics output that can be imported into Microsoft products:

EMF and WMF

CGM

PNG

JPEG and TIFF

GIF and BMP

EPS

HTML (PNG)

RTF

ACTIVEX (RTF)

ACTIVEX (RTF)

ACTXIMG (PNG)

JAVAIMG (PNG)

Note the following:
- The ODS HTML destination generates two files: a PNG file (by default) that contains the graph and an HTML file that enables you to view the graph file.
- The ACTIVEX device is used with the ODS RTF or ODS HTML destination to create an RTF or HTML file that contains code that is consumed by the ActiveX Control.
- The ACTXIMG and JAVAIMG devices generate a PNG file that contains a static graph that is generated by the ACTIVEX and JAVA devices respectively.
- Procedures that do not support the ACTIVEX, ACTXIMG, JAVA, and JAVAIMG devices produce a GIF file when the ACTIVEX, ACTXIMG, JAVA, or JAVAIMG device is used.

Table 8.1 on page 117 provides a brief comparison of these graphics output formats and lists some of the graphics output devices that generate each output type. For detailed information on all of the graphics output devices, see "Overview" on page 67.

Table 8.1 Comparison of the Graphics and Document Types

Type	Advantages and Limitations	Devices
EMF and WMF	Advantages: ☐ Most Windows-based applications recognize the EMF and WMF formats. ☐ Graphs stored in EMF or WMF can usually be edited after they are imported. ☐ Graphs are imported at full size into Office, and can be resized without a loss of quality. Limitations: ☐ The EMF format does not support transparency (see "Working Around the EMF and CGM Transparency Limitation" on page 119). ☐ Only one graph per file is supported.	☐ SASEMF and SASWMF ☐ EMF and WMF
CGM	Advantages: ☐ Graphs stored in CGM files can be edited after they are imported. ☐ The image can be resized without a loss of quality. Limitations: ☐ The format does not support transparency (see "Working Around the EMF and CGM Transparency Limitation" on page 119). ☐ Because the default CGM filter is not installed by default in Microsoft Office, to import CGM files, you must install the CGM filter (see "About the Default CGM Filter for Microsoft Office" on page 120). ☐ Although the CGM format supports multiple images per file, not all versions of Microsoft Office can import more than one image per file (see "About the Default CGM Filter for Microsoft Office" on page 120).	☐ CGMOFML (landscape) ☐ CGMOFMP (portrait)

Type	Advantages and Limitations	Devices
PNG	Advantages: ☐ Designed to display images on the Web. ☐ Uses lossless data compression. ☐ Supports transparency (with the PNGT device). ☐ Can store high-resolution images. ☐ Supports truecolor images. Limitation: cannot be resized without a loss of quality.	☐ PNG (no transparency) ☐ PNG300 (no transparency) ☐ PNGT (transparency) ☐ UPNG (no transparency) ☐ UPNGT (transparency)
JPEG and TIFF	Advantages: ☐ JPEG is widely used for displaying photographs on the Web. ☐ Both can store high-resolution graphics. Limitations: ☐ JPEG uses lossy compression. ☐ The SAS/GRAPH JPEG device supports only 256 colors. ☐ TIFF is not a Web graphics format. ☐ JPEG and TIFF images cannot be resized without a loss of quality.	☐ JPEG ☐ TIFFP (color) ☐ TIFFB (monochrome)
GIF and BMP	Advantages: ☐ GIF supports transparent backgrounds. ☐ GIF can store multiple images per file when it is formatted as an animated GIF. ☐ Both support the IBACK option and the IMAGE annotation function for including logos and other images in the background of the graph. Limitations: ☐ Both formats have a fixed resolution of 96 DPI. ☐ The GIF standard is limited to 256 colors. ☐ Cannot be resized without a loss of quality.	☐ BMP (720x480) ☐ BMP20 (720 480, BMP 2.0) ☐ GIF (800x600) ☐ GIFANIM (1280x1024, multi-image) ☐ UGIF (Universal Printer)

Type	Advantages and Limitations	Devices
EPS	Advantages: ☐ Can contain a combination of vector and bitmap objects. ☐ Can be resized after it is imported into Office 97 or Office 2000. Limitations: ☐ The images should not be edited after they are imported. ☐ Because the system display does not use the PostScript language to render the graph, these graphics might be visible only when printed to a PostScript printer. ☐ Because the preview is created automatically in Office 2002 and later, the image should not be resized after it is imported. ☐ Although this format can store more than one image per file, an EPS file should contain only one image.	☐ UEPS (gray scale) ☐ UEPSC (color) ☐ PSEPSF (gray scale) ☐ PSEPSFA4 (gray scale) ☐ PSLEPSF (gray scale) ☐ PSLEPSFC (color)
HTML	Advantages: ☐ Can store text and graphics. ☐ In Office 2000 and later, and in Microsoft Word in Office 97, the images are loaded into the document automatically when the HTML is imported. Limitation: In Office 97, the images are not loaded into a PowerPoint or Excel document when the HTML is imported. Only the text and tables are imported.	☐ JPEG ☐ GIF and UGIF ☐ ACTIVEX ☐ ACTXIMG and JAVAIMG, which create PNG files ☐ PNG, PNGT, UPNG, and UPNGT
RTF	Advantages: ☐ Designed specifically for sharing documents between word processors. ☐ Can store both text and graphics.	☐ JPEG ☐ ACTIVEX ☐ ACTXIMG and JAVAIMG, which create PNG files ☐ PNG, PNGT, UPNG, and UPNGT ☐ SASEMF and EMF

Working Around the EMF and CGM Transparency Limitation

For the EMF and CGM devices, you can work around the transparency limitation as follows:

☐ For EMF, use the CBACK= or IBACK= graphics options to assign the matching color or image for the graph background. You could instead edit the EMF file after it is imported to remove the default background.

- For CGM, use the CBACK= graphics to assign a matching background color to the CGM file. The CGM devices do not support the IBACK= graphics option or the IMAGE function. To have an image in the document or slide appear as the background of the graph, edit the graph after it is imported to remove the background created by SAS so that the document background shows through.

About the Default CGM Filter for Microsoft Office

To import CGM files in Microsoft Office, you must install the default CGM filter. For information on the CGM filter and how to install it for your version of Microsoft Office, visit the Microsoft Support Web site:

   `http://support.microsoft.com`

Enhancing Your Graphs

You can use various features in SAS/GRAPH that enable you to enhance your graphs. The following table lists some of these features.

Table 8.2 Features that can Enhance Your Graph

Feature in SAS/GRAPH	Reference
Changing the style of the graphic	Chapter 10, "Controlling The Appearance of Your Graphs," on page 133.
Adding annotations to the graph	Chapter 29, "Using Annotate Data Sets," on page 641
Making the graph interactive	Chapter 17, "Creating Interactive Output for ActiveX," on page 453
Adding drill-down links and data tips to the graph	Chapter 27, "Enhancing Web Presentations with Chart Descriptions, Data Tips, and Drill-Down Functionality," on page 595
Animating the graph	Chapter 21, "Generating Web Animation with GIFANIM," on page 519

Importing Your Graphs into Microsoft Office

This section describes how to import SAS/GRAPH graphics and documents into Microsoft Office 2007 products. For instructions on how to import graphics and documents for other versions of Microsoft Office, contact Technical Support.

Importing Graphs into Microsoft Word

To insert a SAS/GRAPH graphics file into a Microsoft Word 2007 document:

1 If you have not already done so, open your Microsoft Word document and position your cursor where you want to insert your graph.

2 Select the **Insert** tab.

3 On the **Insert** tab, click the **Picture** icon in the Illustrations group. The Insert Picture dialog box opens.

4 In the Insert Picture dialog box, select your graphics output file, and then click **Insert**.

To insert a SAS/GRAPH document into a Microsoft Word 2007 document:

1 Do one of the following based on the type of the document you are importing from:
   - If you are importing from an HTML document, open the document in your Web browser.
   - If you are importing from an RTF document, open the document in Microsoft Word.

2 If you have not already done so, open the target document and position your cursor where you want to insert your graph.

3 In the HTML or RTF document, right-click the graph, and then select **Copy** from the pop-up menu.

4 In the target document, right-click in the page area, and then select **Paste** from the pop-up menu.

If the graph you have imported is an ActiveX graph, you can right-click on your graph in your document and change various attributes of your graph using the pop-up menu. For more information on this menu, select **Help ▶ Graph Control Help** from the pop-up menu.

If the graph you have imported is an animated GIF, you must convert the Microsoft Word document to HTML, and then open the HTML version of your document in your Web browser to play the animated GIF.

Importing Graphs into Microsoft Excel

To insert a SAS/GRAPH graphics file into a Microsoft Excel 2007 spreadsheet:

1 If you have not already done so, open your Microsoft Excel spreadsheet.

2 Locate the cell that you want to import your graph to. Resize the cell to accommodate the graph, if necessary.

3 Select the **Insert** tab.

4 In the **Insert** tab, click **Picture** in the Illustrations group. The Insert Picture dialog box appears.

5 In the Insert Picture dialog box, select your graphics output file, and then click **Insert**.

6 Adjust the size of the graph and cell, if necessary.

To insert a SAS/GRAPH document into a Microsoft Excel 2007 spread sheet:

1 Open the SAS/GRAPH document that you want to import from:
   - If the document is an HTML document, open it in your Web browser or Microsoft Word.
   - If the document is an RTF document, open it in Microsoft Word.

2 If you have not already done so, open your Microsoft Excel spreadsheet.

3 Locate the cell that you want to import your graph to. Resize the cell to accommodate the graph, if necessary.

4 In the HTML or RTF document that you are importing from, right-click your graph, and then select **Copy** from the pop-up menu.

5 In your spread sheet, right-click in the cell that you are importing to, and then select **Paste** from the pop-up menu.

6 Adjust the size of the graph and cell, if necessary.

If the graph you have imported is an ActiveX graph, you can right-click on your graph in your spreadsheet and change various attributes of your graph using the pop-up menu. For more information on this menu, select **Help ▶ Graph Control Help** from the pop-up menu.

Importing Graphs into Microsoft PowerPoint

To insert a SAS/GRAPH graphics file into a Microsoft PowerPoint 2007 presentation:

1 If you have not already done so, open your Microsoft PowerPoint presentation.
2 Locate the slide on which you want to insert your graph. Insert a new slide, if necessary.
3 Click the `Insert` tab.
4 In the `Insert` tab, click `Picture` in the Illustrations group. The Insert Picture dialog box appears.
5 In the Insert Picture dialog box, select your graphics output file, and then click `Insert`.
6 Adjust the size and position of the graph, if necessary.

To insert a SAS/GRAPH document into a Microsoft PowerPoint 2007 presentation:

1 Open the SAS/GRAPH document that you want to import from:
   - If the document is an HTML document, open it in your Web browser or Microsoft Word.
   - If the document is an RTF document, open it in Microsoft Word.
2 If you have not already done so, open your Microsoft PowerPoint presentation.
3 Locate the slide on which you want to insert your graph. Insert a new slide, if necessary.
4 In the HTML or RTF document that you are importing from, right-click the graph, and then select `Copy` from the pop-up menu.
5 In your PowerPoint presentation, right-click in the slide that you are importing to, and then select `Paste` from the pop-up menu.
6 Adjust the size and position of the graph, if necessary.

If the graph you have imported is an ActiveX graph, you can change various attributes of your graph dynamically as follows:

1 Right-click your graph, and then select **SAS Graph v9 Object ▶ Edit** to activate the ActiveX Control.
2 Right-click your graph again, and then select an item from the pop-up menu to change one or more attributes of the graph. You can change the chart type, style, and so on, using this menu. For more information on this menu, select **Help ▶ Graph Control Help** from the pop-up menu.
3 To deactivate the ActiveX Control, deselect your graph.

If the graph you have imported is an animated GIF, you must set the PowerPoint mode to Slide Show to play the animated GIF as follows:

1 In the left panel, select the slide that contains your animated GIF.
2 Click the `Slide Show` tab.
3 On the `Slide Show` tab, click `From Current Slide` in the Start Slide Show group.
4 Verify that your animated GIF plays properly.
5 Press the Esc key to exit the Slide Show mode.

CHAPTER

9

Writing Your Graphs to a PDF File

About Writing Your Graphs to a PDF File **123**
Changing the Page Layout **124**
Adding Metadata to Your PDF File **124**
Adding Bookmarks for Your Graphs **124**
Changing the Default Compression Level for Your PDF File **125**
Examples **125**
    Creating a Multipage PDF File with Bookmarks and Metadata **125**
    Creating a PDF/A-1b-Compliant File that Contains Multiple Graphs Per Page **127**
    Creating a Multiple-Page PDF File Using BY-Group Processing **129**
    Creating a Multiple-Page PDF File Using the GREPLAY Procedure **129**

About Writing Your Graphs to a PDF File

You can use the ODS PDF destination to write your graph output to a PDF Version 1.4 file or a PDF file that is compliant with PDF/A-1b standards and can be archived. You can add multiple graphs to your PDF file with one or more graphs per page. You can also add bookmarks, links, and document metadata in your PDF file, and use system options to change the default page layout of your document.

The ODS PDF destination supports the SAS/GRAPH fonts, the TrueType fonts that are installed with the Base SAS product, and the resident PDF fonts. The resident PDF fonts are the Base 14 fonts that are installed by default with the Adobe Acrobat Reader. These fonts include:

    Courier

    Courier/oblique

    Courier/bold

    Courier/bold/oblique

    Helvetica

    Helvetica/oblique

    Helvetica/bold

    Helvetica/bold/oblique

    Times

    Times/italic

    Times/bold

Times/bold/italic

Symbol

ITC Zapf Dingbats

For more information on fonts, see Chapter 11, "Specifying Fonts in SAS/GRAPH Programs," on page 155.

By default, the ODS PDF destination writes your output to a PDF Version 1.4 file. To write your graphs to a PDF file that can be archived, add the PRINTER=PDFA option to your ODS statement. The PDFA Universal Printer shortcut device creates a PDF file that is compliant with PDF/A-1b standards and can be archived. See Chapter 6, "Using Graphics Devices," on page 67 for information on the PDFA Universal Printer shortcut device. See "Creating a PDF/A-1b-Compliant File that Contains Multiple Graphs Per Page" on page 127 for an example of how to create an archivable PDF file.

Changing the Page Layout

Use the following system options to change the page layout for your PDF document:

- ORIENTATION=PORTRAIT | LANDSCAPE | REVERSEPORTRAIT | REVERSELANDSCAPE
- PAPERSIZE="*paper-size*"
- LEFTMARGIN=*value*
- RIGHTMARGIN= *value*
- TOPMARGIN= *value*
- BOTTOMMARGIN=*value*

See *SAS Language Reference: Dictionary* for information on these system options. See "Creating a Multipage PDF File with Bookmarks and Metadata" on page 125 for an example of how to use these system options to change the page layout of a PDF file.

Adding Metadata to Your PDF File

Use the following ODS options to add document metadata to the PDF file:

- AUTHOR="*author-name*"
- KEYWORDS="*word1 word2 ...* "
- SUBJECT="*document-subject*"
- TITLE="*document-title*"

See "Creating a Multipage PDF File with Bookmarks and Metadata" on page 125 for an example of how to add metadata to a PDF file.

Adding Bookmarks for Your Graphs

You can use an ODS PROCLABEL=*label* statement to add bookmarks for your graphs. The PROCLABEL= ODS option specifies the name of the top-level bookmark. The description for each procedure that you run after your ODS PROCLABEL= statement is added as a subtopic under the top-level bookmark that the PROCLABEL= option defines. You can use the DESCRIPTION= option to set the text of the subtopic bookmark for each graph procedure. If you do not specify a description, the default graph description is used. See "Creating a Multipage PDF File with Bookmarks and Metadata" on page 125 for an example.

Changing the Default Compression Level for Your PDF File

You can use the COMPRESS= ODS option to change the default compression level for your PDF file. The COMPRESS= option can be set to an integer value between 0 and 9, which specifies the level of compression. A value of 0 means no compression. The default level is 6.

Examples

This section provides the following examples:

"Creating a Multipage PDF File with Bookmarks and Metadata" on page 125

"Creating a PDF/A-1b-Compliant File that Contains Multiple Graphs Per Page" on page 127

"Creating a Multiple-Page PDF File Using BY-Group Processing" on page 129

"Creating a Multiple-Page PDF File Using the GREPLAY Procedure" on page 129

Creating a Multipage PDF File with Bookmarks and Metadata

Here is an example that creates a multipage PDF file with bookmarks and metadata using RUN-group processing. Each page displays a single graph in the landscape orientation, and is set up for A4 paper with a 1 cm right, left, and bottom margin, and a 2 cm top margin. The PROCLABEL= ODS option is used to set the top-level bookmark for each category of graphs. The DESCRIPTION= option is used with each procedure to set the text of each subheading bookmark.

```
/* Close the LISTING destination */
ods listing close;

/* Reset the graphics options   */
goptions reset=all;

/* Open the PDF destination */
ods pdf style=seaside
   file="MyDoc.pdf"   /* Output filename */
   compress=0         /* No compression */
   /* Add metadata */
   author="J. L. Cho"
   subject="Auto makers"
   title="Car Makers by MPG and Vehicle Type"
   keywords="automobiles cars MPG sedans trucks wagons SUVs";

/* Modify the PDF page properties */
options orientation=LANDSCAPE
   papersize=A4
   leftmargin=1cm
```

```
         rightmargin=1cm
         bottommargin=1cm
         topmargin=2cm;

/* Set the top-level bookmark for the first set of graphs */
ods proclabel="Makes By MPG";

/* Create the first set of graphs */
proc gchart data=sashelp.cars;
   pie Make / name="HighMPG" other=3
      description="High-MPG"; /* Set subheading text */
      title1 "30 MPG or Better";
      where MPG_Highway >= 30;
   run;
   pie Make / name="MedMPG" other=3
      description="Average-MPG"; /* Set subheading text */
      title1 "Between 20 MPG and 29 MPG";
      where MPG_Highway < 30 and MPG_Highway >= 20;
   run;
   pie Make / name="LowMPG" other=3
      description="Low-MPG"; /* Set subheading text */
      title1 "19 MPG or less";
      where MPG_Highway < 20;
   run;
quit;

/* Set the top-level bookmark for the second set of graphs */
ods proclabel="Makes By Type";

/* Create the second set of graphs */
proc gchart data=sashelp.cars;
   pie Make / name="Sedans" other=3
      description="Sedans"; /* Set subheading text */
      title1 "Sedans";
      where Type = "Sedan";
   run;
   pie Make / name="SUVs" other=3
      description="SUVs"; /* Set subheading text */
      title1 "SUVs";
      where Type="SUV";
   run;
   pie Make / name="Trucks" other=3
      description="Trucks"; /* Set subheading text */
      title1 "Trucks";
      where type="Truck";
   run;
   pie Make / name="Wagons" other=3
      description="Wagons"; /* Set subheading text */
      title1 "Wagons";
      where type="Wagon";
   run;
   pie Make / name="Sports" other=3
      description="Sports Cars"; /* Set subheading text */
      title1 "Sports Cars";
```

```
            where type="Sports";
      run;
quit;

/* Close the PDF destination */
ods pdf close;
ods listing;

/* Reset the graphics options */
goptions reset=all;
```

This creates a PDF file with the bookmarks shown in the following display:

The document metadata is displayed on the **Description** tab of the Document Properties dialog box. To open the Document Properties dialog box, type CTRL-D anywhere in the PDF viewer window or right-click in the PDF viewer window, and then select **Document Properties** from the pop-up menu. The following display shows the document metadata that is displayed for this example.

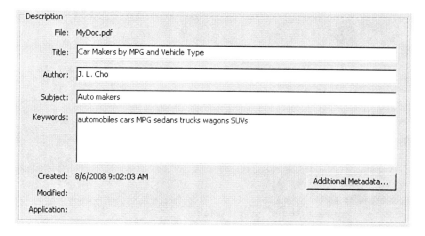

Creating a PDF/A-1b-Compliant File that Contains Multiple Graphs Per Page

Here is an example that creates the PDF file FourVbars.pdf, which contains four graphs on one page and can be archived. The PRINTER=PDFA ODS option is used to create a PDF file that is compliant with PDF/A-1b standards. To create a standard Version 1.4 PDF file, remove the PRINTER=PDFA option from the ODS statement.

```
/* Close the LISTING destination */
ods listing close;

/* Set page options */
options orientation=portrait rightmargin=0.1in leftmargin=0.1in;
goptions reset=all ftext="Helvetica/bold";

/* Open PDF */
ods pdf style=printer
    printer=pdfa          /* Create an archivable PDF */
    file="FourVbars.pdf"  /* Output filename */
    startpage=never;      /* Do not insert a pagebreak after each graph */

/* Create a slide for the graphs */
goptions hsize=0 vsize=0;

proc gslide;
    title1 "1997 Quarterly U.S. Sales By State";
run;

/* Size each graph 4in x 4in */
goptions hsize=4in vsize=4in;
title1;

/* Generate the graphs */
proc gchart data=sashelp.prdsal3;
   /* Create the Q1 graph in the top-left quadrant */
   title2 "First Quarter";
   goptions horigin=0 vorigin=5;
   pie State / sumvar=Actual type=mean;
      where country="U.S.A." AND quarter=1 AND Year=1997;
   run;

   /* Create the Q2 graph in the top-right quadrant */
   goptions horigin=4 vorigin=5;
   title2 "Second Quarter";
   pie State / sumvar=Actual type=mean;
      where country="U.S.A." AND quarter=2 AND Year=1997;
   run;

   /* Create the Q3 graph in the bottom-left quadrant */
   title2 "Third Quarter";
   goptions horigin=0 vorigin=0;
   pie State / sumvar=Actual type=mean;
      where country="U.S.A." AND quarter=3 AND Year=1997;
   run;

   /* Create the Q4 graph in the bottom-right quadrant */
   title2 "Fourth Quarter";
   goptions horigin=4 vorigin=0;
   pie State / sumvar=Actual type=mean;
      where country="U.S.A." AND quarter=4 AND Year=1997;
```

```
        run;
quit;

/* Close PDF and reopen LISTING */
ods pdf close;
ods listing;

/* Reset the graphics options */
goptions reset=all;
```

Creating a Multiple-Page PDF File Using BY-Group Processing

Here is an example that uses BY-group processing to create a multiple-page PDF file that contains one graph per page in the landscape orientation.

```
/* Specify the landscape page orientation */
options orientation=landscape;

/* Close the LISTING destination */
ods listing close;

/* Reset the options */
goptions reset=all;

/* Open the PDF destination */
ods pdf style=statistical;

/* Create our data set by extracting 1994 data from sashelp.prdsale */
/* and sorting by product */
proc sort data=sashelp.prdsale(where=(Year=1994)) out=work.prdsale;
   by product;
run;

/* Generate the graphs */
title1 "1994 Monthly Sales By Product";
proc gchart data=work.prdsale;
   hbar month /sumvar=actual type=sum sum;
      by product;
   run;
quit;

/* Close the PDF destination */
ods pdf close;

/* Reset the graphics options */
goptions reset=all;

/* Open the LISTING destination */
ods listing;
```

Creating a Multiple-Page PDF File Using the GREPLAY Procedure

Here is an example that uses the GREPLAY procedure to create a PDF file that contains four graphs.

```
/* Specify the landscape page orientation */
options orientation=portrait;

/* Close the LISTING destination */
ods listing close;

/* Reset the options and set NODISPLAY */
goptions reset=all nodisplay;

/* Open the PDF destination */
ods pdf style=statistical file="Mygraph.pdf";

/* Create our data set by extracting 1994 data from sashelp.prdsale */
/* and sorting by quarter */
proc sort data=sashelp.prdsale(where=(Year=1994)) out=work.prdsale;
   by quarter;
run;

/* Delete the old GRSEGs */
proc greplay igout=work.gseg nofs;
   delete _all_;
run;

/* Generate the graphs */
proc gchart data=work.prdsale;
   vbar product /sumvar=actual discrete type=mean mean;
      title1 "1994 Q1 Average Sales By Product";
      where quarter=1;
   run;

      title1 "1994 Q2 Average Sales By Product";
      where quarter=2;
   run;

      title1 "1994 Q3 Average Sales By Product";
      where quarter=3;
   run;

      title1 "1994 Q4 Average Sales By Product";
      where quarter=4;
   run;
quit;

/* Replay the graphs to the PDF file */
goptions display;
proc greplay igout=work.gseg nofs;
   replay _all_;
run;
quit;

/* Close the PDF destination */
ods pdf close;

/* Reset the graphics options */
```

```
goptions reset=all;

/* Open the LISTING destination */
ods listing;
```

CHAPTER 10

Controlling The Appearance of Your Graphs

Overview **133**
Style Attributes Versus Device Entry Parameters **134**
About Style Templates **135**
    ODS Destinations and Default Styles **135**
    Recommended Styles **136**
    Examples of Output Using Different Styles **136**
Specifying a Style **139**
    Changing the Current Style by Using the STYLE= Option in ODS Destination Statements **139**
    Changing the Default Style in the SAS Registry **139**
Overriding Style Attributes With SAS/GRAPH Statement Options **140**
Precedence of Appearance Option Specifications **141**
Viewing the List of Styles Provided by SAS **141**
    Using The TEMPLATE Procedure **141**
    Using the Templates Window **141**
Modifying a Style **142**
    Using the TEMPLATE Procedure **142**
    Example: Modifying a Style Element **142**
    Ways to Modify Graph Fonts Or Colors Specified By Styles **143**
    Modifying the GraphFonts And GraphColors Style Elements **143**
Graphical Style Element Reference for Device-Based Graphics **144**
    The GraphColors Style Element **144**
    The GraphFonts Style Element **145**
    Font Specifications In The GraphFonts Style Element **146**
    Style Elements For Use With Device-Based SAS/GRAPH Output **146**
Turning Off Styles **153**
Changing the Appearance of Output to Match That of Earlier SAS Releases **154**

Overview

The appearance of SAS/GRAPH output is determined by ODS styles by default. Along with table and page attributes, ODS styles contain a collection of graphical attributes such as color, marker shape, line pattern, fonts, and so on. Many carefully designed styles that enhance the visual impact of the graphics are shipped with SAS. In addition to creating visually appealing graphics, the styles ensure that different groups of data can be easily distinguished from one another. They also ensure that data of equal importance is given equal visual emphasis.

These styles produce professional-looking graphics without additional code in your SAS programs and without modifying the styles themselves. However, you can use SAS/GRAPH statement options to override specific elements in the styles, or you can modify style elements to create a customized style for yourself or your organization.

Table 10.1 Controlling Graph Appearance

Method	Description	Level of Complexity	Reference
Specify a different style template.	Specify a style template with the STYLE= option to change the appearance of the entire graph. Requires no further modification.	Low	"Changing the Current Style by Using the STYLE= Option in ODS Destination Statements" on page 139
Use appearance options.	Specify an appearance option using SAS/GRAPH procedure options or global statement options to change various aspects of your graph. This method requires modification of your SAS/GRAPH program.	Medium	"Overriding Style Attributes With SAS/GRAPH Statement Options" on page 140
Modify individual style elements.	Specify or change style attributes in order to modify a style element. This requires the use of PROC TEMPLATE style statements.	High	"Modifying a Style" on page 142

You can turn off the use of styles if needed. In this case, the default appearance of your output is controlled by device entry parameters. See "Style Attributes Versus Device Entry Parameters" on page 134 and "Turning Off Styles" on page 153 for more information.

Note: This section covers only device-based graphics. See "Device-Based Graphics and Template-Based Graphics" on page 6. △

Style Attributes Versus Device Entry Parameters

The default appearance of SAS/GRAPH output is determined by either style attributes or device entry parameters, depending on the setting of the GSTYLE system option and on the device that is being used.

By default, the GSTYLE system option is in effect, and the appearance of all SAS/GRAPH output is determined by style attributes. If the NOGSTYLE system option is in effect, then the device entry parameters govern the appearance of SAS/GRAPH output for all devices except the Java and ActiveX devices. The Java and ActiveX devices always use styles to determine appearance. The setting of the GSTYLE system option has no effect on the Java and ActiveX devices.

Table 10.2 The GSTYLE System Option and Default Appearance

Current Device	GSTYLE	NOGSTYLE
Java or ActiveX device	style	style
All other devices	style	device entry parameters

For information on device entries, see "What Is a SAS/GRAPH Device?" on page 68 and "Viewing and Modifying Device Entries" on page 85. See also "Changing the Appearance of Output to Match That of Earlier SAS Releases" on page 154 and "Turning Off Styles" on page 153.

About Style Templates

An ODS style is a collection of named *style elements* that provides specific visual attributes for your graphical and tabular SAS output. Each style element is a named collection of *style attributes* such as background color, text color, marker symbol, line style, font face, font size, as well as many others. Each graphical element of a plot, such as a marker, a bar, a line or a title, derives its visual attributes from a specific style element from the current style.

Note: The style that a destination uses is applied to tabular output as well as graphical output. △

ODS Destinations and Default Styles

Every ODS output destination, except the Document and Output destinations, has a default style associated with it. These styles are tailored for each destination, therefore your output might look different depending on which destination you use. If your program does not specify a style, SAS uses the styles listed in Table 10.3 on page 135.

Table 10.3 Default Style Templates

ODS Destination	Default Style Name
DOCUMENT	Not applicable
LISTING	Listing
OUTPUT	Not applicable
HTML	Default (Styles.Default)
LATEX	Default (Styles.Default)
PRINTER	Printer
RTF	Rtf
Measured RTF	Rtf

The default style for each destination is set in the SAS registry. Changing the style specified in the SAS registry can be a convenient way to apply a company's style to all output sent to all destinations. See "Changing the Default Style in the SAS Registry" on page 139.

Chapter 3, "Getting Started With SAS/GRAPH," on page 39 shows examples of graphs using several styles, including the default styles for the most commonly used destinations. "Examples of Output Using Different Styles" on page 136 shows examples of graphs and tables using the Printer, Rtf, Analysis, and Journal styles.

Recommended Styles

SAS provides a set of styles that have been designed by GUI experts to address the needs of different situations. Table 10.4 on page 136 describes a subset of the styles provided by SAS that are particularly well-suited to displaying graphics.

Table 10.4 Recommended Style Templates

Desired Output	Recommended Styles	Comments
Full Color	Default (Styles.Default)	Gray background, optimized for HTML output
	Analysis	Yellow background
	Statistical	White background, colored fills
	Listing	White background, optimized for color format on white paper
	Printer	White background; serif fonts; optimized for PS and PDF output
	Rtf	Similar to Printer; optimized for RTF output
Black and White	onochromePrinter	Black and white output; patterned fills; optimized for PCL output
	Journal2	Interior filled areas have no color
Gray Scale	Journal	Interior filled areas are gray scale

Note: Certain ODS styles map textures onto graph elements. For the Java devices, these textures can be applied to two-dimensional rectangles only. Therefore, styles with textures cannot be applied to three-dimensional bar and pie charts in Java graphs. △

Chapter 3, "Getting Started With SAS/GRAPH," on page 39 shows examples of graphs using several styles, including the default styles for the most commonly used destinations. "Examples of Output Using Different Styles" on page 136 shows examples of graphs and tables using the Printer, Rtf, Analysis, and Journal styles.

Examples of Output Using Different Styles

Each of the following sets of output was created using a different style. Additional examples of output in Chapter 3, "Getting Started With SAS/GRAPH," on page 39.

Figure 10.1 Output Using the Printer Style

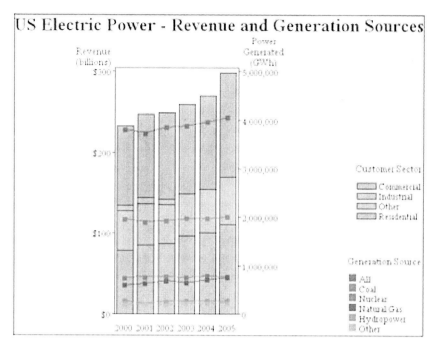

Figure 10.2 Output Using The RTF Style

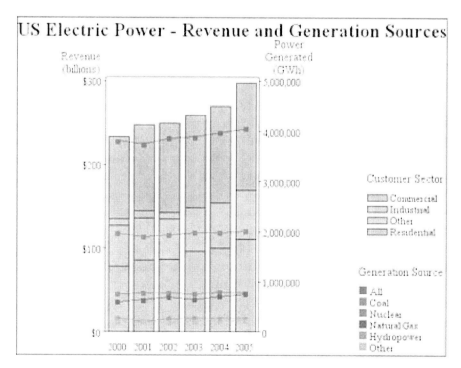

Figure 10.3 Output Using The Analysis Style

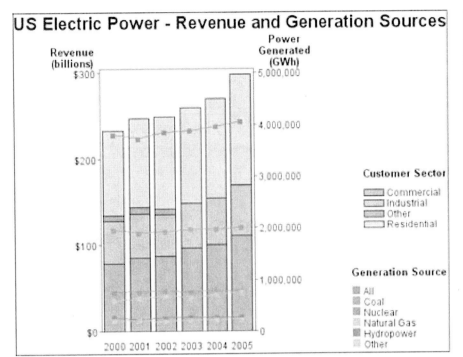

Figure 10.4 Output Using The Journal Style

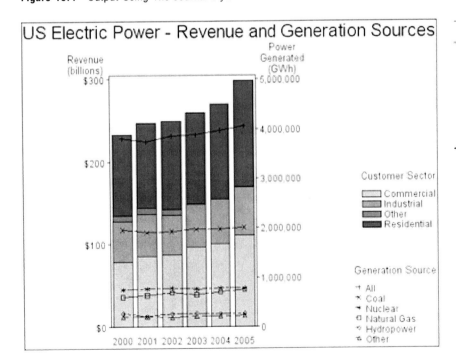

Note: The table in Figure 10.4 on page 138 was sent to the PDF destination. △

Specifying a Style

Changing the Current Style by Using the STYLE= Option in ODS Destination Statements

Changing the current style for an ODS destination is the easiest, simplest way of changing the appearance of your output. Changing the current style requires only the use of the STYLE= option in an ODS destination statement. By specifying only STYLE=*style-definition* in your ODS destination statement, you can create an entirely different appearance for your graphs. For example, you can specify that ODS apply the Styles.Journal style template to all HTML output with one of the following statements:

```
ods html style=styles.journal;
ods html style=journal;
```

This style is applied to all output for that destination until you change or close the destination or start a new SAS session.

Changing the Default Style in the SAS Registry

By default, the SAS registry applies a default style to the output for each ODS destination. The default styles for each destination are listed in Table 10.3 on page 135. To permanently change the default style associated with a destination, you can change the setting of Selected Style in the SAS registry.

CAUTION:
  If you make a mistake when you modify the SAS registry, then your system might become unstable or unusable. See "Managing the SAS Registry" in *SAS Language Reference: Concepts*. △

Note: You many have more than one SAS registry. Each site has a SAS registry in SASHELP. Each directory from which you run SAS has an individual registry in SASUSER. If you run SAS from multiple locations, and you want to change default styles via the SAS registry, you might need to change it in multiple locations. For more information, see "The SAS Registry" in *SAS Language Reference: Concepts*. △

For more information on ODS and the SAS registry, see "Changing SAS Registry Settings for ODS" in *SAS Output Delivery System: User's Guide*.

To permanently change the default style for a particular destination:

1. Select **Solutions ▶ Accessories ▶ Registry Editor**, or issue the command REGEDIT in the SAS command line.

2. Select **ODS ▶ Destinations**.

3. Select the destination that you want to change the default style for.

4. Select **Selected Style**, right-click, and select **Modify**. The Edit String Value window appears.

5. Type the style in the Value Data text box and click **OK**.

Display 10.1 SAS Registry Showing Selected Style Setting

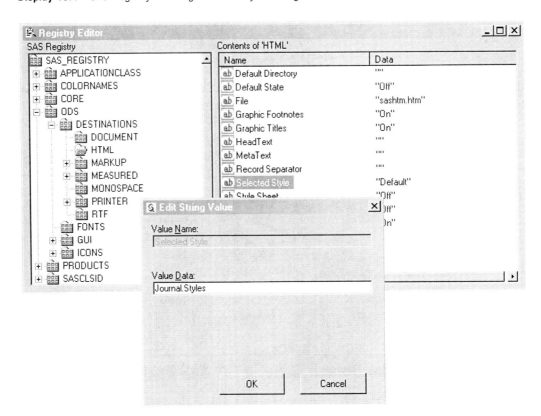

Overriding Style Attributes With SAS/GRAPH Statement Options

By default, the attributes of various elements of the graph are derived from specific style elements (or from device entry parameters if the NOGSTYLE system option is in effect), unless explicitly overridden with procedure or global statement options. For example, you can use the CTITLE= and CTEXT= options in the GOPTIONS global statement to change the color of the text in all of your graphs. You can use the SYMBOL statement to specify colors for markers. The settings remain in effect until you change them or end your SAS session. For information on GOPTIONS, see "GOPTIONS Statement" on page 220 and "Specifying Colors in a GOPTIONS Statement" on page 168. See the examples in Chapter 14, "SAS/GRAPH Statements," on page 197.

Instead of specifying global options, which affect all of your SAS/GRAPH output, you can specify options on specific action statements that affect only the output produced by that statement. Values that you specify on procedure action statements override default style attributes (or device entry parameters) and global options. For an example, see Example 6 on page 1441.

The documentation for each option that overrides a style element includes the name of the style element and attribute. For example, the documentation for the CAXIS= option for the GCHART procedure includes the following style reference information:

CAXIS=

Style reference: Color attribute of the GraphAxisLines element

If you want to change the color of the same graphical elements that are affected by the CAXIS= option by modifying a style, then you need to modify the Color attribute of the GraphAxisLines element. See "Modifying a Style" on page 142 for more information.

Attributes that are used repeatedly might be best specified in an ODS style. However, if you have created a customized style, be aware that you might need to make this style available to anyone that you send your SAS code to.

Attributes that are used only once or occasionally are best specified using SAS/GRAPH statements.

Precedence of Appearance Option Specifications

When you specify options that override style attributes or device parameters, the general order of precedence that SAS/GRAPH uses is as follows:

1 options in a SAS/GRAPH procedure action statement
2 options in AXIS, FOOTNOTE, LEGEND, NOTE, PATTERN, SYMBOL, or TITLE statements
3 graphics options in a GOPTIONS statement

   a color options in the GOPTIONS statement that control specific graph elements such as the background color or title text color
   b the color list specified with the COLORS= option in the GOPTIONS statement

4 attributes specified in the current style or, if the NOGSTYLE option is in effect, device parameters in a device entry for the current device
5 default hardware settings for a device.

SAS/GRAPH uses the first specification it finds in this list. Any exceptions to this rule are noted in the documentation for the specific option as described in "Overriding Style Attributes With SAS/GRAPH Statement Options" on page 140.

Viewing the List of Styles Provided by SAS

You can view the styles that SAS provides using the TEMPLATE procedure or through the Templates window.

Using The TEMPLATE Procedure

To view the list of all styles available, submit the following code:

```
proc template;
    list styles;
run;
```

SAS writes the list of available styles in the Output window.

Using the Templates Window

To view the list of all styles available, follow these steps:

1 Open the Templates window. You can open the Templates window in two ways:
   □ Enter the **odstemplates** command on the SAS command line.
   □ In the Results window, select the Results folder. Right-click and select **Templates** to open the Templates window.

The Templates window contains the item stores `Sasuser.Templat` and `Sashelp.Tmplmst`.

2 Double-click an item store, such as `Sashelp.Tmplmst`, to expand the list of directories where ODS templates are stored. The templates that SAS provides are in the item store `Sashelp.Tmplmst`.

3 Double-click `Styles` to view the list of styles defined in the selected item store.

4 Double-click the style definition that you want to view. For example, the Default style definition is the template store for HTML output. Similarly, the Rtf style definition is the template store for RTF output.

To view the actual style definition, double-click on a style name. The style definition is displayed in the Template Browser window.

Modifying a Style

Using the TEMPLATE Procedure

Within the TEMPLATE procedure, you can use the DEFINE STYLE statement to create a completely new style or you can start from an existing style. When you create styles from existing styles, you can modify the individual style elements.

For complete documentation on using PROC TEMPLATE to modify and create styles, see "TEMPLATE Procedure: Creating a Style Definition" in *SAS Output Delivery System: User's Guide*.

Example: Modifying a Style Element

The style element GraphData1 is defined in the Default style as follows:

```
proc template;
   define style Styles.Default;
       ...more style elements...
   class GraphData1 /
          markersymbol = "circle"
          linestyle = 1
          contrastcolor = GraphColors('gcdata1')
          color = GraphColors('gdata1');
```

You can use the DEFINE STYLE statement in the TEMPLATE procedure to create a new style from the Default style and modify the GraphData1 style element. The following program creates the new style MyStyleDefault, which inherits all of its style elements and style attributes from the Default style, and modifies the GraphData1 style element:

```
proc template;
   define style MyStyleDefault;
   parent=Styles.Default;
   style GraphData1 from GraphData1 /
          markersymbol = "triangle"
          linestyle = 2
          contrastcolor = GraphColors("gcdata1")
```

```
            color = GraphColors("gdata1");
      end;
run;
```

The new GraphData1 uses the same colors as the original GraphData1, but specifies a different marker symbol and line style.

To use the new MyStyleDefault style for HTML output, specify the STYLE= option:

```
ods html style=MyStyleDefault;
```

Ways to Modify Graph Fonts Or Colors Specified By Styles

There are different ways to change the fonts or colors used by a style. Which method you choose depends on how extensively you want to change the font or color specifications used in your output. You can do any of the following:

- Modify a specific style element that controls a specific graphical element. For example, the GraphValueText element specifies the font and color for tick mark values and legend value descriptions. You could change the font or color specified by the GraphValueText element for the Analysis style. Changes to specific style elements affect only the graphical elements they control and affect them in only the styles where you change them. See "Style Elements For Use With Device-Based SAS/GRAPH Output" on page 146 for information on the specific style elements that you can modify.

- Modify the font or color specifications in the GraphFonts or GraphColors style elements for a specific style. The settings specified in GraphFonts and GraphColors are referenced by specific style elements elsewhere in the style. Other style elements that reference the GraphFonts or GraphColors style elements use the modified settings. See "The GraphFonts Style Element" on page 145 and "The GraphColors Style Element" on page 144 for more information. A single change in the specifications in the GraphFonts or GraphColors style elements can potentially change the appearance of several graphical elements and affect output of any style that refers to GraphFonts or GraphColors.

- Modify the font settings for one or more subkeys in the SAS registry. Many styles refer to the font settings in the SAS registry to determine the fonts to use for various graphical elements. Modifying the SAS registry settings changes the fonts used for all styles that refer to the subkeys that you change. See *SAS Output Delivery System: User's Guide* for information on changing SAS registry settings. (Colors used by the styles supplied by the company are not controlled through the SAS registry.)

Modifying the GraphFonts And GraphColors Style Elements

The attributes in the GraphFonts and GraphColors style elements are used as the values for specific style elements elsewhere in the style. In other words, the GraphFonts and GraphColors elements are *abstract elements*. They are used to assign values to other elements.

For example, the GraphFonts element could be defined follows:

```
class GraphFonts
      "Fonts used in graph styles" /
      'GraphDataFont' = ("<sans-serif>, <MTsans-serif> ",7pt)
      'GraphValueFont' = ("<sans-serif>, <MTsans-serif>",9pt)
      'GraphLabelFont' = ("<sans-serif>, <MTsans-serif> ",10pt,bold)
      'GraphFootnoteFont' = ("<sans-serif>, <MTsans-serif>",10pt)
```

```
'GraphTitleFont' = ("<sans-serif>, <MTsans-serif>",11pt,bold);
```

Each attribute, GraphDataFont, GraphValueFont, GraphLabelFont, and so on, defines a list of fonts for use by SAS/GRAPH whenever the corresponding attribute is referenced. These attributes are specified elsewhere in the style as the value of a another font attribute. (For information on the syntax used in the GraphFonts style element, see "Font Specifications In The GraphFonts Style Element" on page 146.)

For example, the GraphValueText element specifies the font and color for tick mark values and legend value descriptions. Suppose the GraphValueText element is defined as follows:

```
class GraphValueText /
       font = GraphFonts('GraphValueFont')
       color = GraphColors('gtext');
```

The font and color for GraphValueText are specified by elements in the GraphFonts and GraphColors style elements.

GraphFonts('GraphValueFont')
  tells SAS/GRAPH to use the font specified by the GraphValueFont attribute in the GraphFonts style element.

GraphColors('gtext')
  tells SAS/GRAPH to use the color specified by the gtext attribute in the GraphColors style element.

To change the font and color for tick mark values and legend value descriptions, you could modify either of the following:

- the FONT= and COLOR= attributes in the GraphValueText element
- the GraphValueFont attribute in the GraphFonts style element and the gtext attribute in the GraphColors style element.

However, because elements in GraphFonts and GraphColors are referred to by other elements in the style, changing the values in GraphFonts and GraphColors result in more extensive changes than modifying a specific style element such as GraphValueText directly. If you modify the GraphValueText element directly, your modifications affect only the items controlled by GraphValueText. If you modify the GraphValueFont or gtext attributes, then your modifications might affect other portions of the graph in addition to tick mark values and legend value descriptions. This list includes pie labels, regression equations, data point labels, bar labels, and graph titles.

The styles supplied with SAS/GRAPH are designed to provide a consistent visual appearance for all graphical elements in your output. Modifying attributes in the GraphFonts or GraphColors elements instead of modifying several specific style elements makes it easier to maintain the consistent appearance in your output.

The tables listed in "Graphical Style Element Reference for Device-Based Graphics" on page 144 describe the portions of SAS/GRAPH output that are affected by elements and attributes defined in the styles.

Graphical Style Element Reference for Device-Based Graphics

The GraphColors Style Element

The GraphColors style element specifies the colors that are used for different categories of graphical elements. Table 10.5 on page 145 lists the style attributes that are defined in the GraphColors style element and the graphical elements that they affect by default.

Table 10.5 GraphColors Attributes For Device-Based Output

GraphColors Attribute[1]	Portion of Graph Affected
gaxis	Axis lines and tick marks
gborderlines	Border around the graph wall, legend border, and borders to complete axis frame
gconnectLine	Line for connecting boxes
gfloor	Graph floor
ggrid	Grid lines
glabel	Axis labels and legend titles
glegend	Background of the legend
goutline	Outlines for data primitives such as bars, pie slices, and boxes
gshadow	Drop shadows used with text
gtext[2]	Graph titles, tick mark values, and legend value descriptions
gwalls	Frame area in two-dimensional graphs and vertical walls in three-dimensional graphs
gdata1–gdata12	Data items; gdata1–gdata12 apply to filled areas; gcdata1–gcdata12 apply to markers and lines
gcdata1–gcdata12	
gramp2cstart	Gradient contours, surfaces, continuous choropleth maps, and continuous block maps when areas are not used
gramp2cend	
gconramp2cstart	Continuous block maps when areas are used
gconramp2cend	

1 Elements in the GraphColors style element that are not included in this table are used with template-based (ODS Graphics) output only. (See "Device-Based Graphics And Template-Based Graphics" in Chapter 1, "Introduction to SAS/GRAPH Software".)

2 The gtext attribute does not affect text that is not rendered as part of the graph. See also "Controlling Titles and Footnotes with Java and ActiveX Devices in HTML Output" in Chapter 13, "Managing Your Graphics With ODS".

The GraphFonts Style Element

The GraphFonts style element specifies the fonts that are used for different categories of graphical elements. Table 10.6 on page 146 lists the style attributes that are defined in the GraphFonts style element and the graphical elements that they affect by default.

Table 10.6 GraphFonts Attributes For Device-Based Output

GraphFonts Attributes*	Portion of Graph Affected
GraphDataFont	Contour labels
GraphValueFont	Axis tick mark labels, legend value description labels, data values in statistics tables, pie labels, regression equations, data point labels, bar labels
GraphLabelFont	Axis labels, legend labels, column headings in statistics tables
GraphFootnoteFont	Footnotes
GraphTitleFont	Titles

* The GraphUnicode and GraphAnnoFont attributes are used with ODS graphics only.

Font Specifications In The GraphFonts Style Element

Font definitions in the GraphFonts style element can refer to registry entries, they can specify a specific font, or they can specify a font family. For example:

```
'GraphLabelFont' = ("<MTsans-serif>, Arial, sans-serif",10pt,bold)
```

<MTsans-serif>
specifies the font family identified by the **MTsans-serif** subkey in the SAS registry. The less than and greater than signs tell SAS that this is the name of a subkey in the SAS registry. Because it is the first font listed, SAS uses this font if possible. To view the font settings in the SAS registry, select **ODS ▸ FONTS** in the SAS registry. See *SAS Output Delivery System: User's Guide* for information on changing SAS registry settings.

Arial
specifies the Arial font family. If SAS cannot find the first font listed, it tries to find the second font listed.

sans-serif
specifies the san-serif font family. If SAS cannot find the specific fonts listed, then it looks for a font in the san-serif font family.

10pt,bold
specifies the weight and style that should be used.

In this example, if the SAS registry entry for the **MTsans-serif** subkey specifies Albany AMT, then SAS/GRAPH first tries to use the Albany AMT 10 point bold font. If it cannot find this font, then it tries to use Arial 10 point bold, and so on.

Note: SAS might not be able to find a specific font unless it is registered with the FONTREG procedure. The fonts provided by SAS are already registered. If you want to add additional fonts, see *SAS Language Reference: Concepts* for information on registering TrueType fonts. See *Base SAS Procedures Guide* for information on the FONTREG procedure. △

Style Elements For Use With Device-Based SAS/GRAPH Output

The style elements listed in the following tables affect SAS/GRAPH output and can be used in styles. These tables list each style element, the portion of the graph it affects or was created to use with, and its attribute values. Attribute values can be changed with PROC TEMPLATE, as described in "Using the TEMPLATE Procedure" on page 142 and "Example: Modifying a Style Element" on page 142. For complete documentation on the style attributes that can be specified in each style element, see "Style Attributes and Their Values" in the section "TEMPLATE Procedure: Creating a Style Definition" in *SAS Output Delivery System: User's Guide*.

Table 10.7 Device-Based Graph Style Elements: General Graph Appearance

Style Element	Portion of Graph Affected	Recognized Attributes	Attribute Values in DEFAULT Style
DropShadowStyle	Used with text types	Color	GraphColors("gshadow")
Graph	Graph size and outer border appearance	OutputWidth	Not set
		OutputHeight	Not set
		BorderColor	Inherited
		BorderWidth	Inherited
		CellPadding	0
		CellSpacing	Inherited
GraphAxisLines	X, Y, and Z axis lines	Color	GraphColors("gaxis")
		LineStyle	1
		LineThickness	1px
GraphBackground	Background of the graph	Transparency	Not set
		BackgroundColor	Colors("docbg")
		Gradient_Direction	Not set
		StartColor	Not set
		EndColor	Not set
		BackgroundImage	Not set
		Image	Not set
		VerticalAlign	Not set
		TextAlign	Not set
GraphBorderLines	Border around graph wall, legend border, borders to complete axis frame	Color	GraphColors("gborderlines")
		LineThickness	1px
		LineStyle	1
GraphCharts	All charts within the graph	Transparency	Not set
		BackgroundColor	Not set
		Gradient_Direction	Not set
		StartColor	Not set
		EndColor	Not set
		BackgroundImage	Not set
		Image	Not set
		VerticalAlign	Not set
		TextAlign	Not set

Style Element	Portion of Graph Affected	Recognized Attributes	Attribute Values in DEFAULT Style
GraphDataText	Text font and color for point and line labels	Font or *font-attributes** Color	GraphFonts("GraphDataFont") Not set GraphColors("gtext")
GraphFloor	3D floor	BackgroundColor Transparency Gradient_Direction StartColor EndColor BackgroundImage Image VerticalAlign TextAlign	GraphColors("gfloor") Not set Not set Not set Not set Not set Not set Not set Not set
GraphFootnoteText	Text font and color for footnotes	Font or *font-attributes** Color	GraphFonts("GraphFootnoteFont") Not set GraphColors("gtext")
GraphGridLines	Horizontal and vertical grid lines drawn at major tick marks	Color LineStyle LineThickness Transparency displayopts	GraphColors("ggrid") 1 1px .5 "Auto"
GraphGridLines	Horizontal and vertical grid lines drawn at major tick marks	Color LineStyle LineThickness Transparency displayopts	GraphColors("ggrid") 1 1px .5 "Auto"
GraphLegendBackground	Background color of the legend	Color Transparency	Colors("glegend") Not set
GraphOutlines	Outline properties for fill areas such as bars, pie slices, and box plots.	Color LineStyle LineThickness	GraphColors("goutlines") 1 1px
GraphTitleText	Text font and color for titles	Font or *font-attributes** Color	GraphFonts("GraphTitleFont") Not set GraphColors("gtext")

Style Element	Portion of Graph Affected	Recognized Attributes	Attribute Values in DEFAULT Style
GraphValueText	Text font and color for axis tick values and legend values	Font or *font-attributes** Color	GraphFonts("GraphValueFont") Not set GraphColors("gtext")
GraphWalls	Vertical walls bounded by axes	Transparency BackgroundColor Gradient_Direction StartColor EndColor BackgroundImage Image	Not set GraphColors("gwalls") Not set Not set Not set Not set Not set

* *Font-attributes* can be one of the following: FONTFAMILY=, FONTSIZE=, FONTSTYLE=, FONTWEIGHT=.

Table 10.8 Style Elements Affecting Device-Based Non-Grouped Graphical Data Representation

Style Element	Portion of Graph Affected	Default Attributes	Attribute Values in DEFAULT Style
ThreeColorAltRamp	Line contours, markers, and data labels with segmented range color response	StartColor NeutralColor EndColor	GraphColors("gconramp3start") GraphColors("gconramp3cneutral") GraphColors("gconramp3end")
ThreeColorRamp	Gradient contours, surfaces, markers, nad data labels with continuous color response	StartColor NeutralColor EndColor	GraphColors("gramp3cstart") GraphColors("gramp3cneutral") GraphColors("gramp3cend")
TwoColorAltRamp	Line contours, markers, and data labels with segmented range color response	StartColor EndColor	GraphColors("gconramp2cstart") GraphColors("gconramp2cend")
TwoColorRamp	Gradient contours, surfaces, markers, and data labels with continuous color response	StartColor EndColor	GraphColors("gramp2cstart") GraphColors("gramp2cend")

Table 10.9 Style Elements Affecting Device-Based Grouped Graphical Data Representation

Style Element	Portion of Graph Affected	Default Attributes	Attribute Values in DEFAULT Style
GraphData1	Primitives related to 1st grouped data items. Color applies to filled areas. ContrastColor applies to markers and lines.	Color	GraphColors("gdata1")
		ContrastColor	GraphColors("gcdata1")
		MarkerSymbol	"Circle"
		LineStyle	1
		MarkerSize	Not set
		LineThickness	Not set
		Gradient_Direction	Not set
		StartColor	Not set
		EndColor	Not set
		BackGroundImage	Not set
		Image	Not set
GraphData2	Primitives related to 2nd grouped data items	Color	GraphColors("gdata2")
		ContrastColor	GraphColors("gcdata2")
		MarkerSymbol	"Plus"
		LineStyle	4
		MarkerSize	Not set
		LineThickness	Not set
		Gradient_Direction	Not set
		StartColor	Not set
		EndColor	Not set
		BackGroundImage	Not set
		Image	Not set
GraphData3	Primitives related to 3rd grouped data items	Color	GraphColors("gdata3")
		ContrastColor	GraphColors("gcdata3")
		MarkerSymbol	"X"
		LineStyle	8
		MarkerSize	Not set
		LineThickness	Not set
		Gradient_Direction	Not set
		StartColor	Not set
		EndColor	Not set
		BackGroundImage	Not set
		Image	Not set

Style Element	Portion of Graph Affected	Default Attributes	Attribute Values in DEFAULT Style
GraphData4	Primitives related to 4th grouped data items	Color	GraphColors("gdata4")
		ContrastColor	GraphColors("gcdata4")
		MarkerSymbol	"triangle"
		LineStyle	5
		MarkerSize	Not set
		LineThickness	Not set
		Gradient_Direction	Not set
		StartColor	Not set
		EndColor	Not set
		BackGroundImage	Not set
		Image	Not set
GraphData5	Primitives related to 5th grouped data items	Color	GraphColors("gdata5")
		ContrastColor	GraphColors("gcdata5")
		MarkerSymbol	"square"
		LineStyle	14
		MarkerSize	Not set
		LineThickness	Not set
		Gradient_Direction	Not set
		StartColor	Not set
		EndColor	Not set
		BackGroundImage	Not set
		Image	Not set
GraphData6	Primitives related to 6th grouped data items	Color	GraphColors("gdata6")
		ContrastColor	GraphColors("gcdata6")
		MarkerSymbol	"Asterisk"
		LineStyle	26
		MarkerSize	Not set
		LineThickness	Not set
		Gradient_Direction	Not set
		StartColor	Not set
		EndColor	Not set
		BackGroundImage	Not set
		Image	Not set

Style Element	Portion of Graph Affected	Default Attributes	Attribute Values in DEFAULT Style
GraphData7	Primitives related to 7th grouped data items	Color	GraphColors("gdata7")
		ContrastColor	GraphColors("gcdata7")
		MarkerSymbol	"Diamond"
		LineStyle	15
		MarkerSize	Not set
		LineThickness	Not set
		Gradient_Direction	Not set
		StartColor	Not set
		EndColor	Not set
		BackGroundImage	Not set
		Image	Not set
GraphData8	Primitives related to 8th grouped data items	Color	GraphColors("gdata8")
		ContrastColor	GraphColors("gcdata8")
		MarkerSymbol	Not set
		LineStyle	20
		MarkerSize	Not set
		LineThickness	Not set
		Gradient_Direction	Not set
		StartColor	Not set
		EndColor	Not set
		BackGroundImage	Not set
		Image	Not set
GraphData9	Primitives related to 9th grouped data items	Color	GraphColors("gdata9")
		ContrastColor	GraphColors("gcdata9")
		MarkerSymbol	Not set
		LineStyle	41
		MarkerSize	Not set
		LineThickness	Not set
		Gradient_Direction	Not set
		StartColor	Not set
		EndColor	Not set
		BackGroundImage	Not set
		Image	Not set

Style Element	Portion of Graph Affected	Default Attributes	Attribute Values in DEFAULT Style
GraphData10	Primitives related to 10th grouped data items	Color	GraphColors("gdata10")
		ContrastColor	GraphColors("gcdata10")
		MarkerSymbol	Not set
		LineStyle	42
		MarkerSize	Not set
		LineThickness	Not set
		Gradient_Direction	Not set
		StartColor	Not set
		EndColor	Not set
		BackGroundImage	Not set
		Image	Not set
GraphData11	Primitives related to 11th grouped data items	Color	GraphColors("gdata11")
		ContrastColor	GraphColors("gcdata11")
		MarkerSymbol	Not set
		LineStyle	2
		MarkerSize	Not set
		LineThickness	Not set
		Gradient_Direction	Not set
		StartColor	Not set
		EndColor	Not set
		BackGroundImage	Not set
		Image	Not set
GraphData12	Primitives related to 12th grouped data items	Color	GraphColors("gdata12")
		ContrastColor	GraphColors("gcdata12")
		MarkerSymbol	Not set
		LineStyle	Not set
		MarkerSize	Not set
		LineThickness	Not set
		Gradient_Direction	Not set
		StartColor	Not set
		EndColor	Not set
		BackGroundImage	Not set
		Image	Not set

Turning Off Styles

To turn off styles, specify the SAS system option NOGSTYLE. To change the setting of the SAS system option from GSTYLE to NOGSTYLE, you can do either of the following:

- Submit the following OPTIONS statement:

```
OPTIONS NOGSTYLE;
```

- Enter **OPTIONS** on the SAS command line, or select **Tools ▸ Options ▸ System** to open the SAS System Options window. Expand **Graphics**, and select **Driver settings**. Right-click on **Gstyle**, select **Modify value**, and select *0=False* as the new value.

Changing the Appearance of Output to Match That of Earlier SAS Releases

SAS/GRAPH 9.2 introduces many new features that significantly change the default appearance of your SAS/GRAPH output. To produce output that looks as if it was produced with previous versions of SAS/GRAPH, do the following:

- Specify the NOGSTYLE system option. This option turns off the use of ODS styles. See "Turning Off Styles" on page 153.
- Specify the FONTRENDERING=HOST_PIXELS system option. This option specifies whether devices that are based on the SASGDGIF, SASGDTIF, and SASGDIMG modules render fonts by using the operating system or by using the FreeType engine. This option applies to certain native SAS/GRAPH devices (see "Device Categories And Modifying Default Output Attributes" on page 72). For example, this option works for GIF, TIFFP, JPEG, and ZPNG devices, but it is not applicable to PNG, SVG, or SASPRT* devices.
- Specify DEVICE=ZGIF on the GOPTIONS statement when you are sending output to the HTML destination.
- In other cases where your application specifies a device, specify a compatible Z device driver, if applicable. See "Devices" on page xvii for more information.

CHAPTER
11

Specifying Fonts in SAS/GRAPH Programs

Introduction: Specifying Fonts in SAS/GRAPH Programs **155**
SAS/GRAPH, System, and Device-Resident Fonts **155**
TrueType Fonts That Are Supplied by SAS **156**
Determining What Fonts Are Available **157**
Default Fonts **157**
Viewing Font Specifications in the SAS Registry **158**
Specifying a Font **159**
  Specifying Font Modifiers (/bold, /italic, and /unicode) **159**
  Using a Registry Subkey **159**
  Specifying International Characters (Unicode Encoding) **159**
  Specifying Special Characters Using Character and Hexadecimal Codes **160**
Methods For Specifying Fonts **163**
  Using SAS/GRAPH Global Statement Options to Specify Fonts **164**
  Using GOPTIONS to Specify Fonts **164**
  Changing The Font Specifications Used By a Style **165**
  Precedence of Font Specifications **165**

Introduction: Specifying Fonts in SAS/GRAPH Programs

SAS/GRAPH provides access to a variety of fonts, or typefaces, to display text and special characters for your graphics output. SAS provides a number of TrueType fonts that you can use in your applications. By default, ODS styles use system fonts, including the TrueType fonts shipped with SAS, for the various titles, labels, and other text in SAS/GRAPH output. You can modify the default fonts by modifying the styles, by specifying graphics options, or by using font options in procedure action statements. You can specify special characters using character codes or hexadecimal codes.

SAS/GRAPH, System, and Device-Resident Fonts

There are three types of fonts that you can use when you generate output with SAS/GRAPH.

SAS/GRAPH fonts
   fonts stored in the SASHELP.FONTS catalog, and fonts created by the user and stored in a GFONT*n* catalog. These fonts can be used only by SAS/GRAPH procedures or other procedures that generate GRSEG output files. Examples of SAS/GRAPH fonts include Swiss, Simulate, and Marker. These fonts are provided for specialized purposes only. For information on these fonts, see Appendix 2, "Using SAS/GRAPH Fonts," on page 1643.

system fonts
: fonts that can be used by any SAS procedure and by other software, such as Microsoft Word. These fonts include TrueType and Type1 fonts. Examples of system fonts include Albany AMT, Monotype Sorts, and Arial. Some system fonts, such as Helvetica, can also be present as device-resident fonts. System fonts are installed on the operating system, and then registered with SAS using the FONTREG procedure. System fonts generally provide the highest quality output. SAS/GRAPH installs and registers a set of TrueType fonts, and it is recommended that you use these fonts whenever possible. See "TrueType Fonts That Are Supplied by SAS" on page 156 for more information.

device-resident fonts
: fonts that are burned into the chips in a device's hardware. These fonts are specific to the device being used and are not portable between devices. Some device-resident fonts, such as Helvetica, can also be present as system fonts.

TrueType Fonts That Are Supplied by SAS

SAS/GRAPH installs and registers a set of TrueType fonts, that are referred to collectively as system fonts. TrueType fonts that are shipped with SAS are listed in Table 11.1 on page 156.

You can use these fonts in your SAS programs by assigning the font name to font options, enclosed in quotes. For example, you can specify the following:

```
goptions ftext="Thorndale AMT";
```

Table 11.1 TrueType Fonts Supplied by SAS

Albany AMT*	Thorndale Duospace WT SC	GungsuhChe
Cumberland AMT*	Thorndale Duospace WT TC	Dotum
Thorndale AMT*	Arial Symbol*	DotumChe
Symbol MT	Times New Roman Symbol*	Gulim
Monotype Sorts	MS PMincho	GulimChe
Monotype Sans WT J	MS Mincho	NSimSun
Monotype Sans WT K	MS PGothic	SimHei
Monotype Sans WT SC	MS UI Gothic	SimSun
Monotype Sans WT TC	Batang	PMingLiU
Thorndale Duospace WT J	BatangChe	MingLiU
Thorndale Duospace WT K	Gungsuh	HeiT

* Albany AMT, Cumberland AMT, Thorndale AMT, Arial Symbol, and Times New Roman Symbol are font families. Normal, bold, italic, and bold italic versions of these fonts are provided.

For more information about using TrueType fonts with SAS/GRAPH, see *SAS Language Reference: Concepts*.

Determining What Fonts Are Available

The fonts listed in Table 11.1 on page 156 are available on all systems where SAS is installed. It is recommended that you use these fonts when possible. Additional system fonts that are available to your application and the methods for determining those fonts depend on the following:

- the operating environment that you are working in
- the device or universal printer that you are using

For more information on determining what fonts are available, see *SAS Language Reference: Concepts* and the SAS documentation for your operating environment.

You can add additional fonts to your system for use by SAS/GRAPH, but all fonts must be registered with the FONTREG procedure. See *Base SAS Procedures Guide* for more information.

All of the fonts that have been registered with the FONTREG procedure are listed in the SAS registry. To view the list of registered fonts, follow these steps:

1 Open the registry editor by either selecting **Solutions ▶ Accessories ▶ Registry Editor** or by issuing the command REGEDIT in the command line.

2 Select **CORE ▶ PRINTING ▶ FREETYPE ▶ FONTS**.

The SAS/GRAPH fonts are available on all systems where SAS/GRAPH is installed, but they are provided primarily for special uses. See Appendix 2, "Using SAS/GRAPH Fonts," on page 1643 for more information.

When you are deciding what font to use, consider all operating environments in which your SAS code will be run. For example, if you specify a font, such as Arial, that is available only on Windows systems, then your output will appear different on other systems. If you specify one of the fonts that is installed with SAS (see Table 11.1 on page 156) then your output will appear the same on all systems. It is recommended that you use system fonts whenever possible.

Default Fonts

Many of the default fonts are specified in the SAS registry. See "Viewing Font Specifications in the SAS Registry" on page 158. The SAS registry is localized, so fonts that are specified by the SAS registry are dependent on your locale.

For most devices, when you are using styles (the GSTYLE system option is in effect), fonts are specified by the current style. Each style specifies fonts for various graph elements such as axis labels, graph titles, tick mark labels, and so on. See "Modifying the GraphFonts And GraphColors Style Elements" on page 143 and "The GraphFonts Style Element" on page 145 for more information about the font specifications in the styles. See "Style Attributes Versus Device Entry Parameters" on page 134 for more information on the GSTYLE system option.

Table 11.2 on page 158 shows the fonts used when styles are not active (the NOGSTYLE system option is in effect).

Table 11.2 Fonts Used By Default When NOGSTYLE System Option Is In Effect

Device	TITLE1	All Other Text
GIF, JPEG, PNG, TIFF, SVG, SASBMP, SASPRTx, GIFANIM, PCL, PS, PDF	Swiss	Font specified by `<MTmonospace>` subkey in the SAS registry
BMP, ZGIF, ZPNG, ZJPEG, ZTIFF in all environments except z/OS EMF and WMF on Windows Display devices for Graph window: WIN, XCOLOR, IBMPCGX	Swiss	Font specified by DMSFont
SASEMF, SASWMF, EMF and WMF in other operating environments	Swiss	Font specified by `<MTmonospace>` subkey in the SAS registry
JAVAMETA	Swiss	Font specified in the **Chartype 1** field in the device entry
CGM, ZPCL, ZPS, ZPDF	Swiss	Font specified by the device entry

The DMS font is controlled by the FONT= option on Windows, by the Xdefaults X resources on UNIX, and by the host display code on z/OS. For more information, see the SAS documentation for your operating environment.

Note: In some cases, SAS/GRAPH can switch to Simulate. When styles are turned off (the NOGSTYLE system option is in effect), the only default font that is scalable is the font used for TITLE1. If the height specified for other fonts is not equal to one, then SAS/GRAPH switches to Simulate. See "The SIMULATE Font" on page 1652 for more information. △

The Java and ActiveX devices ignore the NOGSTYLE system option; they always use styles. When you are using the Java and ActiveX devices (which always use styles), the fonts are determined at run time. The fonts are resolved based on the fonts available on the system where the graph is viewed. When you use the JAVA or ACTIVEX device, the fonts specified by the styles are also specified in the HTML or RTF file that is generated. When the file is viewed, if a font is not available, the font mapper on the system where the file is viewed determines the font that is substituted.

See "Specifying a Font" on page 159 and "Methods For Specifying Fonts" on page 163 for information on how to override default font specifications.

Viewing Font Specifications in the SAS Registry

To view the font settings in the SAS registry, follow these steps:

1 Open the registry editor by either selecting **Solutions ▶ Accessories ▶ Registry Editor** or by issuing the command REGEDIT in the command line.

2 Select **ODS ▶ Fonts**.

Each entry in the registry consists of a name, such as `<MTsans-serif>` or `<MTmonospace>` followed by its value, such as "Albany AMT" or "Cumberland AMT".

Note: The Fonts key contains subkeys that specify which fonts to use based on the locale. △

For more information, see "The SAS Registry" in *SAS Language Reference: Concepts*.

Specifying a Font

To specify a font in your SAS program, include a font name, enclosed in quotes, anywhere fonts are supported. For example, you can specify Thorndale AMT as the font for legend labels as follows:

```
legend label=(font="Thorndale AMT"  "Generation Source");
```

You can change between fonts, specify font modifiers such as **/bold**, and specify special characters. Font names are not case-sensitive. For example, the following FOOTNOTE statement prints E=mc^2.

```
footnote font="Thorndale AMT/bold" "E=mc" font="Albany AMT" "b2"x;
```

Specifying Font Modifiers (/bold, /italic, and /unicode)

To add a modifier such as bold or italic to a font, follow the font name with */modifier*. For example:

```
axis1 value=(font="Cumberland AMT/bold/italic" );
```

SAS/GRAPH recognizes three font modifiers.

/bold or **/bo**
  specifies bold text.

/italic or **/it**
  specifies italic text.

/unicode or **/unic**
  specifies special characters using Unicode code points. See "Specifying International Characters (Unicode Encoding)" on page 159 for more information.

  Note: The **/unicode** modifier is not supported by the Java or ActiveX devices. △

Note: With the ACTIVEX and ACTXIMG devices you can specify only one modifier at a time. Specifying font modifiers is not supported by the JAVA or JAVAIMG devices. △

Note: You cannot specify font modifiers if you specify the font using a registry subkey. △

Using a Registry Subkey

You can specify a font by specifying a registry subkey such as **<MTsans-serif>** or **<MTmonospace>** instead of specifying a font name. For example:

```
title font="<MTsans-serif>" "My Title";
```

The font specified by the **<MTsans-serif>** registry subkey will be used for the title.

The SAS registry is localized. If you specify a font using a registry subkey, the actual font that is used will be the localized value specified in your registry.

See also "Viewing Font Specifications in the SAS Registry" on page 158 and "Font Specifications In The GraphFonts Style Element" on page 146.

Specifying International Characters (Unicode Encoding)

You can use the **/unicode** modifier with a hexadecimal code to print any character in a font. This modifier can be used only with fonts that support Unicode encoding. Most of the TrueType fonts listed in Table 11.1 on page 156 support Unicode encoding.

For example, the following statement uses the **/unicode** modifier and a hexadecimal code (see "Specifying Special Characters Using Character and Hexadecimal Codes" on page 160) to display the symbol for the Euro sign.

```
title "Euro Symbol" font="Albany AMT/unicode" "20ac"x;
```

Unicode Character Code Charts can be found on the Unicode Web site at **http://www.unicode.org/charts**. See also *SAS Language Reference: Concepts* for information on printing international characters.

The Java and ActiveX devices do not support the **/unicode** modifier.

Specifying Special Characters Using Character and Hexadecimal Codes

Some fonts contain characters that are not mapped to the keyboard and cannot be typed directly into a text string. To display these special characters, substitute a character code or a hexadecimal value in the text string. Hexadecimal values are recommended over character codes.

Note: You can also display special characters using *unicode code points*. Unicode code points are specified with the **/unicode** font modifier followed by a hexadecimal value. See "Specifying International Characters (Unicode Encoding)" on page 159 for more information. △

Character codes include the letters, numbers, punctuation marks, and symbols that are commonly found on a keyboard. They are usually associated with symbols or national alphabets. These codes enable you to display the character by specifying the font and using the keyboard character in the text string. For example, on Windows operating environments, to produce the character ζ, you can specify the Symbol MT font and the character code **z** in the text string.

```
title font="Symbol MT" "z";
```

Hexadecimal values are any two-digit hexadecimal numbers enclosed in quotation marks, followed by the letter x. For example, "3D"x. (In double-byte character sets, the hexadecimal values contain four digits, for example, "4E60"x. Unicode characters also contain four digits.)

You display characters with hexadecimal values the same way that you display them with character codes. You specify the font that contains the special character and place the hexadecimal value in the text string. For example, this TITLE statement uses hexadecimal A9 to produce © in the Albany AMT font.

```
title font="Albany AMT" "a9"x;
```

Note: The character code or hexadecimal value associated with characters in a font might be dependent on the key map that is currently being used. Keymaps are not used if the **/unicode** modifier is specified, a symbol font is specified, or NOKEYMAP is specified in the font header. Contact Technical Support if you need assistance with creating or modifying key maps. △

To determine the hexadecimal codes that you need to specify for a specific character, you can use the program shown in Example Code 11.1 on page 161. This program displays 224 characters of a font together with the hexadecimal codes for each character. As shown here, it displays the characters in the Symbol font. You can change the font displayed by this program to any font available on your system. Also, some fonts have many more characters than those displayed by the program below.

Note: Some fonts, such as Albany AMT, display variations due to the national characters for that locale. Symbol fonts, such as Monotype Sorts, are not affected by your locale encoding. For double-byte encodings, the second half of the table might be blank or show small rectangles. △

Example Code 11.1 SAS Program For Displaying Hexadecimal Codes For Special Characters

```
goptions reset=all;

/*****************************************************/
/* Generate the hexadecimal values.  The A values    */
/* do not include 0 and 1 because these values are   */
/* reserved for commands in most hardware fonts.     */
/*****************************************************/

data one;
   do a="2","3","4","5","6","7","8","9","a","b","c","d","e","f";
      do b="0","1","2","3","4","5","6","7","8","9","a","b","c","d","e","f";
         char=input(a||b,$hex3.);
         output;
      end;
   end;
run;

/*****************************************************/
/* Create annotation data set to show the            */
/* hexadecimal values and the corresponding font     */
/* characters underneath the hexadecimal value.      */
/*****************************************************/

data anno;
   length text $2. style $ 25.;
   retain xsys  "3"    ysys   "3"
          tempy 95     x      0
          size  1.5    count  0
          y     0      position "6";
   set one;
   count = count + 1;
   x = x + 4;

   y     = tempy;
   text  = compress(a||b);
   style = "Albany AMT/bold";
   output;

   y        = tempy - 3;
   function = "label";

   /* Modify this statement to use the */
   /* font that you want to display.   */
   style    = "Monotype Sorts";
   text     = char;
   output;

   if int(count/16) = (count/16)
      then do;
```

```
         x = 0;
   tempy = tempy - 6;
         end;
 run;


/********************************************************/
/* Create the table.  The symbol is shown below its     */
/* hexadecimal value.  For example, a circle with       */
/* the number one inside is the hexadecimal value       */
/* AC in the Monotype Sorts system font. To use         */
/* this symbol, specify:                                */
/* font="Monotype Sorts" "AC"x;                         */
/********************************************************/

proc ganno anno=anno;
run;
quit;
```

Figure 11.1 on page 162 and Figure 11.2 on page 163 show the output of the program above for the TrueType fonts Symbol MT and Monotype Sorts.

Figure 11.1 Symbol MT Font

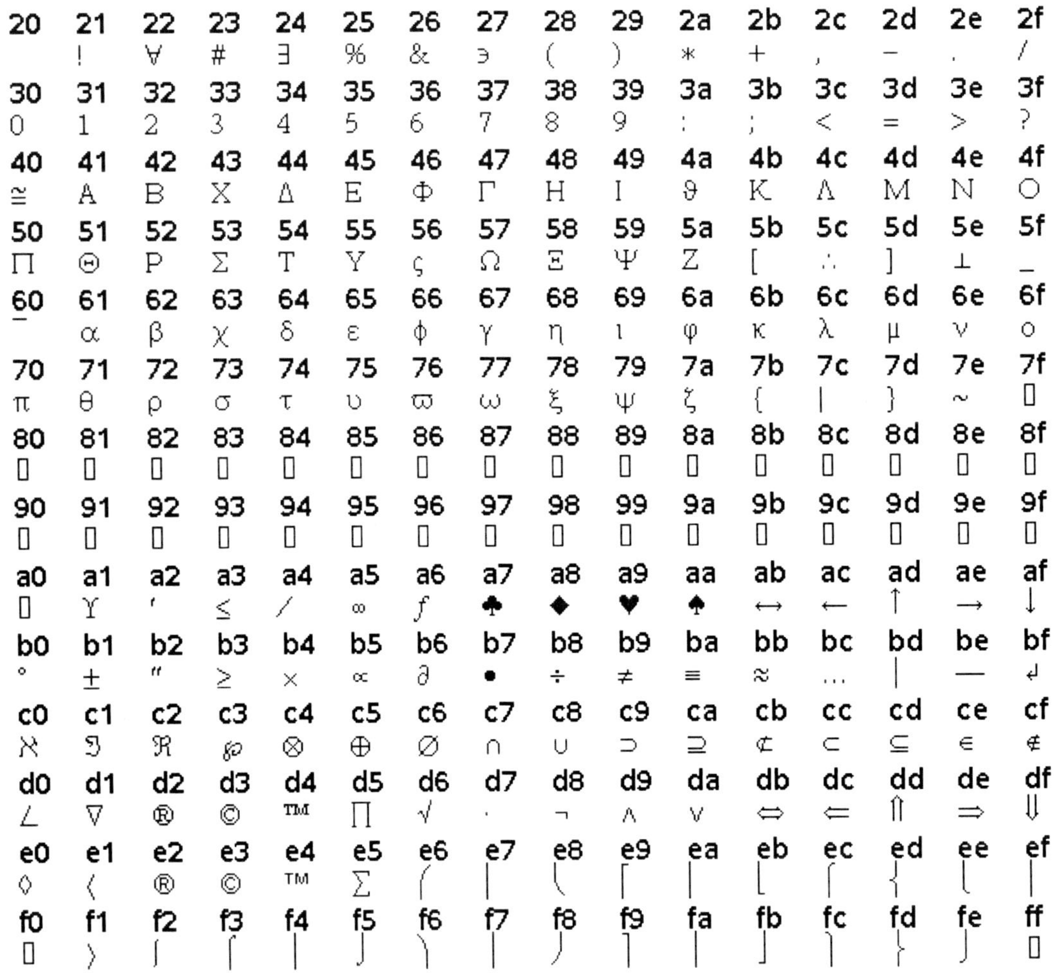

20	21	22	23	24	25	26	27	28	29	2a	2b	2c	2d	2e	2f
	!	∀	#	∃	%	&	∋	(	)	*	+	,	−	.	/
30	31	32	33	34	35	36	37	38	39	3a	3b	3c	3d	3e	3f
0	1	2	3	4	5	6	7	8	9	:	;	<	=	>	?
40	41	42	43	44	45	46	47	48	49	4a	4b	4c	4d	4e	4f
≅	A	B	X	Δ	E	Φ	Γ	H	I	ϑ	K	Λ	M	N	O
50	51	52	53	54	55	56	57	58	59	5a	5b	5c	5d	5e	5f
Π	Θ	P	Σ	T	Y	ς	Ω	Ξ	Ψ	Z	[	∴	]	⊥	_
60	61	62	63	64	65	66	67	68	69	6a	6b	6c	6d	6e	6f
	α	β	χ	δ	ε	φ	γ	η	ι	ϕ	κ	λ	μ	ν	ο
70	71	72	73	74	75	76	77	78	79	7a	7b	7c	7d	7e	7f
π	θ	ρ	σ	τ	υ	ϖ	ω	ξ	ψ	ζ	{	|	}	~	□
80	81	82	83	84	85	86	87	88	89	8a	8b	8c	8d	8e	8f
□	□	□	□	□	□	□	□	□	□	□	□	□	□	□	□
90	91	92	93	94	95	96	97	98	99	9a	9b	9c	9d	9e	9f
□	□	□	□	□	□	□	□	□	□	□	□	□	□	□	□
a0	a1	a2	a3	a4	a5	a6	a7	a8	a9	aa	ab	ac	ad	ae	af
□	ϒ	′	≤	/	∞	f	♣	♦	♥	♠	↔	←	↑	→	↓
b0	b1	b2	b3	b4	b5	b6	b7	b8	b9	ba	bb	bc	bd	be	bf
°	±	″	≥	×	∝	∂	•	÷	≠	≡	≈	…	|	—	↵
c0	c1	c2	c3	c4	c5	c6	c7	c8	c9	ca	cb	cc	cd	ce	cf
ℵ	ℑ	ℜ	℘	⊗	⊕	∅	∩	∪	⊃	⊇	⊄	⊂	⊆	∈	∉
d0	d1	d2	d3	d4	d5	d6	d7	d8	d9	da	db	dc	dd	de	df
∠	∇	®	©	™	∏	√	·	¬	∧	∨	⇔	⇐	⇑	⇒	⇓
e0	e1	e2	e3	e4	e5	e6	e7	e8	e9	ea	eb	ec	ed	ee	ef
◊	⟨	®	©	™	∑	⎛	⎜	⎝	⎡	⎢	⎣	⎧	⎨	⎩	⎪
f0	f1	f2	f3	f4	f5	f6	f7	f8	f9	fa	fb	fc	fd	fe	ff
□	⟩	∫	⌠	⎮	⌡	⎞	⎟	⎠	⎤	⎥	⎦	⎫	⎬	⎭	□

Figure 11.2 Monotype Sorts Font

20	21	22	23	24	25	26	27	28	29	2a	2b	2c	2d	2e	2f
30	31	32	33	34	35	36	37	38	39	3a	3b	3c	3d	3e	3f
40	41	42	43	44	45	46	47	48	49	4a	4b	4c	4d	4e	4f
50	51	52	53	54	55	56	57	58	59	5a	5b	5c	5d	5e	5f
60	61	62	63	64	65	66	67	68	69	6a	6b	6c	6d	6e	6f
70	71	72	73	74	75	76	77	78	79	7a	7b	7c	7d	7e	7f
80	81	82	83	84	85	86	87	88	89	8a	8b	8c	8d	8e	8f
90	91	92	93	94	95	96	97	98	99	9a	9b	9c	9d	9e	9f
a0	a1	a2	a3	a4	a5	a6	a7	a8	a9	aa	ab	ac	ad	ae	af
b0	b1	b2	b3	b4	b5	b6	b7	b8	b9	ba	bb	bc	bd	be	bf
c0	c1	c2	c3	c4	c5	c6	c7	c8	c9	ca	cb	cc	cd	ce	cf
d0	d1	d2	d3	d4	d5	d6	d7	d8	d9	da	db	dc	dd	de	df
e0	e1	e2	e3	e4	e5	e6	e7	e8	e9	ea	eb	ec	ed	ee	ef
f0	f1	f2	f3	f4	f5	f6	f7	f8	f9	fa	fb	fc	fd	fe	ff

Methods For Specifying Fonts

In general, there are four ways to specify fonts. The method you choose depends on how extensively you want to change font specifications used in your program..

- Many procedures support font options that enable you to specify the fonts for certain graph elements. For example, with the GCHART procedure, you can use the FONT= suboption with the PLABEL= option to control the font for the pie slice labels. With the GKPI procedure, you can use the BFONT= option to specify the font for boundary labels. Changes specified using procedure options affect the output of the current invocation of the procedure only.

  For information on the font options that are available for a specific procedure, see the documentation for the procedure.

- You can specify fonts in the AXIS, LEGEND, or SYMBOL global statements. Fonts specified with these statements affect the output of any procedure that references those statements. See "Using SAS/GRAPH Global Statement Options to Specify Fonts" on page 164.

- You can specify fonts in the GOPTIONS statement. The GOPTIONS statement is also a global statement, and specifications in the GOPTIONS statement affect all output in the current SAS session. Using the FTEXT= graphics option is frequently the best solution if you are dealing with any of the following situations.

- You want to specify the fonts only for the current SAS session.
- You want to specify the fonts only for a specific application.
- You do not need all of your output to use the same style.
- You do not want your code to be dependent on registry settings or a customized style. For example, you might want to run your program as a stored process or send it to others who might not have the same registry settings.

See "Using GOPTIONS to Specify Fonts" on page 164.

- If you want all of your output to use the same ODS style, you can create a new style by copying and modifying an existing style and changing the font settings. Your new style can be used for all your ODS output at your site to ensure a consistent appearance. If you always want all of your output to have a specific appearance, then modifying a style might be the best alternative. See "Changing The Font Specifications Used By a Style" on page 165.

Using SAS/GRAPH Global Statement Options to Specify Fonts

Font options on SAS/GRAPH AXIS, LEGEND, and SYMBOL global statements enable you to specify fonts for the following:

- axis labels, reference line labels, and tick mark values
- legend labels and legend value descriptions
- contour line labels and plot point labels

For example, the following statement could be used to label contour lines:

```
symbol value="Deep" font="CUMBERLAND AMT/bold/italic";
```

See Example 2 on page 1116 for an example that uses SYMBOL statements to label contour lines.

As with the options specified in the GOPTIONS statement, options specified with these global statements remain in effect until you change them or until you start a new SAS session.

For specific information on each of the global statements, see Chapter 14, "SAS/GRAPH Statements," on page 197.

Using GOPTIONS to Specify Fonts

The GOPTIONS statement has several options that can be used to specify fonts for your graphs.

- FBY= sets the BY line font in your graphs.
- FTEXT= sets the font for all the text in your graphs.
- FTITLE= sets the font for the first title in your graphs.

For example, to specify Cumberland AMT for all of the text in your graphs, use

```
goptions ftext="Cumberland AMT";
```

Settings specified in the GOPTIONS statement remain in effect until you change them, until you specify **reset=all**, or until you close the SAS session.

If you want most or all of the text in your output to use a single font, specifying this font with the FTEXT= graphics option is frequently the best alternative. Using the

FTEXT= option in the GOPTIONS statement instead of adding font specifications to several procedure action statements in addition to other global statements makes your code easier to maintain.

Note: The FBY= option is not supported by the Java or ActiveX devices. For specific information on the GOPTIONS statement, see "GOPTIONS Statement" on page 220. Information for specific graphics options is in Chapter 15, "Graphics Options and Device Parameters Dictionary," on page 327. △

Note: When you are sending SAS/GRAPH output to the HTML or RTF destinations (MARKUP destinations), titles and footnotes can be rendered as part of your graph image or as part of the HTML or RTF files. Where your titles and footnotes are rendered determines the fonts that are used for them. See "Controlling Titles and Footnotes with Java and ActiveX Devices in HTML Output" on page 194 for information on the GTITLE and GFOOTNOTE destination options and the ODS USEGOPT statement. △

Changing The Font Specifications Used By a Style

There are three ways to change the font specifications used by a style. Which method you choose depends on how extensively you want to change the fonts used in your output.

- You can modify the style element that controls a specific graph element such as graph titles or contour line labels.
- You can modify the abstract font specifications in the GraphFonts class. These font specifications can be referenced in multiple places in a style and affect several graph elements.
- You can modify the font settings in the SAS registry that the styles use to determine the default fonts. Changes to the SAS registry affect the fonts used by all styles that reference the SAS registry entry.

Modifying an existing style to use different fonts might be the best alternative if you need to create a style for all of your company's output. If you only want to change the fonts used in a few applications, then using the GOPTIONS statement is a better alternative.

For information on changing the font specifications used by the styles, see "Ways to Modify Graph Fonts Or Colors Specified By Styles" on page 143.

Precedence of Font Specifications

When SAS/GRAPH is trying to determine the font to use for a specific graph element, it uses the first font that it finds from the following list.

1. Fonts specified on procedure action statement options such as the PLABEL= option in the PIE statement in the GCHART procedure.
2. Fonts specified on the AXIS, LEGEND, or SYMBOL statements.
3. Fonts specified with the GOPTIONS global statement.
4. Default fonts as described in "Default Fonts" on page 157

CHAPTER 12

SAS/GRAPH Colors and Images

Using SAS/GRAPH Colors and Images **167**
Specifying Colors in SAS/GRAPH Programs **168**
    Specifying Colors in a GOPTIONS Statement **168**
    Defining and Using a Color List **169**
        Introduction to the Color Lists **169**
        Using a Device's Color List **169**
        Building a Color List with the GOPTIONS COLORS= Option **169**
    Color-Naming Schemes **170**
        Introduction to Color-Naming Schemes **170**
        RGB Color Codes **171**
        CMYK Color Codes **171**
        HLS Color Codes **172**
        HSV (or HSB) Color Codes **174**
        Gray-Scale Color Codes **175**
        SAS Color Names and RGB Values in the SAS Registry **175**
        Color Naming System Values **176**
        Using the Color Utility Macros **177**
    Processing Limitations For Colors **180**
        Maximum Number of Colors Displayed on a Device **180**
        Replaying Graphs on a Device That Displays Fewer Colors **180**
Specifying Images in SAS/GRAPH Programs **181**
    Image File Types Supported by SAS/GRAPH **181**
    Displaying an Image in a Graph Background **182**
    Displaying an Image in Graph Frame **184**
    Displaying Images on Data Elements **185**
    Displaying Images Using Annotate **187**
    Displaying Images using DSGI **188**
    Disabling and Enabling Image Output **190**

Using SAS/GRAPH Colors and Images

The appearance of SAS/GRAPH output is determined by the current ODS style by default. Styles set the overall appearance of your output, including the colors and fonts that are used. Some styles also add an image to the background of your graphs.

You can turn off the use of styles if needed. In this case, the default appearance of your output is controlled by device entry parameters. See "Style Attributes Versus Device Entry Parameters" on page 134 and "Turning Off Styles" on page 153 for more information.

In either case (using ODS styles or using device parameters), you can override the default colors by specifying options in your SAS/GRAPH program. Whether you are using ODS styles or you have turned styles off, you can the change these colors as described in "Specifying Colors in SAS/GRAPH Programs" on page 168. You can add images to your output as described in "Specifying Images in SAS/GRAPH Programs" on page 181.

Specifying Colors in SAS/GRAPH Programs

SAS/GRAPH enables you to set colors in several ways. You can do any of the following:

- specify colors in procedure action statements for any procedures that create graphics output. For example, the CAXIS= option in the HBAR statement specifies a color for the response and midpoint axis lines. These options are described in the documentation for the individual procedures.
- specify colors in global statements that enhance procedure output: AXIS, FOOTNOTE, LEGEND, PATTERN, SYMBOL, and TITLE. You can also specify colors in the NOTE statement, which is a local statement, not a global statement. See Chapter 14, "SAS/GRAPH Statements," on page 197.
- use options in the GOPTIONS statement that define colors for specific graphics elements. See "Specifying Colors in a GOPTIONS Statement" on page 168.
- define a color list with the GOPTIONS COLORS= option. See "COLORS" on page 340
- specify a different style, modify an existing style, or create a custom style. See Chapter 10, "Controlling The Appearance of Your Graphs," on page 133 for more information on styles.
- modify the color list in the device entry for the device that you want to use. However, the colors listed in the device entry are not used unless styles are turned off. See "Using a Device's Color List" on page 169 and Chapter 38, "The GDEVICE Procedure," on page 1125 for more information.

See "Precedence of Appearance Option Specifications" on page 141 for information on which settings take precedence when colors are set in more than one way.

Specifying Colors in a GOPTIONS Statement

The GOPTIONS statement has several graphics options that set colors for specific graphical elements. These colors are used unless they are overridden by more specific options specified on other global statements or on procedure statements.

Option	Sets the color for
CBACK=	background for graphics output
CBY=	BY lines in graphics output
CPATTERN=	fill patterns
CSYMBOL=	SYMBOL definitions
CTEXT=	all text and the border in graphics output
CTITLE=	border, plus all titles, footnotes, and notes

You can also use the COLORS= option in a GOPTIONS statement to specify a list of colors rather than specific colors for individual graphical elements. Refer to Chapter 15, "Graphics Options and Device Parameters Dictionary," on page 327 for complete information about each of these graphics options.

Defining and Using a Color List

Introduction to the Color Lists

Each device is associated with a list of colors that it can use. This list is defined in the device entry for the device. You can modify this list as needed. However, this device-specific list of colors is not used unless you turn off styles by specifying the NOGSTYLE system option. See "Using a Device's Color List" on page 169.

You can also use the GOPTIONS statement to specify a list of colors for SAS/GRAPH to use instead of the device-specific color list or the colors defined by the current style. Colors specified in the GOPTIONS statement are always used regardless of the setting of the GSTYLE or NOGSTYLE system option. See "Building a Color List with the GOPTIONS COLORS= Option" on page 169 for more information.

The color selected from a color list varies depending on the procedure using the color and graphical element it's drawing. Usually, the first color in the list is used; however, certain procedures can select other colors. For example, if the CAXIS= option is not specified in the GCONTOUR procedure's PLOT statement, the procedure selects the second color from the color list to draw the axes. See the documentation for individual procedures for more information.

Using a Device's Color List

If you specify the NOGSTYLE system option and you do not define a color list with the COLORS= graphics option, then SAS/GRAPH uses the color list from the current device. This color list is found in the device entry of the specified device. The color list might change if you select a different device during a SAS session.

When SAS/GRAPH assigns colors from the current device's color list, this assignment uses some of the colors that you can specify for a graph. The limit on the number of colors that can be used in your output is set by the current device. For example, the PNG device is a true color device and can use up to 16 million different colors. However, the GIF device is limited to 256 colors.

To view, create, or modify a device's color list, use the GDEVICE procedure. See Chapter 38, "The GDEVICE Procedure," on page 1125.

To reset a color list back to the default color list, for the current device driver, specify the COLORS= option without specifying any colors.

```
goptions colors=;
```

Building a Color List with the GOPTIONS COLORS= Option

To build a color list, use the COLORS= option in the GOPTIONS statement. A color list specified with the COLORS= option overrides the color list of the current device. Building a color list is useful for selecting a subset of colors in a specific order for graphics output. For example, to ensure that the colors red, green, and blue are available in that order, you can specify any of the following:

```
goptions colors=(red green blue);
goptions colors=(CXFF0000 CX00FF00 CX0000FF);
goptions colors=(medium_red medium_green medium_blue);
```

You can specify colors in any color-naming schemes described in "Color-Naming Schemes" on page 170. Each value specified in a color list must be one of the following:

- a valid color name, not to exceed 64 characters
- a valid color code, not exceed eight characters

Note: The COLORS= graphics option provides only a default lookup table. Any time you explicitly select any other colors in your SAS/GRAPH program, those colors are used to draw the graphical elements for which you have specified them. △

See "COLORS" on page 340 for more information.

Color-Naming Schemes

Introduction to Color-Naming Schemes

The valid color-naming schemes are as follows:

- RGB (red green blue)
- CMYK (cyan magenta yellow black)
- HLS (hue lightness saturation)
- HSV (hue saturation brightness), also called HSB
- Gray scale
- SAS color names (from the SAS Registry)
- SAS Color Naming System (CNS)

Table 12.1 on page 170 shows examples of each color-naming scheme.

Table 12.1 Examples of Specifying Colors

Color-Naming Scheme	Example
RGB	`COLORS=(cx98FB98 cxDDA0DD cxFFDAB9 cxDB7093 cxB0E0E6)`
CMYK	`COLORS=("FF00FF00" "00FFFF00" "FFFFFF00")`
HLS	`COLORS=(H14055FF H0F060FF H0B485FF H07880FF)`
HSV	`COLORS=(V0F055FF v010FFFF v03BFFFF v12C55E8)`
Gray Scale	`COLORS=(GRAY4F GRAY6D GRAY8A GRAYC3)`
SAS Registry Colors	`COLORS=(palegreen plum peachpuff palevioletred powderblue)`
CNS Color Names	`COLORS=("very light purplish blue" "light vivid green" "medium strong yellow" "dark grayish green")`

You can also mix color-naming schemes in the same statement, for example:

```
goptions colors=(cxEE0044 "vivid blue" darkgreen);
```

Note: Hardware characteristics of your output device might cause some colors with different color definitions to appear the same. The same color is likely to appear different on different devices and might not appear correctly on some devices. To determine whether your device supports a specific color-naming scheme, refer to your graphics device documentation. △

Each of the color-naming schemes supported by SAS/GRAPH has its advantages and disadvantages based on how the output is used. For example, if you are creating a report that will be viewed online only, then specifying colors using the RGB naming scheme or the SAS color names defined in the registry might produce better results. If you are creating a report for publishing in printed form, you might want to use the CMYK color-naming scheme.

The color utility macros enable you to create colors for a specific color-naming scheme. These macros convert color values between color-naming schemes. See "Using the Color Utility Macros" on page 177.

Note: Invalid color names, such as a misspelled color name, are mapped to gray, and a NOTE is issued to the SAS log. A valid color name that is not supported by the current device is mapped to the closest color that is supported by the device. △

RGB Color Codes

The RGB color-naming scheme is usually used to define colors for a display screen. This color-naming scheme is based on the properties of light. With RGB color codes, a color is defined by its red, green, and blue components. Individual amounts of each color are added together to create the desired color. All the colors combined together create white. The absence of all color creates black.

Color names are in the form CX*rrggbb*, where the following is true:

- CX indicates to SAS that this is an RGB color specification.
- *rr* is the red component.
- *gg* is the green component.
- *bb* is the blue component.

The components are given as hexadecimal numbers in the range 00 through FF (0% to 100%), where lower values are darker and higher values are lighter. This scheme allows for up to 256 levels of each color component (over 16 million different colors).

Table 12.2 Examples of RGB Color Values

Color	RGB Value
red	CXFF0000
green	CX00FF00
blue	CX0000FF
white	CXFFFFFF
black	CX000000

Any combination of the color components is valid. Some combinations match the colors produced by predefined SAS color names. See "Using the SAS Registry to Control Color" in *SAS Language Reference: Concepts* for information on viewing the RGB combinations that match predefined SAS color names.

CMYK Color Codes

CMYK is a color-naming scheme used in four-color printing. CMYK is based on the principles of objects reflecting light. Combining equal values of cyan, magenta, and yellow produces process black, which might not appear as pure black. The black component (K) of CMYK can be used to specify the level of blackness in the output. A lack of all colors produces white, when the output is printed on white paper.

To specify the colors from a printer's Pantone Color Look-Up Table, you can use the CMYK color-naming scheme. Specify colors in terms of their cyan, magenta, yellow, and black components. Color names are of the form *ccmmyykk*, where the following is true:

- *cc* is the cyan component.
- *mm* is the magenta component.
- *yy* is the yellow component.
- *kk* is the black component.

The components are given as hexadecimal numbers in the range 00 through FF, where higher values are darker and lower values are brighter. This scheme allows for up to 256 levels of each color component. Quotation marks are required when the color value starts with a number instead of a letter.

Table 12.3 Examples of CMYK Color Values

Color	CMYK Value
red	00FFFF00
green	FF00FF00
blue	FFFF0000
white	00000000
process black (using cyan, magenta, and yellow ink)	FFFFFF00
pure black (using only black ink)	000000FF

Note: You can specify a CMY value by making the *kk*, the color's black component, zero (00). △

CMYK color specifications are for devices that support four colors. If a CMYK color is used on a three-color device, the device processes the color specification. The resulting colors might not be as expected. Different CMYK colors might map to the same device color because a four-color space supports more colors than a three-color space.

HLS Color Codes

The HLS color-naming scheme follows the Tektronix Color Standard illustrated in Figure 12.1 on page 174. To make the HLS color model consistent with the HSV coordinate system, Tektronix places blue at zero degrees. With the HLS color naming-scheme, you specify colors in terms of hue, lightness, and saturation levels. HLS color names are of the form H*hhhllss*, where the following is true:

- H indicates that this is an HLS color specification.
- *hhh* is the hue component.
- *ll* is the lightness component.
- *ss* is the saturation component.

The components are given as hexadecimal numbers. The hue component has the range of 000 through 168 hexadecimal (168 hexadecimal is equivalent to 360 decimal). Both the lightness and saturation components are hexadecimal and scaled to a range of 0 to 255 expressed with values of 00 through FF (0% to 100%). Thus, they provide 256 levels for each component.

Table 12.4 Examples of HLS Color Codes

Color	HLS Color Code
red	H07880FF
green	H0F080FF
blue	H00080FF
light gray	H000BB00
white*	H*xxx*FF00, such as H000FF00
black*	H*xxx*0000 such as H0000000

* When the saturation is set to 00, the color is a shade of gray that is determined by the lightness value. Therefore, white is defined as H*xxx*FF00 and black as H*xxx*0000, where *xxx* can be any hue.

Figure 12.1 Tektronix Color Standard

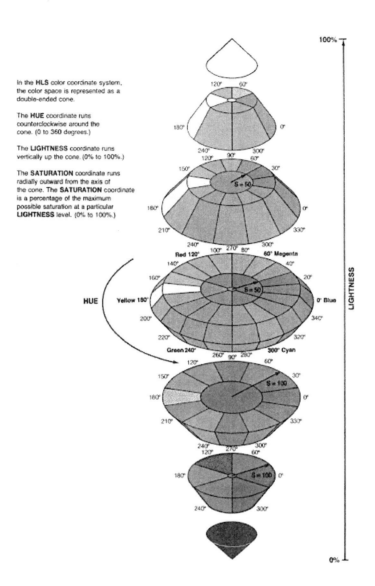

HSV (or HSB) Color Codes

Specify the HSV color-naming scheme in terms of hue, saturation, and value (or brightness) components. HSV color names are of the form V*hhhssvv*, where the following is true:

- V indicates that this is an HSV color specification.
- *hhh* is the hue component.
- *ss* is the saturation component.
- *vv* is value or brightness component.

The components are given as hexadecimal numbers. The hue component has the range of 000 through 168 hexadecimal (168 hexadecimal is equivalent to 360 decimal). Both the saturation and value (brightness) components are hexadecimal, scaled to a range of 0 to 255, and expressed with values of 00 through FF. Thus, they provide 256 levels for each component.

Table 12.5 Examples of HSV (or HSB) Color Codes

Color	HSV Color Code
red	V000FFFF
green	V078FFFF
blue	V0F0FFFF
light gray*	V*xxx*00BB such as V07900BB
white*	V*xxx*00FF such as V07900FF
black*	V*xxx*00000 such as V0790000

* When the saturation is set to 00, the color is a shade of gray. The value component determines the intensity of gray level. The *xxx* can be any hue.

Gray-Scale Color Codes

Specify the lightness or darkness of gray using the word GRAY and a lightness value. Gray-scale color codes are of the form GRAY*ll*. The value *ll* is the lightness of the gray and is given as a hexadecimal number in the range 00 through FF. This scheme allows for 256 levels on the gray scale.

Note: GRAY, without a lightness value, is a SAS color name defined in the SAS registry (see "SAS Color Names and RGB Values in the SAS Registry" on page 175). Its value is CX808080. Invalid color specifications are mapped to GRAY. △

Table 12.6 Examples of Gray-Scale Color Codes

Color	Gray-Scale Color Codes	RGB equivalent
white	GRAYFF	CXFFFFFF
light gray	GRAYC0	CXC0C0C0
dark gray	GRAY40	CX404040
black	GRAY00	CX000000

SAS Color Names and RGB Values in the SAS Registry

SAS provides, in the SAS Registry, a set of color names and RGB values that you can use to specify colors. These color names and RGB values are common to most Web browsers. You can specify the name itself or the RGB value associated with that color name. To view the color names as associated RGB values that are defined in the registry, submit the following code;

```
proc registry list
  startat="COLORNAMES";
run;
```

SAS prints the output in the SAS log.

You can also create your own color values by adding them to the registry. For more information on viewing and modifying the list of color names, see "Using the SAS Registry to Control Color" in *SAS Language Reference: Concepts*.

Color Naming System Values

With CNS, you specify a color value by specifying lightness, saturation, and hue, in that order, using the terms shown in the following table.

Table 12.7 Color Naming System Values

Lightness	Saturation	Hue
Black	Gray	Blue
Very Dark	Grayish	Purple
Dark	Moderate	Red
Medium	Strong	Orange/Brown
Light	Vivid	Yellow
Very Light		Green
White		

Follow these rules when you are determining the CNS color name:

- The lightness values black and white should not be used with saturation or hue values.
- If not specified, medium is the default lightness value and vivid is the default saturation value.
- Gray is the only saturation value that can be used without a hue.
- Unless the color you want is black, white, or some form of gray, you must specify at least one hue.

One or two hue values can be used in the CNS color name. When using two hue values, the hues must be adjacent to each other in the following list: *blue*, *purple*, *red*, *orange/brown*, *yellow*, *green*, and then returning to *blue*. When two hues are used, the resulting color is a combination of both colors. Use the suffix **ish** to reduce the effect of a hue when two hues are combined. Reddish purple is less red than red purple. If you are using a color with an **ish** suffix, this color must precede the color without the **ish** suffix.

Color names can be written in the following ways:

- without space separators between words
- with an underscore to separate words
- with a space to separate words, enclosed in quotation marks

For example, the following are all valid color specifications:

- verylightmoderatepurplishblue
- very_light_moderate_purplish_blue
- "very light moderate purplish blue"

Note: If a CNS color name is also a color name in the SAS Registry, the SAS Registry color value takes precedence. Some CNS color names and color names in the SAS Registry have different color values. To use a CNS color value when the color name is also in the SAS Registry, do the following:

- Include a space to separate the words.
- Enclose the entire color name in quotation marks.

Using the Color Utility Macros

The color utility macros enable you to define colors for a specific color-naming scheme and convert color values between color-naming schemes.

The %COLORMAC macro contains several subcomponent macros that can be used to construct and convert color values for the different color-naming schemes supported by SAS. The %HELPCLR macro provides information about the %COLORMAC subcomponent macros. The following table shows information displayed in your SAS log when you call the %HELPCLR macro from the command line.

Table 12.8 Using the %HELPCLR macro

Use...	To...
%HELPCLR;	List the color utility macro names with help information.
%HELPCLR(ALL);	Display the short descriptions and examples for each of the color utility macros.
%HELPCLR(*macroname*);	Obtain a short description and an example of a specific color utilities macro. Replace *macroname* with the name of the color utility macro you are interested in.

When the color utility macros are invoked, the calculated color value is directed to the SAS log. The calculated color can also be used to perform in-place substitutions in the code.

Table 12.9 %CMY(*cyan, magenta, yellow*);

Description	Usage Example

Replace *cyan, magenta, yellow* with numeric values to create an RGB color value. The numeric values that are used in place of *cyan, magenta, yellow* indicate the percentage of each color to be included in the RGB value.	Entering the following code into your Program Editor:
	`%COLORMAC;`
	`data _null_;`
	`put "%CMY(100,0,100)";`
	`run;`
	
	Returns the RGB value CX00FF00 which is green.

Table 12.10 %CMYK(*cyan, magenta, yellow, black*);

Description	Usage Example

Replace *cyan, magenta, yellow, black* with numeric values to create a CMYK color value. The numeric values that are used in place of *cyan, magenta, yellow, black* indicate the percentage of each color to include in the CMYK color value. See "CMYK Color Codes" on page 171 for more information on the color value produced by using this macro.	Entering the following code into your Program Editor:
	`%COLORMAC;`
	`data _null_;`
	`put %CMYK(0,46,16,31);`
	`run;`
	
	Returns the CMYK value 0075294F which is purple.

Note: In the PUT statement, %CMYK(*cyan, magenta, yellow, black*), should not be placed in quotations. △

Table 12.11 %CNS (*colorname*);

Description	Usage Example

Replace *colorname* with a color-naming scheme color name to create an HLS color value. See "HLS Color Codes" on page 172 for more information on HLS color values. For more information on valid color-naming scheme color names see "Color Naming System Values" on page 176 or enter the following into the command-line of the Program Editor:
`%HELPCLR(CNS);`
	Entering the following code into your Program Editor:
	`%COLORMAC;`
	`data _null_;`
	`put "%CNS(GRAYISH REDDISH PURPLE)";`
	`run;`
	
	Returns the HLS value H04B8040 which is grayish reddish purple.

Note: The %CNS macro accepts only CNS color names where a space is used to separate the words in the color name. △

Table 12.12 %HLS(*hue, lightness, saturation*);

Description	Usage Example

Replace *hue, lightness, saturation* with numeric values to create an HLS color value. *Hue* should be replaced with any value from 0 to 360. *Lightness* and *saturation* indicate a percentage to be included in the HLS color values. See "HLS Color Codes" on page 172 for more information.	Entering the following code into your Program Editor:
	`%COLORMAC;`
	`data _null_;`
	`put "%HLS(0,50,100)";`
	`run;`
	Returns the HLS value H00080FF which is blue.

Table 12.13 %HSV(*hue, saturation, value*);

Description	Usage Example

Replace *hue, saturation, value* with numeric values to create an HLS value from HSV components. *Hue* should be replaced with any value from 0 to 360. *Saturation* and *value* (brightness) indicate a percentage to be included in the HLS color value. See "HSV (or HSB) Color Codes" on page 174 and "HLS Color Codes" on page 172 for more information.	Entering the following code into your Program Editor:
	`%COLORMAC;`
	`data _null_;`
	`put "%HSV(0,100,75)";`
	`run;`
	Returns the HSV value V000FFBF which is dark red.

Table 12.14 %RGB(*red, green, blue*);

Description	Usage Example

Replace *red, green, blue* with numeric values to create an RGB color value from RGB color components. The numeric values that are used in place of *red, green, blue* indicate the percentage of each color to be included in the RGB color value. See "RGB Color Codes" on page 171 for more information.	Entering the following code into your Program Editor:
	`%COLORMAC;`
	`data _null_;`
	`put "%RGB(100,100,0)";`
	`run;`
	Returns the RGB value CXFFFF00 which is yellow.

Table 12.15 %HLS2RGB(*hls*);

Description	Usage Example

Replace *hls* with an HLS color value to create an RGB color value. See "HLS Color Codes" on page 172 and "RGB Color Codes" on page 171 for more information.	Entering the following code into your Program Editor:
	`%COLORMAC;`
	`data _null_;`
	`put "%HLS2RGB(H04B8040)";`
	`run;`
	Returns the RGB value CX9F5F8F which is grayish reddish purple.

Table 12.16 %RGB2HLS(*rgb*);

Description	Usage Example

Replace *rgb* with an RGB color value to create an HLS color value. See "RGB Color Codes" on page 171 and "HLS Color Codes" on page 172 for more information.	Entering the following code into your Program Editor:
	`%COLORMAC;`
	`data _null_;`
	`put "%RGB2HLS(CX9F5F8F)";`
	`run;`
	
	Returns the HLS value H04C7F40 which is grayish reddish purple.

Note: Round-trip conversions using the HLS2RGB and RGB2HLS macros might produce ultimate output values that differ from the initial input values. For example, converting CXABCDEF (a light blue) using %RGB2HLS produces H14ACDAD. Converting this value back to RGB using %HLS2RGB returns CXAACCEE. While not identical, the colors are very similar on the display, and when printed. △

For additional information on color-naming schemes. See *Effective Color Displays: Theory and Practice* by David Travis and *Computer Graphics: Principles and Practice* by Foley, van Dam, Feiner, and Hughes.

Processing Limitations For Colors

Using colors in SAS/GRAPH is limited by the number of colors that you can use in one graph and by the capabilities of your device.

Maximum Number of Colors Displayed on a Device

The number of colors that you can display is limited by the graphics output device. If you create a graph with more colors than the device can display, the colors are mapped to an existing color for display. You might also receive a note in the SAS log telling you when a color is mapped to another color, along with the name of the replacement color.

If your device can support 16 million colors, it might not let you use all of them at once. The MAXCOLORS device parameter tells SAS/GRAPH the maximum number of colors it can display simultaneously. MAXCOLORS is the number of foreground colors plus the background color. If you use more than the number of colors set by the MAXCOLORS device parameter, the excess colors are remapped.

Note: The MAXCOLORS device parameter defaults to the number of colors that the basic model of each graphics device supported can display. If your graphics device can display more colors than the base model, use the PENMOUNTS= graphics option to specify the number of colors your graphics device can display. You can also use the GDEVICE procedure to modify the value of the MAXCOLORS device parameter. △

Replaying Graphs on a Device That Displays Fewer Colors

You can use the GREPLAY procedure to display previously created graphs. Sometimes you might need to replay the graphs on a device that cannot display as many colors as the device on which the graph was originally developed. Use the CMAP statement (see "CMAP" on page 339) to control some of the remapping.

When you replay graphs on devices that display fewer colors than are in the graph, two situations can cause problems:

- Colors are specified that the device does not support.
- More colors are specified than the device can display at one time.

If you specify colors on a device that does not support the colors requested, the colors are remapped to gray. A note is issued to the SAS log telling you when a color is mapped gray.

The number of colors that your device can display affects the actual colors displayed. If your graphics output device can create a maximum of 64 distinct colors, and your graph contains 256 colors, then the 65th through the 256th color specifications are remapped to the colors specified in the current style. If the NOGSTYLE system option is in effect, the colors are remapped to the device's available colors and might not display as the color you specify.

You can use the TARGETDEVICE= graphics option to preview the way a graph is going to look on a different device. Set the device entry name of the device driver to this graphics option. The graph is displayed as close as possible to the display when the other device is used.

Note: When you use the TARGETDEVICE= graphics option, SAS/GRAPH uses the color list of the target device as the default color list; any color that you explicitly use is displayed when you preview the graph, although the color might be mapped by the target device. Refer to "TARGETDEVICE" on page 424 for complete information about the TARGETDEVICE= graphics option. △

Specifying Images in SAS/GRAPH Programs

SAS/GRAPH enables you to display images as part of your graph. You can place an image in the background area of a graph, in the backplane of graphs that support frames, or on the bars of two-dimensional bar charts. You can also apply images at specified graph-coordinate positions using the Annotate facility or the DATA Step Graphics Interface (DSGI).

The images you add to your graphs can be SAS files or external files, in a range of image formats.

Image File Types Supported by SAS/GRAPH

For displaying images in your graphs, SAS/GRAPH supports the image file types shown in the following table.

File Type	Description
BMP (Microsoft Windows Device Independent Bitmap)	supports color-mapped and true color images stored as uncompressed or run-length encoded. BMP was developed by Microsoft Corporation for storing images under Windows 3.0.
DIB (Microsoft Windows Device Independent Bitmap)	see the description of BMP.
GIF (Graphics Interchange Format)	supports only color-mapped images. GIF is owned by CompuServe, Inc.
JPEG (Joint Photographic Experts Group)	supports compression of images with the use of JPEG File Interchange Format (JFIF) software. JFIF software is developed by the Independent Joint Photographic Experts Group.

File Type	Description
PBM (Portable Bitmap Utilities)	supports gray, color, RGB, and bitmap files. The Portable Bitmap Utilities is a set of free utility programs that were primarily developed by Jeff Poskanzer.
PCD (Photo CD)	Kodak Photo CD format which supports multiple image resolutions.
PCX (PC Paintbrush)	supports bitmap, color-mapped, and true color images. PCX and PC Paintbrush are owned by Zsoft Corporation.
PNG (Portable Network Graphic)	supports truecolor, gray-scale, and 8-bit images.
TGA (Targa)	supports true color images. Targa is owned by Truevision, Inc.
TIFF (Tagged Image File Format)	internally supports a number of compression types and image types, including bitmap, color-mapped, gray-scale, and true color. TIFF was developed by Aldus Corporation and Microsoft Corporation.
XBM (X Window Bitmaps)	supports bitmap images only. XBM is owned by MIT X Consortium.
XWD (X Window Dump)	supports all X visual types (bitmap, color-mapped, and true color.) XWD is owned by MIT X Consortium.

Displaying an Image in a Graph Background

To place an image on the graph background, use the IBACK= option in a GOPTIONS statement. Specify either the path to the image file in quotation marks or a fileref that has been defined to point to the image file as follows:

```
goptions iback="external-image-file" | fileref;
```

For example, the following program creates a pie chart with a background image:

```
goptions reset=all
         htitle=1.25
         colors=(cx7c95ca cxde7d6f  cx66ada0
                 cxb689cd cxa9865b cxbabc5c)
         iback="external-image-file";
title "Projected Automobile Sales";
data sales;
    input Month Amount;
    informat month monyy.;
    datalines;
jan08 200
feb08 145
mar08 220
apr08 180
may08 155
```

```
jun08 250
;
proc sort;
   by month;
proc gchart;
   format month monname8.;
   pie month / discrete freq=amount value=inside
               noheading coutline=black;
run;
quit;
```

Because the default value for the IMAGESTYLE= graphics option is TILE, the image is copied as many times as needed to fill the background area.

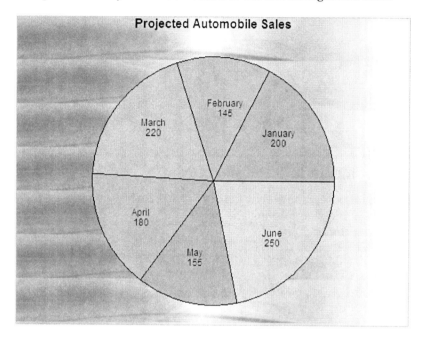

You can specify IMAGESTYLE=FIT in the GOPTIONS statement to stretch the image so that a single image fits within the entire background area.

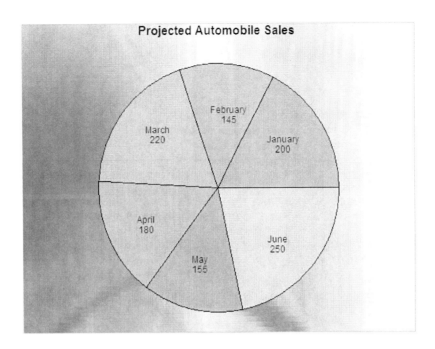

Displaying an Image in Graph Frame

Procedure action statements that support the IFRAME= support frames, which are the backplanes behind the graphs. The backplane is the area within the graph axes. To place an image on the backplane of a graph, specify the IFRAME= option in the procedure action statement that generates the graph. On the IFRAME= option, specify either the path to the image file in quotation marks or a fileref that has been defined to point to the image file:

```
iframe=fileref | "external-image-file";
```

For example, the following program creates a vertical bar chart and adds an image to the graph frame:

```
goptions reset=all htitle=1.25 colors=(yellow cxde7d6f);
title "Projected Automobile Sales";
data sales;
    input Month Amount;
    informat month monyy.;
  datalines;
jan08 200
feb08 145
mar08 220
apr08 180
may08 155
jun08 250
;
proc sort;
   by month;
proc gchart;
   format month monname8.;
   vbar month / discrete freq=amount inside=freq
                coutline=black  iframe="external-image-file";
run;
```

```
quit;
```

Because the default value for the IMAGESTYLE= graphics option is TILE, the image is copied as many times as needed to fill the frame area.

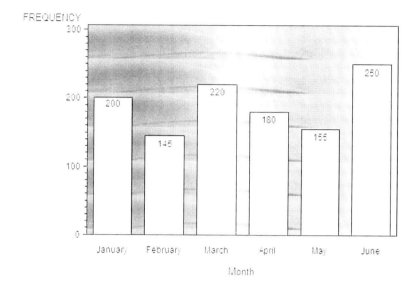

You can specify IMAGESTYLE=FIT in the GOPTIONS statement to stretch the image so that a single image fits within the entire frame area.

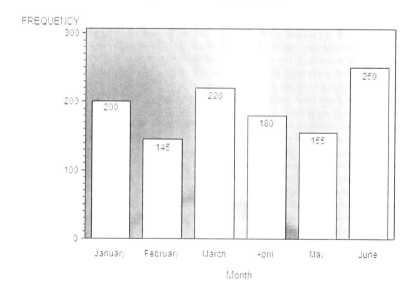

Displaying Images on Data Elements

You can place images on the bars in two-dimensional bar charts generated by the GCHART HBAR or VBAR statements. You can also place images on the bars in three-dimensional bar charts if you are using the ACTIVEX device.

On the IMAGE= option of the PATTERN statement, specify either the path to the image file in quotation marks or a fileref that has been defined to point to the image file.

```
pattern image=fileref | "external-image-file";
```

By default, the image is tiled on the bar, which means that the image is copied as many times as needed to fill each bar. Specify IMAGESTYLE=FIT in the PATTERN statement to stretch the image as needed to fill each bar.

```
pattern image="external-image-file" imagestyle=fit;
```

To tile subsequent images, reset the PATTERN statement or by specify IMAGESTYLE=TILE.

Note: Images are supported on bar charts generated by the HBAR and VBAR statements. If an image is specified on a PATTERN statement that is used with another type of chart, then the PATTERN statement is ignored and default pattern rotation is affected. If you submit a PIE statement when an image has been specified in the PATTERN= option, the default fill pattern is used for the pie slices. Each pie slice displays the same fill pattern. △

The following example places an image on the bars of a vertical bar chart:

```
goptions reset=all htitle=1.25 colors=(yellow cxde7d6f);
title "Projected Automobile Sales";
data sales;
    input Month Amount;
    informat month monyy.;
   datalines;
jan08 200
feb08 145
mar08 220
apr08 180
may08 155
jun08 250
;
proc sort;
   by month;
pattern1 image="external-image-file";
proc gchart;
   format month monname8.;
   vbar month / discrete freq=amount inside=freq
             coutline=black;
 run;
quit;
```

The image is tiled to fill each bar.

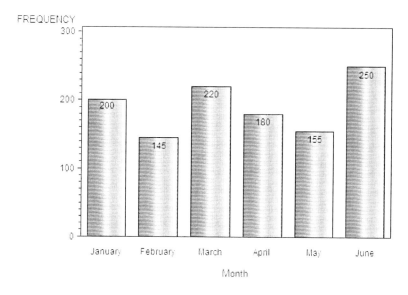

If the PATTERN IMAGESTYLE=FIT option is used, the image is stretched to fill each bar.

```
pattern=fileref | "external-image-file" imagestyle=fit;
```

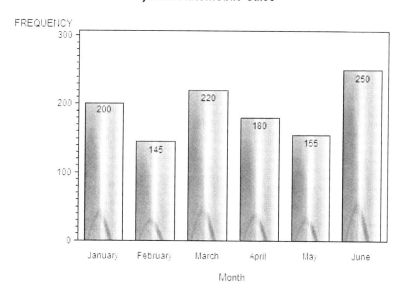

Displaying Images Using Annotate

The Annotate facility enables you to display an image at the coordinate location that you specify with the X and Y variables. To display an image, do the following:

- Specify the image file in quotation marks on the IMGPATH variable.
- Set the image coordinates with the X and Y variables.
- Specify the IMAGE function.

One corner of the image is located by the current X and Y position. The opposite corner is located by the X and Y variables associated with the IMGPATH variable.

```
goptions reset=all border htitle=1.25
    hsize=5.5in vsize=4.2in;
data my_anno;
    length function $8;
    xsys="3"; ysys="3"; when="a";
function="move";  x=55;  y=55;  output;
function="image"; style="fit"; imgpath="external-image-file";
        x=x+15;  y=y+18;  output;
run;
title1 "GMAP with Annotated Image";
proc gmap data=maps.us map=maps.us anno=my_anno;
    id state;
    choro state/
    levels=1
    nolegend
    statistic=freq;
run;
quit;
```

The `style="fit"` variable on the IMAGE function stretches the image as needed to fill the area.

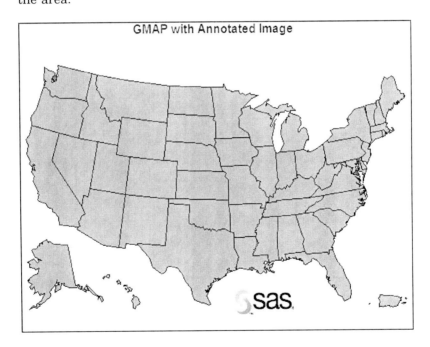

To tile the image to fill the area, set the STYLE variable equal to `"tile"`.

Displaying Images using DSGI

Using the DATA Step Graphics Interface (DSGI), you can display an image in a designated position. To display an image, specify the file specification for the image file in quotation marks on the GDRAW('IMAGE',...) function.

This code displays the image in the screen coordinates (20, 20) to (40, 40). The last parameter, FIT, indicates how to display the image.

```
rc=gdraw("image", "external-image-file", 20, 20, 40, 40, "fit");
```

"Image File Types Supported by SAS/GRAPH" on page 181 shows the supported image file formats.

```
goptions reset=all
    ftext="Albany AMT/bold" htitle=1.25
    hsize=5.5in vsize=4.2in;
title "DSGI with Image";
data image;
    rc=ginit();
    rc=graph("clear");
    rc=gdraw("image","external-image-file",
            5, 5, 90, 90,"tile");
    rc=graph("update");
    rc=gterm();
run;
quit;
```

If you specify the TILE keyword for the GDRAW('IMAGE',...) function, the image is copied as many times as needed to fill the specified area.

```
rc=gdraw("image","external-image-file",
        5, 5, 90, 90,"tile");
```

DGSI with Image

If you specify the FIT keyword for the GDRAW('IMAGE',...) function, the image is stretched to fit within the entire area.

```
rc=gdraw("image","external-image-file",
        5, 5, 90, 90,"fit");
```

DGSI with Image

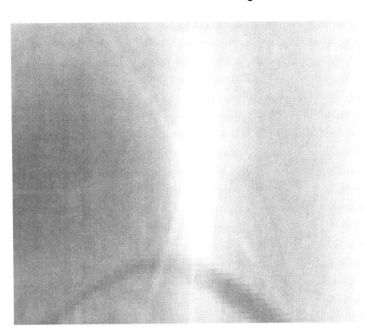

Disabling and Enabling Image Output

The NOIMAGEPRINT graphics option disables image output without removing code from your SAS/GRAPH program. It is useful for printing output without images.

```
goptions noimageprint;
```

To enable image output, reset the GOPTIONS statement or specify the IMAGEPRINT graphics option.

```
goptions imageprint;
```

CHAPTER

13

Managing Your Graphics With ODS

Introduction **191**
Managing ODS Destinations **191**
Specifying a Destination **192**
ODS Destination Statement Options **192**
ODS and Procedures that Support RUN-Group Processing **194**
Controlling Titles and Footnotes with Java and ActiveX Devices in HTML Output **194**
　Controlling Where Titles and Footnotes are Rendered **194**
　Controlling the Text Font, Size, and Color **195**
　Using Graphics Options with ODS (USEGOPT) **195**

Introduction

The Output Delivery System (ODS) manages all output created by procedures and enables you to display the output in a variety of forms, such as HTML, PDF, and RTF. The ODS destination statements provide options for control of many relevant features.

Managing ODS Destinations

ODS supports multiple destinations for procedure output. The most frequently used destinations are LISTING, HTML, RTF, and PDF, although many more destinations are available.

ODS destinations can be open or closed. When a destination is open, ODS can send output to it, and when a destination is closed, ODS cannot send output to it. You can have several destinations open at the same time, and SAS will send output to each destination. The LISTING destination is open by default.

An open destination always uses system resources. It is best to close any destinations if you do not need the output from that destination.

Note: For more information on ODS destinations, see *SAS Output Delivery System: User's Guide.* △

The following table lists the ODS destinations and the default type of output that results from each destination.

Table 13.1 Relevant Destination Table

Destinations	Results	Default Style	Default ImgFmt	Default DPI
DOCUMENT	ODS document	N/A	N/A	N/A
LISTING	SAS output listing	Listing	PNG	100
OUTPUT	SAS data set	N/A	N/A	N/A
HTML	HTML file for online viewing	Default	PNG	100
LATEX [1]	LaTeX file	Default	PostScript	200
PRINTER	printable output in one of three different formats: PCL, PDF, or PS (PostScript)	Printer for PDF and PS, monochromePrinter for PCL	Embedded PNG	150
RTF	output written in Rich Text Format for use with Microsoft Word 2000	RTF	Embedded PNG	200
Measured RTF		RTF	Embedded	200

[1] LATEX is an experimental tagset. Do not use this tagset in production jobs.

Specifying a Destination

To generate output from SAS, a valid ODS destination must be open. By default, the LISTING destination is open. You can use an ODS destination statement, such as ODS HTML, to open a different destination. You can also specify options, such as the HTML filename or the path to an output directory, on the ODS destination statement.

ODS *destination* *<option(s)>;*

The options available vary with the destination that is specified.

ODS Destination Statement Options

There are several destination statement options that you can use to control where your files or graphics should be written, as well as specifying a different style, and specifying the appropriate image resolution in DPI for your output images. For example, the following ODS HTML statement:

- opens the HTML destination
- specifies that images be written to the directory `C:\myfiles\images`
- specifies that the path to the images is specified as `http://www.sas.com/images/image-filename` in the HTML file

- specifies that other output files (for example, the HTML file) be written to the directory `C:\myfiles\`
- specifies that the name of the initial HTML file that is displayed is `barGraph.htm`
- changes the style to Analysis.

```
ods html path="c:\myfiles\"
         gpath="c:\myfiles\images" (url="http://www.sas.com/images/")
         body="barGraph.htm"
         style=analysis;
```

The following ODS HTML statement specifies that the output is sent to the HTML destination. Because it does not specify either the PATH= or GPATH= options, all output is sent to the default SAS folder.

```
ods html body="barGraph.html";
```

The HTML output is written to the file specified by the BODY= option, `barGraph.html`. At start up, the SAS current folder is the same directory in which you start your SAS session. If you are running SAS with the windowing environment in the Windows operating system, then the current folder is displayed in the status bar at the bottom of the main SAS window.

If you do not specify a filename for your output, then SAS provides a default file that is determined by the ODS destination. This file is saved in the SAS current folder. You can check the SAS log to verify the name of the file in which your output is saved.

For complete documentation on ODS destinations, see *SAS Output Delivery System: User's Guide*.

Options that you might want to specify on ODS destination statements are the following:

GPATH= *location* (URL= 'Uniform-Resource-Locator' | NONE)
: specifies the location for all graphics output that is generated while the destination is open. You can specify an external file or a fileref. You can use the URL= suboption to specify a URL that is used in links and references to output files. The GPATH= option is valid for the Listing destination and the Markup family of destinations. If the GPATH option is not specified, the images are written to the location specified by the PATH option. For complete documentation on GPATH= option, see the ODS LISTING statement and the ODS MARKUP statement in *SAS Output Delivery System: User's Guide*.

PATH= *location* (URL= 'Uniform-Resource-Locator' | NONE)
: specifies the location of an external file or a SAS catalog for all markup files. You can specify an external file or a fileref. You can use the URL= suboption to specify a URL that is used in links and references to output files. The PATH= option is valid for the RTF, Measured RTF, and Markup family of destinations. If the PATH option is not specified, images are written to the current working directory. For complete documentation on PATH= option, see the ODS LISTING statement, ODS MARKUP statement, or TAGSET.RTF statement in *SAS Output Delivery System: User's Guide*.

DPI=
: specifies the image resolution in DPI for the output images sent to PRINTER family destinations. The default value for the PRINTER destination is 150. For complete documentation on the DPI= option, see the valid ODS PRINTER statement in *SAS Output Delivery System: User's Guide*.

STYLE=
style-definition
specifies a style to be used for the output. Each ODS destination has a default style for the formatting of output. The style specifies a collection of visual attributes that are used for the rendering of the output. The STYLE= option is valid for all ODS destinations except the Document destination and the Output destination. For complete documentation on the STYLE= option, see the ODS statements in *SAS Output Delivery System: User's Guide*. For more information on using the STYLE= option with SAS/GRAPH output, see Chapter 10, "Controlling The Appearance of Your Graphs," on page 133.

Note: If you specify the PATH= or GPATH= options, the directory name that you specify is used to refer to images that are generated as part of your output. For example, if you are sending output to the HTML destination, and you specify `path="C:\myfiles\"`, then all HTML image tags use that path to refer to your images:

```
<img src="C:\myfiles\myoutput.png">
```

If your browser implements strict security regarding access to local files, you might have problems viewing the images. You can avoid these problems by specifying the URL= suboption. △

ODS and Procedures that Support RUN-Group Processing

When you use ODS, it is wise to specify a QUIT statement at the end of every procedure that supports RUN-group processing. If you end every procedure step explicitly, rather than waiting for the next PROC or DATA step to end it for you, then the QUIT statement clears the selection list, and you are less likely to encounter unexpected results.

Controlling Titles and Footnotes with Java and ActiveX Devices in HTML Output

When you use ODS to send your graphs to an HTML destination, you can choose whether titles and footnotes are rendered as part of the HTML body file, as they are with tabular output, or the graphical image that appears in the Web page.

Where titles and footnotes are rendered determines how you control their font, size, and color.

Controlling Where Titles and Footnotes are Rendered

Where titles and footnotes are rendered depends on the device driver that you are using and on the setting of the ODS statement options GTITLE and GFOOTNOTE.

For the JAVA, JAVAIMG, ACTIVEX, and ACTXIMG device drivers, titles and footnotes are always rendered as part of the HTML body file. The GTITLE and GFOOTNOTE options are ignored for these drivers.

For all other devices, the GTITLE and GFOOTNOTE options determine where the titles and footnotes are rendered. The default settings, GTITLE and GFOOTNOTE, render titles and footnotes as part of the graphic image. If you want titles and footnotes to appear within the HTML body file and not as part of the graphical image, you must specify the NOGTITLE or NOGFOOTNOTE option, as in the following example.

```
/* direct titles and footnotes to the HTML file */
ods html body="filename.htm" nogtitle nogfootnote;
```

If the title or footnote is being output through an ODS markup destination (such as HTML) and the corresponding ODS option NOGTITLE or NOGFOOTNOTE is specified, then the title or footnote is rendered in the body of the HTML file rather than in the graphic itself. Specifying NOGTITLE or NOGFOOTNOTE results in increasing the amount of space allowed for the procedure output area, which can result in increasing the size of the graph. Space that would have been used for the title or footnote is devoted instead to the graph. You might need to be aware of this possible difference if you are using annotate or map coordinates.

Controlling the Text Font, Size, and Color

When you use ODS to send graphics to an HTML destination, and titles and footnotes are rendered as part of the HTML body file instead of the graphic image, then SAS looks for information about how to format titles and footnotes in the following order:

1 SAS looks for options on the TITLE and FOOTNOTE statement. For example, you can specify BOLD, ITALIC, FONT=, or HEIGHT= options on these statements.

2 SAS looks for global options such as CTEXT= and FTITLE= in the GOPTIONS statement. For more information, see "Using Graphics Options with ODS (USEGOPT)" on page 195.

3 SAS looks for information specified in the current style.

When titles and footnotes are rendered as part of the graphic image, SAS looks first for options on the TITLE and FOOTNOTE statement and then for options in the GOPTIONS statement. When titles and footnotes are rendered as part of the graphic image, you do not need to specify the ODS USEGOPT statement.

When titles and footnotes are rendered as part of the body of the HTML file, font sizes that are specified as a percentage are interpreted as a percentage of the size specified by the current style. When titles and footnotes are rendered as part of the image, fonts sizes that are specified as a percentage are interpreted as a percentage of graphics output area. For more information about specifying fonts and font sizes, refer to

- "FTEXT" on page 363 and "FTITLE" on page 363
- "HTEXT" on page 385 and "HTITLE" on page 385
- "GUNIT" on page 378
- "TITLE, FOOTNOTE, and NOTE Statements" on page 279.

Using Graphics Options with ODS (USEGOPT)

When you use ODS to send graphics to an HTML destination, and titles and footnotes are rendered as part of the HTML body file instead of the graphic image, ODS does not recognize the settings for the following graphics options unless you also specify the ODS USEGOPT statement:

- CTEXT=
- CTITLE=
- FTEXT=
- FTITLE=
- HTEXT=
- HTITLE=

For example, the following code generates two graphs. The title for the first graph uses the text color and font as defined by the current style (ASTRONOMY). The title for the second graph uses the font size and color specified by the HTITLE and CTEXT options.

```
ods html file="myout.htm" style=astronomy;
goptions reset=all dev=activex htitle=8 ctext="black";

ods nousegopt;
title "My title";
footnote "My footnote";
proc gchart data=sashelp.class;
   pie age / discrete legend;
run;

ods usegopt;
   pie age / discrete legend;
run;

quit;
ods nousegopt;
ods html close;
```

While ODS USEGOPT is in effect, the settings for these graphics options affect all of the titles and footnotes rendered by ODS. To turn off the use of these graphics option settings for non-graphic output, specify the ODS NOUSEGOPT statement.

The default setting is ODS NOUSEGOPT.

CHAPTER
14

SAS/GRAPH Statements

Overview **197**
    AXIS Statement **198**
    BY Statement **216**
    FOOTNOTE Statement **220**
    GOPTIONS Statement **220**
    LEGEND Statement **225**
    NOTE Statement **238**
    ODS HTML Statement **239**
    PATTERN Statement **240**
    SYMBOL Statement **252**
    TITLE, FOOTNOTE, and NOTE Statements **279**
Example 1. Ordering Axis Tick Marks with SAS Date Values **294**
Example 2. Specifying Logarithmic Axes **297**
Example 3. Rotating Plot Symbols Through the Color List **299**
Example 4. Creating and Modifying Box Plots **302**
Example 5. Filling the Area between Plot Lines **304**
Example 6. Enhancing Titles **307**
Example 7. Using BY-group Processing to Generate a Series of Charts **309**
Example 8. Creating a Simple Web Page with the ODS HTML Statement **313**
Example 9. Combining Graphs and Reports in a Web Page **315**
Example 10. Creating a Bar Chart with Drill-Down Functionality for the Web **321**
    Details **325**
    Building an HREF value **325**
    Creating an image map **326**
    Referencing SAS/GRAPH Output **326**

Overview

SAS/GRAPH programs can use some of the SAS language statements that you typically use with the Base SAS procedures or with the DATA step, such as LABEL, WHERE, and FORMAT. These statements are described in the *SAS Language Reference: Dictionary*.

In addition, SAS/GRAPH has its own set of statements that affect only graphics output generated by the SAS/GRAPH procedures and the graphics facilities Annotate and DSGI. Most of these statements are *global statements*. That is, they can be specified anywhere in your program and remain in effect until explicitly changed or canceled. These are the SAS/GRAPH global statements:

    AXIS
        modifies the appearance, position, and range of values of axes in charts and plots.

FOOTNOTE
: adds footnotes to graphics output. This statement is like the TITLE statement and is described in that section.

GOPTIONS
: submits graphics options that control the appearance of graphics elements by specifying characteristics such as colors, fill patterns, fonts, or text height. Graphics options can also temporarily change device settings.

LEGEND
: modifies the appearance and position of legends generated by procedures that produce charts, plots, and maps.

NOTE
: adds text to the graphics output. This statement is an exception because it is not global but local, meaning that it must be submitted within a procedure. Otherwise, the NOTE statement is like the TITLE statement and is described in that section.

PATTERN
: controls the color and fill of patterns assigned to areas in charts, maps, and plots.

SYMBOL
: specifies the shape and color of plot symbols as well the interpolation method for plot data. It also controls the appearance of lines in contour plots.

TITLE
: adds titles to graphics output. The section describing the TITLE statement includes the FOOTNOTE and NOTE statements.

The above statements are described in this chapter, as well as the following two Base language statements that have a special effect when used with SAS/GRAPH procedures:

BY
: processes data according to the values of a classification (BY) variable and produces a separate graph for each BY-group value. This statement is not a global statement. It must be specified within a DATA step or a PROC step.

ODS HTML
: generates one or more files written in Hypertext Markup Language (HTML). If you use it with SAS/GRAPH procedures, you can specify one of the device drivers GIF, ACTIVEX, or JAVA. ACTIVEX and JAVA are available only with GCHART, GCONTOUR, GMAP, GPLOT, and G3D. With the GIF device driver, the graphics output is stored in GIF files. With the ACTIVEX device driver, graphics output is stored as XML input to ActiveX controls. With the JAVA device driver, graphics output is stored as XML input to Java applets. The HTML files that are generated reference the graphics output. When viewed with a Web browser, the HTML files can display graphics and non-graphics output together on the same Web page.

For more information on the BY, LABEL, OPTIONS, and WHERE statements in Base SAS software, see *SAS Language Reference: Dictionary*.

AXIS Statement

Controls the location, values, and appearance of the axes in plots and charts.

Used by: GCHART, GBARLINE, GCONTOUR, GPLOT, GRADAR, G3D procedures

Restriction: For the G3D procedure, the AXIS statement is supported by the JAVA and ActiveX devices only.

Type: Global

Syntax

AXIS<1...99> <*options*>;
   option(s) can be one or more options from any or all of the following categories:
- axis scale options:
      INTERVAL=EVEN | UNEVEN | PARTIAL
      LOGBASE=*base* | E | PI
      LOGSTYLE=EXPAND | POWER
      ORDER=(*value-list*)
- appearance options:
      COLOR=*axis-color*
      LENGTH=*axis-length* <*units*>
      NOBRACKETS
      NOPLANE
      OFFSET=(<*n1* ><,*n2* >)<*units* > | (<*n1*<*units*>><,*n2*<*units* >>)
      ORIGIN=(<*x*><,*y* >)<*units*> | (<*x*<*units* >><,*y*<*units*>>)
      STAGGER
      STYLE=*line-type*
      WIDTH=*thickness-factor*
- tick mark options:
      MAJOR=(*tick-mark-suboption(s)*)| NONE
      MINOR=(*tick-mark-suboption(s)*)| NONE
- text options:
      LABEL=(*text-argument(s)*)| NONE
      REFLABEL=(*text-argument(s)*)| NONE
      SPLIT="*split-char*"
      VALUE=(*text-argument(s)*)| NONE

Description

AXIS statements specify the following characteristics of an axis:
- the way the axis is scaled
- how the data values are ordered
- the location and appearance of the axis line and the tick marks
- the text and appearance of the axis label and major tick mark values

AXIS definitions are used only when they are explicitly assigned by an option in a procedure that produces graphs with axes.

Figure 14.1 on page 200 illustrates the terms associated with the various parts of axes.

Figure 14.1 Parts of Axes

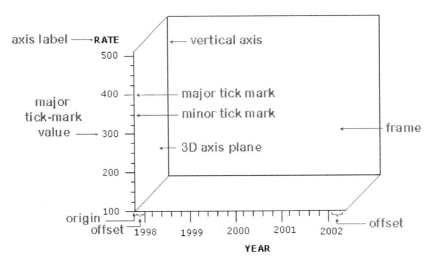

Options

When the syntax of an option includes *units*, use one of these:

CELLS	character cells
CM	centimeters
IN	inches
PCT	percentage of the graphics output area
PT	points

If you omit *units*, a unit specification is searched for in this order:

1. The GUNIT= option in a GOPTIONS statement
2. the default unit, CELLS.

COLOR=*axis-color*
  specifies the color for all axis components (the axis line, all tick marks, and all text) unless you include a more explicit AXIS statement color specification. The following table lists the SAS/GRAPH statement options that can be used to override the COLOR= specification. The table also lists the name of the style reference associated with each of the options.

Table 14.1

Option	Graph Element	Style Reference
AXIS statement:		
LABEL= (COLOR=*color*)	axis label	GraphLabelText
REFLABEL= (COLOR=*color*)	reference-line labels	
VALUE= (COLOR=*color*)	major tick mark values	GraphValueText

Option	Graph Element	Style Reference
calling procedure:		
CTEXT=	all axis text (AXIS label and major tick mark value descriptions)	GraphLabelText
CAXIS=	axis line and major and minor tick marks	GraphAxisLines

If you omit all color options, the AXIS statement looks for a color specification in this order:

1 The CTEXT= graphics option in a GOPTIONS statement.

2 If the CTEXT= option is not used, the color of all axis components is the color of the default style.

Alias: C=

Featured in: "Example 1. Ordering Axis Tick Marks with SAS Date Values" on page 294

INTERVAL=EVEN | UNEVEN | PARTIAL

The INTERVAL option affects the LOGBASE option in the AXIS statement. Specifying the option INTERVAL=UNEVEN and LOGBASE=10, permits non-base10 values to be specified for the ORDER option, while retaining a logarithmic scale for the axis.

Note: PARTIAL is an alias for UNEVEN. They have the same effect. △

Restriction: Not supported by Java and ActiveX

LABEL=(*text-argument(s)*) | NONE

modifies an axis label. *Text-argument(s)* defines the appearance or the text of an axis label, or both. NONE suppresses the axis label. *Text-argument(s)* can be one or more of these:

"text-string"
provides up to 256 characters of label text. By default, the text of the axis label is either the variable name or a previously assigned variable label. Enclose each string in quotes. Separate multiple strings with blanks.

text-description-suboption
modifies a characteristic such as the font, color, or size of the text string(s) that follows it. *Text-description-suboption* can be

ANGLE=*degrees*

COLOR=*text-color*

FONT=*font* | NONE

HEIGHT=*text-height* <*units* >

JUSTIFY=LEFT | CENTER | RIGHT

ROTATE=*degrees*

See "Text Description Suboptions" on page 210 for a complete description.
Specify as many text strings and text description suboptions as you want, but enclose them all in one set of parentheses.

Style Reference: Color attribute of the GraphLabelText style element

Featured in: "Example 1. Ordering Axis Tick Marks with SAS Date Values" on page 294, "Example 2. Specifying Logarithmic Axes" on page 297 , and "Example 7. Using BY-group Processing to Generate a Series of Charts" on page 309

Restriction: Partially supported by Java and ActiveX.

LENGTH=*axis length* <*units* >
specifies the length of the axis in number of units. If you request a length that cannot fit the display, a warning message is written to the log and your graph may produce unexpected results.

This option is not supported by the GRADAR Procedure.

Style Reference: Color attribute of the GraphLabelText graph element.

Restriction: Not supported by Java.

Featured in: "Example 2. Specifying Logarithmic Axes" on page 297 and "Example 9. Combining Graphs and Reports in a Web Page" on page 315 .

LOGBASE=*base* | E | PI
scales the axis values logarithmically according to the value specified. *Base* must be greater than 1. The number of minor tick marks is a function of the logbase, and is calculated as the logbase minus 2. For example, if logbase=10, there are 8 minor tick marks. If logbase=2, then there are no minor tick marks. Because the value of logbase=e (2.718281828) is so close to 2, it also results in no minor tick marks. How the values are displayed on the axis depends on the LOGSTYLE= option. For example, LOGBASE=10 with the default LOGSTYLE=EXPAND generates an axis like the one in Figure 14.2 on page 202.

Figure 14.2 Axis Generated with LOGBASE=10 and LOGSTYLE=EXPAND

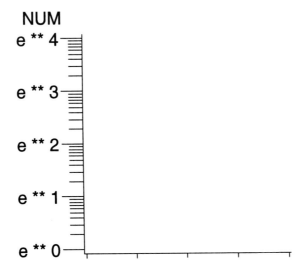

This option is not supported by the GRADAR Procedure.

Featured in: "Example 2. Specifying Logarithmic Axes" on page 297

Restriction: Not supported by Java

LOGSTYLE=EXPAND | POWER
specifies whether the values displayed on the logarithmic axis are the values of the base or the values of the power. LOGSTYLE= is meaningful only when you use LOGBASE=.

LOGSTYLE=EXPAND specifies that the values displayed are the values of the base raised to successive powers and that the minor tick marks are logarithmically placed. For example, if the base is 10, the values displayed are 10, 100, 1000, 10000, and so on. The default is LOGSTYLE=EXPAND. This statement generates an axis like the one in part (a) of Figure 14.3 on page 203:

```
axis logbase=10 logstyle=expand;
```

LOGSTYLE=POWER specifies that the values displayed are the powers to which the base is raised (for example, 1, 2, 3, 4, 5, and so on). For example, this statement generates an axis like the one in part (b) of Figure 14.3 on page 203:

```
axis logbase=10 logstyle=power;
```

Figure 14.3 Axes Generated with the LOGSTYLE=option

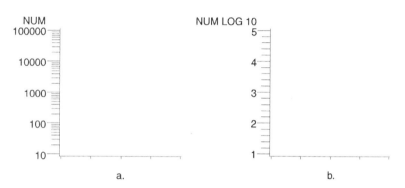

If you use the ORDER= option with a logarithmic axis, the values specified by the ORDER= option must match the style specified by the LOGSTYLE= option. For example, if you specify a logarithmic axis with a base of 2 and you want to display the first five expanded values, use this statement:

```
axis logbase=2 logstyle=expand
     order=(2 4 8 16 32);
```

If you use LOGSTYLE=POWER, the values in the ORDER= option must represent the powers to which the base is raised, as in this example:

```
axis logbase=2 logstyle=power order=(1 2 3 4 5);
```

If the values that are specified by ORDER= do not match the type of values specified by LOGSTYLE=, the request for a logarithmic axis is ignored.

This option is not supported by the GRADAR Procedure.

Featured in: "Example 2. Specifying Logarithmic Axes" on page 297

Restriction: Not supported by Java

MAJOR=(*tick-mark-suboption(s)*)| NONE
modifies the major tick marks. *Tick-mark-suboption(s)* defines the color, size, and number of the major tick marks. NONE suppresses all major tick marks, although the values represented by those tick marks are still displayed.
Tick-mark-suboption can be

   COLOR=*tick-color*

   HEIGHT=*tick-height* <*units* >

   NUMBER=*number-of-ticks*

WIDTH=*thickness-factor*
See "Tick Mark Description Suboptions" on page 214 for complete descriptions. List all suboptions and their values within the parentheses.

AXIS definitions assigned to the group axis of a bar chart by the GAXIS= option ignore MAJOR= because the axis does not use tick marks.

Note: By default, tick marks are now placed at three intervals on the spokes of a GRADAR chart. They are placed at the minimum value, maximum value, and at one value in between. The tick marks on the 12 o'clock spoke are also labeled by default.

HEIGHT is not supported by Java or ActiveX. WIDTH is not supported by Java. △

Featured in: "Example 1. Ordering Axis Tick Marks with SAS Date Values" on page 294 , "Example 2. Specifying Logarithmic Axes" on page 297, and "Example 7. Using BY-group Processing to Generate a Series of Charts" on page 309

Restriction: Partially supported by Java and ActiveX

MINOR=(*tick-mark-suboption(s)*) | NONE
modifies the minor tick marks that appear between major tick marks.
Tick-mark-suboption(s) defines the color, number, or size of the minor tick marks. NONE suppresses all minor tick marks. *Tick-mark-suboption* can be

COLOR=*tick-color*

HEIGHT=*tick-height* <*units* >

NUMBER=*number-of-ticks*

WIDTH=*thickness-factor*
See "Tick Mark Description Suboptions" on page 214 for complete descriptions. List all suboptions and their values within the parentheses.

AXIS definitions assigned to the group axis of a bar chart by the GAXIS= option ignore MINOR= because the axis does not use tick marks.

This option is not supported by the GRADAR Procedure.

HEIGHT is not supported by Java or ActiveX.

Featured in: "Example 1. Ordering Axis Tick Marks with SAS Date Values" on page 294, "Example 2. Specifying Logarithmic Axes" on page 297, and "Example 7. Using BY-group Processing to Generate a Series of Charts" on page 309

Restriction: Partially supported by Java and ActiveX

NOBRACKETS
suppresses the printing of group brackets drawn around the values on the group axis in a bar chart. NOBRACKETS applies only to the group axis of bar charts.

This option is not supported by the GRADAR Procedure.

See also: GROUP= on page 1025 and GAXIS= on page 1025

Restriction: Not supported by Java and ActiveX

NOPLANE
removes either the horizontal or vertical three-dimensional axis plane in bar charts produced by the HBAR3D and VBAR3D statements. NOPLANE affects only the axis to which the AXIS statement applies.

To remove selected axis elements such as lines, values or labels, use specific AXIS statement options. To remove all axis elements except the three-dimensional planes use the NOAXIS option in the procedure. To remove the backplane, use the NOFRAME option in the procedure.

This option is not supported by the GRADAR Procedure.

Featured in: "Example 7. Using BY-group Processing to Generate a Series of Charts" on page 309.

OFFSET=(<n1><,n2>)<units > | (<n1<units>><,n2<units>>)
   specifies the distance from the first and last major tick marks or bars to the ends of the axis line.
   The value of (n1) is the distance from the beginning (origin) of the axis line to the first tick mark or middle of the first bar. The value of (n2) is the distance from the end of the axis line to the last tick mark or middle of the last bar.
   On a horizontal axis, the (n1) offset is measured from the left end of the axis line and the (n2) offset is measured from the right end. On a vertical axis, the (n1) offset is measured up from the bottom of the axis line and the (n2) offset is measured down from the top of the line.
   To specify the same offset for both *n1* and *n2*, use one value, with or without a following comma. For example, either option sets both *n1* and *n2* to 4 centimeters:

   ```
   offset=(4 cm)
   offset=(4 cm,)
   ```

   To specify different offsets, use two values, with or without a comma separating them. For example:

   ```
   offset=(4 cm, 2 cm)
   ```

   To specify only the second offset, use only one value preceded by a comma. This option offsets the last major tick mark or bar three centimeters from the right-hand end of the axis line:

   ```
   offset=(,3 cm)
   ```

   You can specify *units* for the *n1,n2* pair or for the individual offset values.
   This option is not supported by the GRADAR Procedure.
   Featured in: "Example 1. Ordering Axis Tick Marks with SAS Date Values" on page 294
   Restriction: Not supported by Java

ORDER=(*value-list*)
   specifies the order in which data values appear on the axis. The values specified by the ORDER= option are the major tick mark values. You can modify the appearance of these values with the VALUE= option.
   The way you specify *value-list* depends on the type of variable:
   □ For numeric variables, *value-list* is either an explicit list of values or a starting and an ending value with an interval increment, or a combination of both forms:

   n <...*n*>

   n TO *n* <BY *increment*>

   n<...*n*> TO *n* <BY *increment* > <*n* <...*n* > >

   If a numeric variable has an associated format, the specified values must be the unformatted values.

   Values must be listed in either ascending or descending order. By default the increment value is 1. You can use a negative integer for *increment* to specify a value list in descending order. In all forms, multiple *n* values can be separated by blanks or commas. Here are some examples:

   ```
   order=(2 4 6)
   order=(6,4,2)
   order=(2 to 10 by 2)
   order=(50 to 10 by -5)
   ```

   If the specified range is not evenly divisible by the increment value, the highest value displayed on the axis is the last incremental value below the ending value for the range. For example, this value list produces a maximum axis value of 9:

```
order=(0 to 10 by 3)
```

□ *For character variables*, *value-list* is a list of unique character values enclosed in quotes and separated by blanks:

"*value-1*" <..."*value-n*">

If a character variable has an associated format, the specified values must be the *formatted* values for PROC GCHART and the *unformatted* values for PROC GPLOT.

Character values can be specified in any order, but the character strings must match exactly the variable values in case and spelling. For example,

```
order=("Paris" "London" "Tokyo")
```

Observations can be inadvertently excluded if entries in the *value-list* are misspelled or if the case does not match exactly.

□ *For date and time values*, *value-list* can have the following forms:

"*SAS-value*"*i* <..."*SAS-value*"*i*>

"*SAS-value*"*i* TO "*SAS-value*"*i* <BY *interval*>

"*SAS-value*"*i*
is any SAS date, time, or datetime value described for the SAS functions INTCK and INTNX. Enclose the value in quotes and specify one of the following for *i*:

D	date
T	time
DT	datetime

interval
is one of the valid arguments for the INTCK or INTNX functions. These are the default intervals:

DAY	default interval for date
SECOND	default interval for time
DTSECOND	default interval for datetime

These value lists use SAS date and time values:

```
order=("25MAY98"d "04JUL98"d "07SEP98"d)
order=("01JUL97"d to "01AUG97"d)
order=("01JUL97"d to "01JAN98"d by week)
order=("9:25"t to "11:25"t by minute)
order=("04JUN97:12:00:00"dt to
       "10JUN97:12:00:00"dt by dtday)
```

With SAS date and time values, use a FORMAT statement so that the tick mark values have an understandable form. For more information on SAS date and time values, see the *SAS Language Reference: Dictionary*.

With any type of *value-list*, specifying values that are not distributed uniformly or are not in ascending or descending order, generates a warning message in the SAS log. The specified values are spaced evenly along the axis even if the values are not distributed uniformly.

Using the ORDER= option to restrict the values displayed on the axis can result in clipping. For example, if the data range is 1 to 10 and you specify ORDER=(3

TO 5), only the data values from 3 to 5 appear on the plot or chart. For charts, the omitted values are still included in the statistic calculation.

Note: Values out of range do not always produce a warning message in the SAS log. △

CAUTION:
The ORDER= option does not calculate midpoint values; as a result it is not interchangeable with the MIDPOINTS= option in the GCHART procedure. △

You can use the ORDER= option to specify the order in which the midpoints are displayed on a chart, but do not use it to calculate midpoint values. Make sure that the values you specify match the midpoint values that are calculated either by default by the GCHART procedure or by the MIDPOINTS= option. For details, see the description of the MIDPOINTS= option for the appropriate statement in Chapter 36, "The GCHART Procedure," on page 989.

The ORDER= option overrides the suboption NUMBER= described in "Tick Mark Description Suboptions" on page 214.

The ORDER= option is not valid with the ASCENDING, DESCENDING, and NOZEROS options used with the bar chart statements in the GCHART procedure.

This option is not supported by the GRADAR procedure.

Note: The Java applet supports the ORDER= option for numeric axes, but does not support the ORDER= option for categorical, character, midpoint, or group axes.

The ActiveX control supports only simple order lists. Non-uniform interval values, such as dates, are not supported. Only maximum and minimum values are supported with a default interval of one day. △

Featured in: "Example 1. Ordering Axis Tick Marks with SAS Date Values" on page 294, "Example 5. Filling the Area between Plot Lines" on page 304, and "Example 7. Using BY-group Processing to Generate a Series of Charts" on page 309

Restriction: Partially supported by Java and ActiveX

ORIGIN=(<x><,y>)<units> | (<x<units>><,y<units>>)

specifies the *x* coordinate and the *y* coordinate of the origin of the axis. The origin of the horizontal axis is the left end of the axis, and the origin of the vertical axis is the bottom of the axis. The ORIGIN= option explicitly positions the axis anywhere on the graphics output area.

If you specify only one value, with or without a comma following it, only the *x* coordinate is set to that value. For example, this specification sets *x* to 4 centimeters:

```
origin=(4 cm,)
```

If you specify two values, with or without a comma separating them, the first value sets the *x* coordinate and the second value sets the *y* coordinate:

```
origin=(2 pct, 4 pct)
```

If you specify one value preceded by a comma, only the *y* coordinate is set to that value, as shown here:

```
origin=(,3 pct)
```

You can specify *units* for the *x,y* pair or for the individual coordinates.

This option is not supported by the GRADAR Procedure.

Restriction: Not supported by Java and ActiveX

REFLABEL=(*text-argument(s)*) | NONE

creates and defines the appearance of a reference-line label. *Text-argument(s)* defines the appearance or the text of the label, or both. NONE suppresses the reference-line label. *Text-argument(s)* can be one or more of these:

"text-string"
> provides up to 256 characters of label text. By default, a reference line does not have a label. Enclose each string in quotes. Separate multiple strings with blank spaces. The strings are applied to the reference lines specified by the VREF or HREF option.

text-description-suboption
> modifies a characteristic such as the font, color, or size of the text string(s) that follows it. *Text-description-suboption* can be
>
> ANGLE=*degrees*
> AUTOREF
> COLOR=*text-color*
> FONT=*font* | NONE
> HEIGHT=*text-height* <*units*>
> JUSTIFY=LEFT | CENTER | RIGHT
> POSITION=TOP | MIDDLE | BOTTOM
> ROTATE=*degrees*
> T=*n*
>
> See "Text Description Suboptions" on page 210 for a complete description.

Specify as many text strings and text description suboptions as you want, but enclose them all in one set of parentheses.

REFLABEL is not supported by the GRADAR Procedure.

Style Reference: Font and Color attributes of the GraphLabelText element

Restriction: Not supported by Java and ActiveX

STAGGER
> offsets the axis values on a horizontal axis. This option is useful when values overlap on an axis. When specifying the Java and ActiveX devices, the STAGGER option must sometimes be used in conjunction with the ORDER statement.

SPLIT="*split-char*"
> specifies the split character that the AXIS statement uses to break axis values into multiple lines. *Split-char* can be any character value that can be specified in a SAS character variable. The split character must be embedded in the variable values in the data set or in an associated format. When the AXIS statement encounters the split character, it automatically breaks the value at that point and continues on the next line. For example, suppose the data set contains the value **Berlin, Germany**, and you specify SPLIT=",". The value would appear on the axis as follows:

```
Berlin
Germany
```

Note that the split character itself is not displayed.

Axis values specified with VALUE= do not use the split character. For example, suppose you specify this statement:

```
axis1 split="," value=(tick=1 "December, 1999");
```

The value appears on the axis on one line as **December, 1999**. However, any other axis values containing a comma honors the split character.

This option is not supported by the GRADAR Procedure.

Featured in: "Example: Creating Bar Charts with Drill-Down for the Web" on page 618

Restriction: Not supported by Java and ActiveX

STYLE=*line-type*
: specifies a line type for the axis line. Valid values for *line-type* are 0 through 46. If you specify STYLE=0, the axis line is not drawn. The default is 1, a solid line.

    Note: In order for the axis *line* to be altered by the STYLE= option, the NOFRAME option must also be set. If only the STYLE=option is set, the axis *frame* is modified. △

    Note: See also: Figure 14.22 on page 277 for examples of the available line types. △

    Style Reference: Line style attribute of the GraphAxisLine element

VALUE=(*text-argument(s)*) | NONE
: modifies the major tick mark values. That is, this option modifies the text that labels the major tick marks on the axis. *Text-argument(s)* defines the appearance or the text of a major tick mark value, or both. NONE suppresses the major tick mark values, although the major tick marks are still displayed. *Text-argument(s)* can be one or more of these:

    "*text-string*"
    : provides up to 256 characters of text for the major tick mark value. By default, the value is either the variable value or an associated format value. Enclose each string in quotes and separate multiple strings with blanks.

        Specified text strings are assigned to major tick marks in order. If you specify only one text string, only the first tick mark value changes, and all the other tick mark values display the default. If you specify multiple strings, the first string is the value of the first major tick mark, the second string is the value of the second major tick mark, and so on. For example, to change default tick mark values 1, 2, and 3 to `First`, `Second`, and `Third`, use this option:

        ```
        value=("First" "Second" "Third")
        ```

        Note: Although the VALUE= option changes the text displayed at a major tick mark, it does not affect the actual value represented by the tick mark. To change the tick mark values, use the ORDER= option. Also note that with the Java or ActiveX devices, it is necessary to use the ORDER= option to ensure that the same number of tick marks are displayed as are with graphics rendered with the other device drivers. For example, specify ORDER=(1 to 12) to ensure that tick marks for all twelve months are displayed.

        To change the value of midpoints in bar charts produced with the GCHART procedure, use the MIDPOINTS= option in the procedure. △

    text-description-suboption
    : modifies a characteristic such as the font, color, or size of the text string(s) that follows it. *Text-description-suboption* can be

        ANGLE=*degrees*

        COLOR=*text-color*

        FONT=*font* | NONE

        HEIGHT=*text-height* <*units* >

        JUSTIFY=LEFT | CENTER | RIGHT

        ROTATE=*degrees*

        TICK=*n*.

        For a complete description, see "Text Description Suboptions" on page 210.

Place text description suboptions before the text strings they modify. Suboptions not followed by a text string affect the default values. To specify and describe the text for individual values or to produce multi-line text, use the TICK= suboption.

Specify as many text strings and text description suboptions as you want, but enclose them all in one set of parentheses.

Note: If an end user viewing a graph in the Java applet or ActiveX control zooms in on a particular part of a graph for which the VALUE= option is specified, the values are not readjusted in coordination with the zooming. △

Style Element: Color attribute of the GraphLabelText graph element

Featured in: "Example 2. Specifying Logarithmic Axes" on page 297, "Example 7. Using BY-group Processing to Generate a Series of Charts" on page 309, and "Example 9. Combining Graphs and Reports in a Web Page" on page 315

Restriction: Partially supported by Java

WIDTH=*thickness-factor*
> specifies the thickness of the axis line. Thickness increases directly with the value of *thickness-factor*. By default, WIDTH=1.
>
> *Note:* In order for the axis *line* to be altered by the WIDTH= option, the NOFRAME option must also be set. If only the WIDTH=option is set, the axis *frame* is modified.
>
> Java does not support the WIDTH option. ActiveX ignores the WIDTH option for the vertical axis of an AXIS statement with GPLOT and GCONTOUR. △
>
> **Style Reference:** LineThickness attribute of the GraphAxisLines element
>
> **Featured in:** "Example 1. Ordering Axis Tick Marks with SAS Date Values" on page 294
>
> **Restriction:** Not supported by Java and partially supported by ActiveX

Text Description Suboptions

Text description suboptions are used by the LABEL=, REFLABEL=, and VALUE= options to change the color, height, justification, font, and angle of either default text or specified text strings. See the LABEL= option on page 201, the REFLABEL= option on page 207, and the VALUE= option on page 209.

ANGLE=*degrees*
A=*degrees*
> specifies the angle of the *baseline* with respect to the horizontal. A positive value for *degrees* moves the baseline counterclockwise; a negative value moves it clockwise. By default, ANGLE=0 (horizontal) unless the text is automatically angled or rotated to avoid overlapping. .
>
> *Note:* Changing the angle of a vertical axis-label can result in the label being positioned above the graph when using the Java or ActiveX device drivers. △
>
> **Alias:** A=
>
> **Restriction:** Partially supported by Java
>
> **See also:** the ROTATE= suboption on page 213
>
> **Featured in:** "Example: Creating Bar Charts with Drill-Down for the Web" on page 618

AUTOREF
> automatically labels each reference line on an axis with the response value at the reference line's position. The AUTOREF option is used only with the REFLABEL=

option. The automatic labels are applied only to reference lines that do not have specific labels assigned to them. For example, the following option uses the response-axis value as the label for every reference line except the second reference line, which is assigned the label *two*:

```
reflabel=(autoref t=2 "two")
```

Note, however, that if you simultaneously request automatic labeling with a PLOT or BUBBLE statement (using the AUTOHREF or AUTOVREF option), then the automatic labeling can write on top of the custom label you specified using the AXIS statement. You must ensure that your custom labels specified using the AXIS statement are not at the same position as automatic labels requested with a different statement.

Restriction: Not supported by Java, ActiveX, and GIF

COLOR=*text-color*
   specifies the color for the text. If you omit the COLOR= suboption, a color specification is searched for in this order:
   1 the CTEXT= option for the procedure
   2 the CTEXT= option in a GOPTIONS statement
   3 the color of the default style.

   Alias: C=

FONT=*font* | NONE
   specifies the font for the text. See Chapter 11, "Specifying Fonts in SAS/GRAPH Programs," on page 155 for details on specifying *font*. If you omit FONT=, a font specification is searched for in this order:
   1 the FTEXT= option in a GOPTIONS statement
   2 the default style font, NONE.

   Alias: F=

   Restriction: Partially supported by Java

HEIGHT=*text-height* <*units*>
   specifies the height of the text characters in number of units. By default, HEIGHT=1 CELL. If you omit the HEIGHT= option, a text height specification is searched for in this order:
   1 the HTEXT= option in a GOPTIONS statement
   2 the default style value, 1.

   Alias: H=

JUSTIFY=LEFT | CENTER | RIGHT
   specifies the alignment of the text. The default depends on the option with which it is used and the text it applies to.
      □ With the LABEL= option:
         □ for a left vertical axis label, the default is JUSTIFY=RIGHT
         □ for a right vertical axis label, the default is JUSTIFY=LEFT
         □ for a horizontal axis label, the default is JUSTIFY=CENTER.

      □
         With the REFLABEL= option:
         □ for a reference line that intersects a vertical axis, the default is JUSTIFY=CENTER. RIGHT places the text string on the right end of the line, CENTER places the text string in the middle of the line, and LEFTplaces the text string to the left of the line.

- for a reference line that intersects a horizontal axis, the default is JUSTIFY=RIGHT for all procedures except the BAR statement in GBARLINE. For the BAR statement in GBARLINE the default is JUSTIFY=LEFT. RIGHT places the text string just to the right of the line, CENTER is centered on top of the line, and LEFT places the text string just to the left of the line.
- With the VALUE= option:
  - for numeric variables on a vertical axis, the default is JUSTIFY=RIGHT
  - for character variables on a vertical axis, the default is JUSTIFY=LEFT
  - for all variables on a horizontal axis, the default is JUSTIFY=CENTER.

Note: With output using Java and ActiveX, text justification is relative to the text string, not the tick mark. For example, left justification means that the left end of the text string is justified with respect to the drawing location, as well as other strings in a multiline label. Because the text is left justified with respect to the drawing location and not the tick mark, the text string can be placed to the right of a tick mark. △

You can use the JUSTIFY= option to print multiple lines of text by repeating the JUSTIFY= option before the text string for each line. You can also use JUSTIFY= to specify multi-line text at specified major tick marks. For example, this statement produces an axis label and major tick mark values like those shown in Figure 14.4 on page 212.

```
axis label=("Current" justify=c
            "Sales Projections")
      value=(tick=1 "JAN" justify=c "1997"
             tick=2 "FEB" justify=c "1997"
             tick=3 "MAR" justify=c "1997"
             tick=4 "APR" justify=c "1997"
             tick=5 "MAY" justify=c "1997");
```

Figure 14.4 The JUSTIFY= suboption

Specify additional suboptions before any string.

Alias: J=L | C | R

Restriction: Not supported by Java

See also: the suboption TICK= on page 213

POSITION=TOP | MIDDLE | BOTTOM
specifies the position of a reference-line label relative to the reference line. The default is TOP for both vertical and horizontal reference lines. The POSITION= option is available only on the REFLABEL= option.
- For horizontal reference lines, TOP places the label just above the reference line, MIDDLE places the label on the reference line, and BOTTOM places the label just below the reference line.

□ For vertical reference lines, TOP places the label at the top end of the reference line, MIDDLE places the label in the middle of the line, and BOTTOM places the label at the bottom end of the line.

Restriction: Not supported by Java and ActiveX

ROTATE=*degrees*
specifies the angle at which each character of text is rotated with respect to the baseline of the text string. A positive value for *degree* rotates the character counterclockwise; a negative value moves it clockwise. By default, ROTATE=0 (parallel to the baseline) unless the text is automatically angled or rotated to avoid overlapping.

Alias: R=*degrees*

Restriction: Partially supported by Java

See also: the suboption ANGLE= on page 210

TICK=*n*
specifies the *n* reference line or tick mark value. Used only with the REFLABEL= option or the VALUE= option. If neither one is specified, then the TICK= option is ignored.

□ With the REFLABEL= option, the TICK= option specifies the *n*th reference line. It is used to limit modifications to individual reference lines when there are multiple reference lines on an axis. For example, the following option changes the color of only the third reference line's label and leaves all other reference-line labels unchanged:

```
reflabel=(autoref t=3 color=red)
```

Suboptions that *precede* the TICK= option affect all the reference-line labels on an axis. Suboptions that *follow* the TICK= option affect only the specified line's label. For example, the following option assigns the color green to all the reference-line labels on an axis, but left-justifies only the third reference line's label:

```
reflabel(c=green "one" "two" t=3  j=left "three")
```

For the options to be applied to a text string, they must precede the quoted string. In the following option, the `j=left` is ignored because it follows the string:

```
reflabel(c=green "one" "two" t=3 "three" j=left)
```

□

Note: The Java and ActiveX device drivers do not support the REFLABEL option. △

With the VALUE= option, the TICK= option specifies the *n*th major tick mark value. It is used to designate the tick mark value whose text and appearance you want to modify. For example, the following option changes the color of only the third tick mark value and leaves all others unchanged:

```
value=(tick=3 color=red)
```

Suboptions that precede the TICK= option affect all the major tick mark values. Suboptions that follow the TICK= option affect only the specified value. For example, the following option makes all the major tick mark values four units high and colors all of them blue except for the third one, which is red:

```
value=(height=4 color=blue tick=3 color=red)
```

Alias: T=*n*

Using Text Description Suboptions

Text description suboptions affect all the strings that follow them unless the suboption is changed or turned off. If the value of a suboption is changed, the new value affects all the text strings that follow it. Consider this example:

```
label=(font=swiss height=4 "Weight"
       justify=right height=3 "(in tons)")
```

FONT=SWISS applies to both `Weight` and `(in tons)`. HEIGHT=4 affects `Weight`, but is respecified as HEIGHT=3 for `(in tons)`. JUSTIFY=RIGHT affects only `(in tons)`.

Tick Mark Description Suboptions

Tick mark description suboptions are used by the MAJOR= and the MINOR= options to change the color, height, width, and number of the tick marks to which they apply. See the MAJOR= and MINOR= options.

COLOR=*tick-mark-color*
: colors the tick marks. If you omit the COLOR= suboption, a color specification is searched for in this order:

    1. the COLOR= option in the AXIS statement
    2. the CAXIS= option for the procedure
    3. the color of the default style.

    Alias: C=*tick-mark color*

HEIGHT=*tick-height* <*units*>
: specifies the height of the tick mark. The defaults for the HEIGHT= suboption depend on the option with which it is used:

    □ With the MAJOR= option the default height .5 CELLS.

    □ With the MINOR= option the default height .25 CELLS.

    If you specify a negative number, tick marks are drawn inside the axis.

    Alias: H=*tick-height* <*units*>

    Restriction: Not supported by Java and ActiveX.

NUMBER=*number-of-ticks*
: specifies the number of tick marks to be drawn. With the MAJOR= option, *number-of-ticks* must be greater than 1. With the MINOR= option, *number-of-ticks* must be greater than 0.

    With the MAJOR= option, the NUMBER= suboption can be overridden by a major tick mark specification in the procedure, which in turn can be overridden by the ORDER= option.

    With the MINOR= option, the NUMBER= suboption can be overridden by a minor tick mark specification in the procedure.

    The NUMBER= option is not valid with logarithmic axes.

    Alias: N=*number-of-ticks*

WIDTH=*thickness-factor*
: specifies the thickness of the tick mark, where *thickness-factor* is a number. Thickness increases directly with *thickness-factor*. By default, WIDTH=1.

    Style Reference: LineThickness attribute of the GraphAxisLines element.

    Alias: W=*thickness-factor*

Restriction: Partially supported by Java

Using the AXIS Statement

AXIS statements can be defined anywhere in your SAS program. They are global and remain in effect until redefined, canceled, or until the end of your SAS session. AXIS statements are not applied automatically, and must be explicitly assigned by an option in the procedure that uses them.

You can define up to 99 different AXIS statements. If you define two AXIS statements of the same number, the most recently defined statement replaces the previously defined statement of the same number. An AXIS statement without a number is treated as an AXIS1 statement.

Cancel individual AXIS statements by defining an AXIS statement of the same number without options (a null statement):

```
axis4;
```

Canceling one AXIS statement does not affect any other AXIS definitions. To cancel all current AXIS statements, use the RESET= option in a GOPTIONS statement:

```
goptions reset=axis;
```

Specifying RESET=GLOBAL or RESET=ALL cancels all current AXIS definitions as well as other settings.

To display a list of current AXIS definitions in the LOG window, use the GOPTIONS procedure with the AXIS option:

```
proc goptions axis nolist;
run;
```

Assigning AXIS Definitions

AXIS definitions must always be explicitly assigned by the appropriate option in the statement that generates the graph. The following table lists the procedures and statements that generate axes, the type of axis, and the statement option that assigns an AXIS definitions to that axis:

Procedure	Statement that generates an axis	Type of axis	Option that assigns an AXIS definition
GBARLINE	BAR | PLOT	midpoint axis	MAXIS=
		response axis	RAXIS=
GCHART	HBAR | VBAR	group axis	GAXIS=
		midpoint axis	MAXIS=
		response axis	RAXIS=
GCONTOUR	PLOT	horizontal axis	HAXIS=
		vertical axis	VAXIS=
GPLOT	PLOT	horizontal axis	HAXIS=
		vertical axis	VAXIS=
GRADAR	CHART	star axis	STARAXIS=

Some types of axes cannot use certain AXIS statement options:
- Group and midpoint axes ignore the LOGBASE=, MAJOR=, and MINOR= options.
- Midpoint, horizontal and vertical axes ignore the NOBRACKETS option.

BY Statement

Processes data and orders output according to the BY group.

Used by: GAREABAR, GCHART, GBARLINE, GCONTOUR, GMAP, GPLOT, GRADAR, GREDUCE, G3D, G3GRID procedures

Syntax

BY<DESCENDING> *variable*
   <...<DESCENDING> *variable-n*>
   <NOTSORTED>;

Description

The BY statement divides the observations from an input data set into groups for processing. Each set of contiguous observations with the same value for a specified variable is called a *BY group*. A variable that defines BY groups is called a *BY variable* and is the variable that is specified in the BY statement. When you use a BY statement, the graphics procedure performs the following operations:
- processes each group of observations independently
- generates a separate graph or output for each BY group
- automatically adds a heading called a *BY line* to each graph identifying the BY group represented in the graph
- adds BY statement information below the Description field of the catalog entry.

By default, the procedure expects the observations in the input data set to be sorted in ascending order of the BY variable values.

Note: The BY statement in SAS/GRAPH is essentially the same as the BY statement in Base SAS: however, the effect on the output is different when it is used with SAS/GRAPH procedures. △

Required Arguments

variable
   specifies the variable that the procedure uses to form BY groups. You can specify more than one variable. By default, the procedure expects observations in the data set to be sorted in ascending order by all the variables that you specify or to be indexed appropriately.

Options

DESCENDING
   indicates that the data set is sorted in descending order by the specified variable. The option affects only the variable that immediately follows the option name, and must be repeated before every variable that is not sorted in ascending order. For example, this BY statement indicates that observations in the input data set are arranged in descending order of VAR1 values and ascending order of VAR2 values:

```
by descending var1 var2;
```

This BY statement indicates that the input data set is sorted in descending order of both VAR1 and VAR2 values:

```
by descending var1 descending var2;
```

NOTSORTED
specifies that observations with the same BY value are grouped together, but are not necessarily sorted in alphabetical or numeric order. The observations can be grouped in another way, for example, in chronological order.

NOTSORTED can appear anywhere in the BY statement and affects all variables specified in the statement. NOTSORTED overrides DESCENDING if both appear in the same BY statement.

The requirement for ordering or indexing observations according to the values of BY variables is suspended when you use the NOTSORTED option. In fact, the procedure does not use an index if you specify NOTSORTED. For NOTSORTED, the procedure defines a BY group as a set of contiguous observations that have the same values for all BY variables. If observations with the same value for the BY variables are not contiguous, the procedure treats each new value it encounters as the first observation in a new BY group and creates a graph for that value, even if it is only one observation.

Preparing Data for BY-Group Processing

Unless you specify the NOTSORTED option, observations in the input data set must be in ascending numeric or alphabetic order. To prepare the data set, either sort it with the SORT procedure using the same BY statement that you plan to use in the target SAS/GRAPH procedure or create an appropriate index on the BY variables.

If the procedure encounters an observation that is out of the proper order, it issues an error message.

If you need to group data in some other order, you can still use BY-group processing. To do so, process the data so that observations are arranged in contiguous groups that have the same BY-variable values and specify the NOTSORTED option in the BY statement.

For an example of sorting the input data set, see "Example 7. Using BY-group Processing to Generate a Series of Charts" on page 309 .

Controlling BY Lines

By default, the BY statement prints a BY line above each graph that contains the variable name followed by an equal sign and the variable value. For example, if you specify BY SITE in the procedure, the default heading when the value of SITE is **London** would be SITE=London.

Suppressing the BY line   To suppress the entire BY line, use the NOBYLINE option in an OPTION statement or specify HBY=0 in the GOPTIONS statement. See "Example 7. Using BY-group Processing to Generate a Series of Charts" on page 309.

Suppressing the name of the BY variable   To suppress the variable name and the equal sign in the heading and leave only the BY value, use the LABEL statement to assign a null label ("00"X) to the BY variable. For example, this statement assigns a null label to the SITE variable:

```
label site="00"x;
```

Controlling the appearance of the BY line   To control the color, font, and height of the BY lines, use the following graphics options in a GOPTIONS statement:

CBY=*BY-line-color*
:   specifies the color for BY lines.

FBY=*font*
:   specifies the font for BY lines.

HBY=*n<units>*
:   specifies the height for BY lines.

See Chapter 15, "Graphics Options and Device Parameters Dictionary," on page 327 for a complete description of each option.

Naming the Catalog Entries

The catalog entries generated with BY-group processing always use incremental naming. This means that the first entry created by the procedure uses the base name and subsequent entries increment that name. The base name is either the default entry name for the procedure (for example, GPLOT) or the name specified with the NAME= option in the action statement. Incrementing the base name automatically appends a number to each subsequent entry (for example, GPLOT1, GPLOT2, and so on). See also "Specifying the Catalog Name and Entry Name for Your GRSEGs" on page 100. For an example of incremented catalog names, see "Example 9. Combining Graphs and Reports in a Web Page" on page 315.

Using the BY Statement

This section describes the following:

- the effect of BY-group processing on the GCHART, GMAP, and GPLOT procedures
- the interaction between BY-group and RUN-group processing
- the requirements for using BY-group processing with the Annotate facility
- how to include BY information in titles, notes, and footnotes
- how patterns and symbols are assigned to BY-groups
- the effect of using BY-group processing with the ODS HTML statement

For additional information on any of these topics, refer to the appropriate chapter.

With the GCHART Procedure   When you use BY-group processing with the GCHART procedure, you can do the following tasks:

- With the BLOCK, HBAR, and VBAR statements, you can use the PATTERNID=BY option to assign patterns according to BY groups. With PATTERNID=BY, each BY group uses a different PATTERN definition, but all bars or blocks within a BY group use the same pattern. For further information, see "Example: PATTERN and SYMBOL Definitions with BY Groups in the GCHART Procedure" on page 220.
- With the BLOCK statement, you can use the BLOCKMAX= option to produce the same block-height scaling in all block charts in a BY group.
- With the HBAR or VBAR statement, you can use the RAXIS= option to produce the same response axis scaling in all horizontal or vertical bar charts in a BY group.

With the PIE and STAR statements, the effect of a BY statement is similar to that of the GROUP= option, except that the GROUP= option enables you to put more than one graph on a single page while the BY statement does not. Do not use a BY variable as the group variable in STAR or PIE statements.

With the GMAP Procedure   By default, BY-group processing affects both the map data set and the response data set. This means that you get separate, individual output for each map area common to both data sets. For example, if the map data set REGION contains six states and the response data set contains the same six states, and you specify BY STATE in the GMAP procedure, you get six graphs with one state on each graph.

If you use the ALL option in the PROC GMAP statement and you also use the BY statement, you get one output for each map area in the response data set, but that output displays all the map areas in the map data set. Only one map area per output contains response data information; the others are empty. For example, if you create a block map using the data sets REGION and SALES, specify BY STATE, and include the ALL option in the PROC GMAP statement, you get six graphs with six states on each graph. One state per graph has a block; the remaining five are empty. The UNIFORM option applies colors and heights uniformly across all BY-groups.

With the GPLOT Procedure   You can use the UNIFORM option in the PROC GPLOT statement to produce the same axis scaling for all graphs in a BY group. By default, the range of the axes can vary from graph to graph, but UNIFORM forces the scaling to be the same for all graphs generated by the procedure.

The UNIFORM option applies colors and heights uniformly across all BY-groups.

With the RUN Groups   If you use the BY statement with a procedure that processes data and supports RUN-group processing (the GCHART, GMAP, and GPLOT procedures), then each time you submit an action statement or a RUN statement you get a separate graph for each value of the BY variable. For example, each of these two RUN-groups produces a separate plot for every value of the BY variable SITE:

```
/* first run group*/
proc gplot data=sales;
   title1 "Sales Summary";
   by site;
   plot sales*model_a;
run;

   /* second run group */
   plot sales*model_b;
run;
quit;
```

The BY statement stays in effect for every subsequent RUN group until you submit another BY statement or exit the procedure. Variables in subsequent BY statements replace any previous BY variables.

You can also turn off BY-group processing by submitting a null BY statement (BY;) in a RUN group, but when you do this, the null BY statement turns off BY-group processing *and* the RUN group generates a graph.

For more information, see "RUN-Group Processing" on page 56.

With the Annotate Facility   If a procedure that is using BY-group processing also specifies annotation with the ANNOTATE= option in the PROC statement, the same annotation is applied to every graph generated by the procedure.

If you specify annotation with the ANNOTATE= option in the action statements for a procedure, the BY-group processing is applied to the Annotate data set. In this way, you can customize the annotation for the output from each BY group by including the BY variable in the Annotate data set and by using each BY-variable value as a condition for the annotation to be applied to the output for that value.

With TITLE, FOOTNOTE, and NOTE Statements   TITLE, FOOTNOTE, and NOTE statements can automatically include the BY variable name, BY variable values, or BY lines in the text they produce. To insert BY variable information into the text strings used by these statements, use the #BYVAR, #BYVAL, and #BYLINE substitution options. For an example, see "Example 7. Using BY-group Processing to Generate a Series of Charts" on page 309.

With PATTERN and SYMBOL Definitions   By default, when using a BY statement, the graph for each BY group uses the same patterns or symbols in their defined order. For example, if the BY variable contains four values and there are two response levels for each BY value, the PATTERN1 and PATTERN2 or SYMBOL1 and SYMBOL2 statements are used for each graph. Each BY-group starts over with PATTERN1 or SYMBOL1. The UNIFORM option in the GMAP procedure changes this behavior.

Example: PATTERN and SYMBOL Definitions with BY Groups in the GCHART Procedure   The GCHART procedure, when used with SYMBOL or PATTERN definitions, assigns the symbols or patterns in order to each BY group. For example, if the BY variable REGION has four values—`East`, `North`, `South`, and `West`—the patterns are assigned to the BY-groups in this order:

1 PATTERN1 is assigned to `East`
2 PATTERN2 is assigned to `North`
3 PATTERN3 is assigned to `South`
4 PATTERN4 is assigned to `West`.

If you create sets of graphs from several data sets containing the variable REGION, and if you want the same pattern assigned to the same region each time, you must be sure that REGION always has the same four values. Otherwise, the patterns may not be the same across graphs. For example, if the value `North` is missing from the data, the patterns are assigned as follows:

1 PATTERN1 is assigned to `East`
2 PATTERN2 is assigned to `South`
3 PATTERN3 is assigned to `West`.

In this case, `South` is assigned pattern 2 instead of pattern 3 and `West` is assigned pattern 3 instead of pattern 4. To avoid this, include the value `North` for the variable REGION, but assign it a missing value for all other variables.

FOOTNOTE Statement

Writes up to 10 lines of text at the bottom of the graph.

See: "TITLE, FOOTNOTE, and NOTE Statements" on page 279

Syntax

FOOTNOTE<1...10> <*text-argument(s)*>;

GOPTIONS Statement

Temporarily sets default values for many graphics attributes and device parameters used by SAS/GRAPH procedures.

Used by: all statements and procedures in a SAS session

Syntax

GOPTIONS <*options-list*>;
  options-list can be one or more options from any or all of the following categories:
  ☐ reset option
      RESET=ALL | GLOBAL | *statement-name* | (*statement-name(s)*)
  ☐ options that affect the appearance of the display area and the graphics output
      ASPECT=*scaling-factor*
      ALTDESC | NOALTDESC
      AUTOSIZE=ON | OFF | DEFAULT
      BORDER | NOBORDER
      CELL | NOCELL
      GSIZE=*lines*
      HORIGIN=*horizontal-offset* <IN | CM>
      HPOS=*columns*
      HSIZE=*horizontal-size* <IN | CM>
      IBACK= *fileref* | "*external-file*"
      IMAGESTYLE = TILE | FIT
      IMAGEPRINT | NOIMAGEPRINT
      ROTATE=LANDSCAPE | PORTRAIT
      ROTATE | NOROTATE
      TARGETDEVICE=*target-device-entry*
      VORIGIN=*vertical-offset* <IN | CM>
      VPOS=*rows*
      VSIZE=*vertical-size* <IN | CM>
      XMAX=*width* <IN | CM>
      XPIXELS=*width-in-pixels*
      YMAX=*height* <IN | CM>
      YPIXELS=*height-in-pixels*
  ☐ options that affect color
      CBACK=*background-color*
      CBY=*BY-line-color*
      COLORS=<(*colors-list* | NONE)>
      CPATTERN=*pattern-color*
      CSYMBOL=*symbol-color*
      CTEXT=*text-color*
      CTITLE=*title-color*
      PENMOUNTS=*active-pen-mounts*
      PENSORT | NOPENSORT
  ☐ options that control font selection or text appearance
      CHARTYPE=*hardware-font-chartype*
      FASTTEXT | NOFASTTEXT
      FBY=*BY-line-font*

FCACHE=*number-fonts-open*
FONTRES=NORMAL | PRESENTATION
FTEXT=*text-font*
FTITLE=*title-font*
FTRACK=LOOSE | NONE | NORMAL | TIGHT | TOUCH | V5
HBY=*BY-line-height* *<units>*
HTEXT=*text-height* *<units>*
HTITLE=*title-height* *<units>*
RENDER=APPEND | DISK | MEMORY | NONE | READ
RENDERLIB=*libref*
SIMFONT=*software-font*

☐ options that set defaults for procedures and global statements
GUNIT=*units*
INTERPOL=*interpolation-method*
OFFSHADOW=(*x* *<units>*, *y* *<units>* | (*x,y*) *<units>*
V6COMP | NOV6COMP

☐ image animation options
DELAY=*delay-time*
DISPOSAL=NONE | BACKGROUND | PREVIOUS | UNSPECIFIED
INTERLACED | NONINTERLACED
ITERATION=*iteration-count*
TRANSPARENCY | NOTRANSPARENCY

☐ options that affect how your SAS/GRAPH program runs
DISPLAY | NODISPLAY
ERASE | NOERASE
GWAIT=*seconds*
GRAPHRC | NOGRAPHRC
IMAGEPRINT | NOIMAGEPRINT
PCLIP | NOPCLIP
POLYGONCLIP | NOPOLYGONCLIP

☐ options that control how output is sent to devices or files
ADMGDF | NOADMGDF
DEVADDR=*device-address*
DEVICE=*device-entry*
DEVMAP=*device-map-name* | NONE
EXTENSION="*file-type*"
FILECLOSE=DRIVERTERM | GRAPHEND
FILEONLY | NOFILEONLY
GACCESS=*output-format* | "*output-format* > *destination*"
GEND="*string*" <..."*string-n*">
GEPILOG="*string*" <..."*string-n*">
GOUTMODE=APPEND | REPLACE
GPROLOG="*string*" <..."*string-n*">
GPROTOCOL=*module-name*
GSFLEN=*record-length*

GSFMODE=APPEND | PORT | REPLACE
GSFNAME=*fileref*
GSFPROMPT | NOGSFPROMPT
GSTART="*string*" <..."*string-n*">
HANDSHAKE=HARDWARE | NONE | SOFTWARE | XONXOFF
KEYMAP=*map-name* | NONE
POSTGEPILOG="*string*"
POSTGPROLOG="*string*"
PREGEPILOG="*string*"
PREGPROLOG="*string*"
PROMPTCHARS="*prompt-chars-hex-string*"X

☐ options that specify hardware capabilities of the device
CHARACTERS | NOCHARACTERS
CIRCLEARC | NOCIRCLEARC
DASH | NODASH
DASHSCALE=*scaling-factor*
FILL | NOFILL
FILLINC=0...9999
LFACTOR=*line-thickness-factor*
PIEFILL | NOPIEFILL
POLYGONFILL | NOPOLYGONFILL
SYMBOL | NOSYMBOL

☐ options that control printer hardware features
AUTOCOPY | NOAUTOCOPY
AUTOFEED | NOAUTOFEED
BINDING=DEFAULTEDGE | LONGEDGE | SHORTEDGE
COLLATE | NOCOLLATE
DUPLEX | NODUPLEX
GCOPIES=(<*current-copies*><,*max-copies*>)
PAPERDEST=*bin*
PAPERFEED=*feed-increment* <IN | CM>
PAPERLIMIT=*width* <IN | CM>
PAPERSIZE="*size-name*" | (*width,height*)
PAPERSOURCE=*tray*
PAPERTYPE="*type-name*"
PPDFILE=*fileref* | "*external-file*"
REPAINT=*redraw-factor*
REVERSE | NOREVERSE
SPEED=*pen-speed*
UCC="*control-characters-hex-string*"X

☐ options that interact with the operating environment
DRVINIT="*system-command(s)*"
DRVTERM="*system-command(s)*"
PREGRAPH="*system-command(s)*"
POSTGRAPH="*system-command(s)*"

PROMPT | NOPROMPT
- options for mainframe systems
    GCLASS=*SYSOUT-class*
    GDDMCOPY=FSCOPY | GSCOPY
    GDDMNICKNAME=*nickname*
    GDDMTOKEN=*token*
    GDEST=*destination*
    GFORMS="*forms-code*"
    GWRITER="*writer-name*"
    TRANTAB=*table* | *user-defined-table*

Description

The GOPTIONS statement specifies values for *graphics options*. Graphics options control characteristics of the graph, such as size, colors, type fonts, fill patterns, and symbols. If GOPTIONS are specified, they override the default style. In addition, they affect the settings of device parameters, which are defined in the device entry. Device parameters control such characteristics as the appearance of the display, the type of output produced, and the destination of the output.

The GOPTIONS statement enables you to change these settings temporarily, either for a single graph or for the duration of your SAS session. You can use the GOPTIONS statement to do the following tasks:

- override default values for graphics options that control either graphics attributes or device parameters for a single graph or for an entire SAS session
- reset individual graphics options or all graphics options to their default values
- cancel definitions for AXIS, FOOTNOTE, PATTERN, SYMBOL, and TITLE statements

To change device parameters permanently, you must use the GDEVICE procedure to modify the appropriate device entry or to create a new one. See Chapter 38, "The GDEVICE Procedure," on page 1125 for details.

To review the current settings of all graphics options, use the GOPTIONS procedure. See Chapter 44, "The GOPTIONS Procedure," on page 1319 for details.

Options

See Chapter 15, "Graphics Options and Device Parameters Dictionary," on page 327 for a complete description of all graphics options used by the GOPTIONS statement.

Using the GOPTIONS Statement

GOPTIONS statements are global and can be located anywhere in your SAS program. However, for the graphics options to affect the output from a procedure, the GOPTIONS statement must execute before the procedure.

With the exception of the RESET= option, graphics options can be listed in any order in a GOPTIONS statement. The RESET= option should be the first option in the GOPTIONS statement.

A graphics option remains in effect until you either specify the option in another GOPTIONS statement, or use the RESET= option to reset the values, or end the SAS session. When a session ends, the values of the graphics options return to their default values.

Graphics options are additive; that is, the value of a graphics option remains in effect until the graphics option is explicitly changed or reset or until you end your SAS session. Graphics options remain in effect even after you submit additional GOPTIONS statements specifying different options.

To reset an individual option to its default value, submit the option without a value (a null graphics option.) You can use a comma (but it is not required) to separate a null graphics option from the next one. For example, this GOPTIONS statement sets the values for background color, text height, and text font:

```
goptions cback=blue htext=6 pct ftext=albany;
```

To reset only the background color specification to the default and keep the remaining values, use this GOPTIONS statement:

```
goptions cback=;
```

To reset all graphic options to their default values, specify RESET=GOPTIONS:

```
goptions reset=goptions;
```

Alternatively, you can use RESET=ALL, but it also cancels any global statement definitions in addition to resetting all graphics options to default values.

Graphics Option Processing

You can control many graphics attributes through statement options, graphics options, device parameters, or a combination of these. SAS/GRAPH searches these places to determine the value to use, stopping at the first place that gives it an explicit value:

1 statement options

2 the value of the corresponding graphics option

3 the value of a device parameter found in the catalog entry for your device driver

Note: Not every graphics attribute can be set in all three places. See the statement and procedure chapters for the options that can be used with each. △

Some graphics options are supported for specific devices or operating environments only. See the SAS Help facility for SAS/GRAPH or the SAS companion for your operating environment for more information.

LEGEND Statement

Controls the location and appearance of legends on two-dimensional plots, contour plots, maps, and charts.

Used by: GAREABAR, GCHART, GBARLINE, GCONTOUR, GMAP, GPLOT procedures
Type: Global

Syntax

LEGEND<1...99> <*options*>;
   option(s) can be one or more options from any or all of the following categories:
      ☐ appearance options
          ACROSS=*number-of-columns*
          CBLOCK=*block-color*
          CBORDER=*frame-color*

CFRAME=*background-color*
CSHADOW=*shadow-color*
DOWN=*number-of-rows*
FRAME
FWIDTH=*thickness-factor*
REPEAT=1 | 2 | 3
ROWMAJOR | COLMAJOR
SHAPE=BAR(*width,height*) <*units*> | LINE(*length*) <*units*> |
  SYMBOL(*width,height*) <*units*>

□ position-options
  MODE=PROTECT | RESERVE | SHARE
  OFFSET=(<*x* ><,*y* >)<*units* > | (<*x* <*units* >><,*y* <*units* >>)
  ORIGIN=(<*x* ><,*y* >)<*units* > | (<*x* <*units* >><,*y* <*units* >>)
  POSITION=(<BOTTOM | MIDDLE | TOP> <LEFT | CENTER | RIGHT>
    <INSIDE | OUTSIDE>)

□ text-options
  LABEL=(*text-argument(s)*) | NONE
  ORDER=(*value-list*)
  VALUE=(*text-argument(s)*) | NONE

Description

LEGEND statements specify the characteristics of a legend but do not create legends. The characteristics are as follows:

□ the position and appearance of the legend box
□ the text and appearance of the legend label
□ the appearance of the legend entries, including the size and shape of the legend values
□ the text of the labels for the legend values

LEGEND definitions are not automatically applied when a procedure generates a legend. Instead, they must be explicitly assigned with a LEGEND= option in the appropriate procedure statement.

The following figure illustrates the terms associated with the various parts of a legend.

Figure 14.5 Parts of a Legend

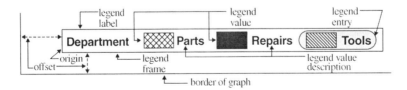

Options

When the syntax of an option includes *units*, use one of these:

CELLS   character cells
CM   centimeters

IN	inches
PT	points
PCT	percentage of the graphics output area

Note: The Java applet does not support CM, IN, or PT. △

If you omit *units*, a unit specification is searched for in this order:

1 GUNIT= in a GOPTIONS statement

2 the default unit, CELLS

ACROSS=*number-of-columns*
specifies the number of columns to use for legend entries. If there are multiple rows and columns in a legend, use the ROWMAJOR and COLMAJOR options to specify the arrangement of legend entries. Specify the ROWMAJOR option to arrange entries (from lowest to highest) starting from left to right, and then top to bottom. Specify the COLMAJOR option to arrange entries starting from top to bottom, and then left to right.

Featured in: "Example 8. Creating a Simple Web Page with the ODS HTML Statement" on page 313

See also: ROWMAJOR, COLMAJOR

CBLOCK=*block-color*
generates and colors a three-dimensional block effect behind the legend. The size and position of the block are controlled by the graphics option OFFSHADOW=(*x,y*).
The CBLOCK= and CSHADOW= options are mutually exclusive. If both are present, SAS/GRAPH software uses the last one specified. The CBLOCK= option is usually used in conjunction with the FRAME, CFRAME=, or CBORDER= options.
The Java applet treats the CBLOCK option like the CSHADOW option.

See also: The OFFSHADOW="OFFSHADOW" on page 394 graphics option and "Creating Drop Shadows and Block Effects" on page 238

Restriction: Not supported by Java.

CBORDER=*frame-color*
draws a colored frame around the legend. This option overrides the FRAME option. CBORDER= can be used in conjunction with the CFRAME= option.

Style Reference: Color attribute of the GraphBorderLines graph element

CFRAME=*background-color*
specifies the background color of the legend. This option overrides the FRAME option. If both the CFRAME= and FRAME= options are specified, only the solid background produced by the CFRAME= option is displayed. The CFRAME= option can be used in conjunction with the CBORDER= option.

Style Reference: Color attribute of the GraphLegendBackground graph element

CSHADOW=*shadow-color*
generates and colors a drop shadow behind the legend. The size and position of the shadow is controlled by the graphics option OFFSHADOW=(*x,y*).
The CSHADOW= and CBLOCK= options are mutually exclusive. If both are present, SAS/GRAPH uses the last one specified. The CSHADOW= option is usually specified in conjunction with the FRAME, CFRAME=, or CBORDER= options.

See also: the OFFSHADOW="OFFSHADOW" on page 394 graphics option and "Creating Drop Shadows and Block Effects" on page 238.

DOWN=*number-of-rows*
> specifies the number of rows to use for legend entries. If there are multiple rows and columns in a legend, use the ROWMAJOR and COLMAJOR options to specify the arrangement of legend entries. Specify the ROWMAJOR option to arrange entries (from lowest to highest) starting from left to right, and then top to bottom. Specify the COLMAJOR option to arrange entries starting from top to bottom, and then left to right. The ROWMAJOR option is the default.

FRAME
> draws a frame around the legend. The color of the frame is the first color in the color list.

FWIDTH=*thickness-factor*
> specifies the thickness of the frame, where *thickness-factor* is a number. The thickness of the line increases directly with *thickness-factor*. By default, FWIDTH=1.
>
> **Restriction:** Not supported by Java and ActiveX

LABEL=(*text-argument(s)*) | NONE
> modifies a legend label. *Text-argument(s)* defines the appearance or the text of a legend label, or both. NONE suppresses the legend label. By default, the text of the legend label is either the variable name or a previously assigned variable label (except in the case of GPLOT with OVERLAY. In that case the default label is "PLOT"). *Text-argument(s)* can be one or more of these:
>
> *"text-string"*
>> provides up to 256 characters of label text. Enclose each string in quotes. Separate multiple strings with blanks.
>
> *text-description-suboption*
>> modifies a characteristic such as the font, color, or size of the text strings that follows it. *Text-description-suboption* can be as follows:
>>
>> COLOR=*text-color*
>>
>> FONT=*font* | NONE
>>
>> HEIGHT=*text-height* <*units*>
>>
>> JUSTIFY=LEFT | CENTER | RIGHT
>>
>> POSITION=(<BOTTOM | MIDDLE | TOP> <LEFT | CENTER | RIGHT>)
>
> *Note:* The Java applet does not support the POSITION= suboption—it draws legend labels at the top-left of the legend. Also, it does not support multiple values for the JUSTIFY= suboption (only the first is honored). The ActiveX control supports the POSITION= option but does not support multiple values for the JUSTIFY suboption (only the first is honored). △
>
> See "Text Description Suboptions" on page 233 for complete descriptions. Specify as many text strings and text description suboptions as you want, but enclose them all in one set of parentheses.
>
> **Style Reference:** Color attribute of the GraphLabelText graph element
>
> **Featured in:** "Example 3. Rotating Plot Symbols Through the Color List" on page 299 and "Example 8. Creating a Simple Web Page with the ODS HTML Statement" on page 313
>
> **Restriction:** Partially supported by Java and ActiveX

MODE=PROTECT | RESERVE | SHARE
> specifies whether the legend is drawn in the procedure output area or whether legend elements can overlay other graphics elements. MODE= can take one of these values:

PROTECT	draws the legend in the procedure output area, but a *blanking area* surrounds the legend, preventing other graphics elements from being displayed in the legend. (A blanking area is a protected area in which no other graphics elements are displayed.)
RESERVE	takes space for the legend from the procedure output area, thereby reducing the amount of space available for the graph. If MODE=RESERVE is specified in conjunction with OFFSET=, the legend can push the graph off the graphics output area. RESERVE is valid only when POSITION=OUTSIDE. If POSITION=INSIDE is specified, a warning is issued and MODE= value is changed to PROTECT.
SHARE	draws the legend in the procedure output area. If the legend is positioned over elements of the graph itself, both graphics elements and legend elements are displayed.

By default, MODE=RESERVE unless POSITION=INSIDE. In this case, the default changes to MODE=PROTECT.

See also: "Positioning the Legend" on page 237

Restriction: Not supported by Java and ActiveX

OFFSET=(<x><,y>)<units> | (<x <units>><,y <units>>)

specifies the distance to move the entire legend; x is the number of units to move the legend right (positive numbers) or left (negative numbers), and y is the number of units to move the legend up (positive numbers) or down (negative numbers).

To set only the x offset, specify one value, with or without a following comma:

```
offset=(4 cm,)
```

To set both the x and y offset, specify two values, with or without a comma separating them:

```
offset=(2 pct, 4 pct)
```

To set only the y offset, specify one value preceded by a comma:

```
offset=(,-3 pct)
```

The OFFSET= option is usually used in conjunction with the POSITION= option to adjust the position of the legend. Moves are relative to the location specified by the POSITION= option, with OFFSET=(0,0) representing the initial position. You can also apply the OFFSET= option to the default legend position.

The OFFSET= option is unnecessary with the ORIGIN= option since the ORIGIN= option explicitly positions the legend and requires no further adjustment. However, if you specify both options, the OFFSET= values are added to the ORIGIN= values,l and the LEGEND is positioned accordingly.

See also: "Positioning the Legend" on page 237 and the option POSITION= on page 230

Restriction: Not supported by Java and ActiveX

ORDER=(*value-list*)
: selects or orders the legend values that appear in the legend. The way you specify *value-list* depends on the type of variable that generates the legend:

    □ For numeric variables, *value-list* is either an explicit list of values, or a starting and an ending value with an interval increment, or a combination of both forms:

    n <...*n*>

    n TO *n* <BY *increment*>

    n <...*n*> TO *n* <BY *increment*> <*n* <...*n*>>

    If a numeric variable has an associated format, the specified values must be the *unformatted* values.

    □ For character variables, *value-list* is a list of unique character values enclosed in quotes and separated by blanks:

    "*value-1*" <..."*value-n*">

    If a character variable has an associated format, the specified values must be the *formatted* values.

    For a complete description of *value-list*, see the option ORDER= on page 205 in the AXIS statement.

    Even though the ORDER= option controls whether a legend value is displayed and where it appears, the VALUE= option controls the text that the legend value displays.

    Restriction: Not supported by Java and ActiveX

ORIGIN=(<*x*><,*y*>)<*units*> | (<*x* <*units* >><,*y* <*units*>>)
: specifies the *x* coordinate and the *y* coordinate of the lower-left corner of the legend box. The ORIGIN= option explicitly positions the legend anywhere on the graphics output area. It is possible to run a legend off the page or overlay the graph.

    To set only the *x* coordinate, specify one value, with or without a following comma:

    ```
    origin=(4 cm,)
    ```

    To set both the *x* and *y* coordinates, specify two values, with or without a comma separating them:

    ```
    origin=(2 pct, 4 pct)
    ```

    To set only the *y* coordinate, specify one value preceded by a comma:

    ```
    origin=(,3 pct)
    ```

    The ORIGIN= option overrides the POSITION= option if both are used. Although using the OFFSET= option with the ORIGIN= option is unnecessary, if the OFFSET= option is also specified, it is applied after the ORIGIN= request has been processed.

    See also: "Positioning the Legend" on page 237

    Restriction: Not supported by Java and ActiveX

POSITION=(<BOTTOM | MIDDLE | TOP> <LEFT | CENTER | RIGHT> <OUTSIDE | INSIDE>)
: positions the legend on the graph. Values for POSITION= are

OUTSIDE or INSIDE	specifies the location of the legend in relation to the axis area.
BOTTOM or MIDDLE or TOP	specifies the vertical position.

LEFT or   specifies the horizontal position.
CENTER or
RIGHT

By default, POSITION=(BOTTOM CENTER OUTSIDE). You can change one or more settings. If you supply only one value the parentheses are not required. If you specify two or three values and omit the parentheses, SAS/GRAPH accepts the first value and ignores the others.

Once you assign the initial legend position, you can adjust it with the OFFSET= option.

The ORIGIN= option overrides the POSITION= option. The value of the MODE= option can affect the behavior of the POSITION= option.

Note: The Java applet defaults to BOTTOM-CENTER and supports all possible combinations of BOTTOM | MIDDLE | TOP with LEFT | CENTER | RIGHT except for MIDDLE-CENTER (which would overwrite the map.) The Java applet does not support INSIDE for positioning. △

See also: OFFSET= option on page 229 and MODE= option on page 228

Restriction: Partially supported by Java

REPEAT=1 | 2 | 3

Use the REPEAT= option to specify how many times the plot symbol is repeated in the legend. Valid values are 1 to 3, with 3 being the default.

ROWMAJOR | COLMAJOR

specifies the arrangement of legend entries when there are multiple rows and multiple columns. Specify the ROWMAJOR option (the default) to arrange entries (from lowest to highest) starting from left to right, and then top to bottom. Specify the COLMAJOR option to arrange the entries starting from top to bottom, and then left to right.

See also: ACROSS=, DOWN=

SHAPE=BAR(*width*<*units*>,*height*<*units*>) <*units*> | LINE(*length*) <*units*> | SYMBOL(*width*<*units*>,*height*<*units*>) <*units*>

specifies the size and shape of the legend values displayed in each legend entry. The SHAPE= value you specify depends on which procedure generates the legend.

BAR(*width,height*)<*units*>

is used with the GCHART and GMAP procedures, with the GPLOT procedure if you use the AREAS= option, and with the GCONTOUR procedure if you use the PATTERN option. Each legend value is a bar of the specified width and height. By default, *width* is 5, *height* is 0.8, and *units* are CELLS. You can specify *units* for the *width,height* pair or for the individual coordinates.

LINE(*length*) <*units*>

is used with the GPLOT and GCONTOUR procedures. Each legend value is a line of the length you specify. Plotting symbols are omitted from the legend values. By default, *length* is 5 and *units* are CELLS. You can specify *units* for *length*.

SYMBOL(*width<units>*,*height<units>*) *<units>*
   is used with the GPLOT procedure. Each legend value (*not* each symbol) is the width and height you specify. For example, this specification produces legend values like the ones in Figure 14.6 on page 232(a):

```
shape=symbol(.5,.5)
```

   This specification produces legend values like the ones in Figure 14.6 on page 232(b):

```
shape=symbol(2,.5)
```

Figure 14.6 Legend Values Produced with SHAPE= SYMBOL

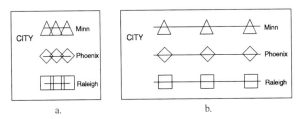

   By default, *width* is 5, *height* is 1, and *units* are CELLS. You can specify *units* for the *width*,*height* pair or for the individual coordinates.
   Restriction: Not supported by Java and ActiveX

VALUE=(*text-argument(s)*)| NONE
   modifies the legend value descriptions. *Text-argument(s)* defines the appearance or the text of the value descriptions. By default, value descriptions are the values of the variable that generates the legend or an associated format value. Numeric values are right-justified and character values are left-justified.
   NONE suppresses the value descriptions although the legend values (bars, lines, and so on) are still displayed. (NONE is not supported by Java or ActiveX). *Text-argument(s)* can be one or more of these:

"*text-string*"
   provides up to 256 characters of text for the value description. Enclose each string in quotes. Separate multiple strings with blanks.
   Specified text strings are assigned to the legend values in order. If you submit only one string, only the first legend entry uses the value of that string. If you specify multiple strings, the first string is the text for the first entry; the second string is the text for the second entry; and so on. For example, this specification produces legend entries like those shown in Figure 14.7 on page 232:

```
value=("1986" "1987" "1988")
```

Figure 14.7 Specifying Value Descriptions with the VALUE= Option

text-description-suboption
    modifies a characteristic such as the font, color, or size of the text string(s) that follows it. *Text-description-suboption* can be as follows:
        COLOR=*text-color*
        FONT=*font* | NONE
        HEIGHT=*text-height* <*units*>
        JUSTIFY=LEFT | CENTER | RIGHT
        TICK=*n*
    See "Text Description Suboptions" on page 233 for complete descriptions. Place text description suboptions before the text strings they modify. Suboptions not followed by a text string affect the default values. To specify and describe the text for individual values or to produce multi-line text, use the TICK= suboption.

Specify as many text strings and text description suboptions as you want, but enclose them all in one set of parentheses.

To order or select legend entries, use the ORDER= option.

See also: "Text Description Suboptions" on page 233 and the option ORDER= on page 230

Restriction: Partially supported by Java and ActiveX

Text Description Suboptions

Text description suboptions are used by the LABEL= and VALUE= options to change the color, height, justification, font, and angle of either default text or specified text strings. See the LABEL= suboption on page 228 and the VALUE= suboption on page 232.

    COLOR=*text-color*
        specifies the color of the text. If you omit the COLOR= suboption, a color specification is searched for in this order:
        1 the CTEXT= option for the procedure
        2 the CTEXT= option in a GOPTIONS statement
        3 the color of the default style
        Alias: C=*text-color*

    FONT=*font* | NONE
        specifies the font for the text. See Chapter 11, "Specifying Fonts in SAS/GRAPH Programs," on page 155 for information on specifying fonts. If you omit the FONT= suboption, a font specification is searched for in this order:
        1 the FTEXT= option in a GOPTIONS statement
        2 the default style font, NONE
        Alias: F=*font* | NONE

    HEIGHT=*text-height* <*units*>
        specifies the height of the text characters in the number of units. By default, HEIGHT=1 CELL. If you omit the HEIGHT= suboption, a text height specification is searched for in this order:
        1 the HTEXT= option in a GOPTIONS statement
        2 the height specified by the default style
        Alias: H=*text-height* <*units*>

    JUSTIFY=LEFT | CENTER | RIGHT
        specifies the alignment of the text. The default for character variables is JUSTIFY=LEFT. The default for numeric variables is JUSTIFY=RIGHT.

Associating a character format with a numeric variable does not change the default justification of the variable.

You can use the JUSTIFY= suboption to print multiple lines of text by repeating the suboption before the text string for each line. For example, this statement produces a legend label and value descriptions like those shown in Figure 14.8 on page 234:

```
legend label=(justify=c "Distribution"
              justify=c "Centers")
   value=(tick=1 justify=c "Portland,"
                 justify=c "Maine"
          tick=2 justify=c "Paris,"
                 justify=c "France"
          tick=3 justify=c "Sydney,"
                 justify=c "Australia");
```

Figure 14.8 Specifying Multiple Lines of Text with the JUSTIFY= Suboption

Specify additional suboptions before any string.

See also: the suboption TICK= on page 235.

Alias: J=L | C | R

POSITION=(<BOTTOM | MIDDLE | TOP> <LEFT | CENTER | RIGHT>)
places the legend label in relation to the legend entries. The POSITION= suboption is used only with the LABEL= option. By default, POSITION=LEFT.

The parentheses are not required if only one value is supplied. If you specify two or three values and omit the parentheses, SAS/GRAPH accepts the first value and ignores the others.

Figure 14.9 on page 235 shows some of the ways the POSITION= suboption affects a multiple-line legend label in which the entries are stacked in a column (ACROSS=1). This figure uses a label specification such as the following:

```
label=("multi-"
        justify=left "line"
        justify=left "label"
        position=left)
```

In this specification, the POSITION= suboption specifies the default value, LEFT, which is represented by the first legend in the figure. The POSITION= value is indicated above each legend. The default justification is used unless you also use the JUSTIFY= suboption.

Figure 14.9 Using the POSITION= Suboption with Multiple-line Legend Labels

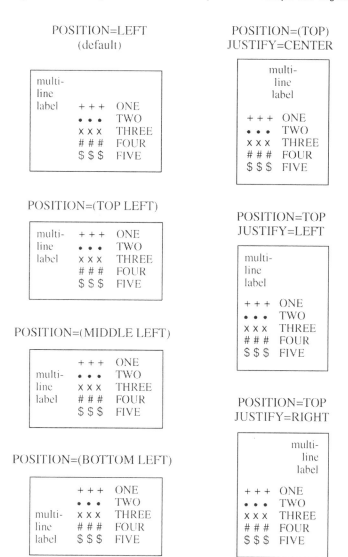

POSITION
In addition, specifying POSITION=RIGHT mirrors the effect of POSITION=LEFT, and specifying POSITION=BOTTOM mirrors the effect of POSITION=TOP.

Restriction: Not supported by Java. Partially supported by ActiveX.

TICK=n
specifies the nth legend entry. The TICK= suboption is used only with the VALUE= option to designate the legend entry whose text and appearance you want to modify. For example, to change the text of the third legend entry to **Minneapolis**, specify the following code:

```
value=(tick=3 "Minneapolis")
```

The characteristics of all other value descriptions remain unchanged.

If you use the TICK= suboption when you designate text for one legend entry, you must also use it when you designate text for any additional legend entries. For example, this option changes the text of both the second and third legend entries:

```
value=(tick=2 "Paris" tick=3 "Sydney")
```

If you omitted TICK=3, the text of the second legend entry would be **ParisSydney**.

Text description suboptions that *precede* the TICK= suboption affect all the value descriptions for the legend unless the same suboption (with a different value) follows a TICK= specification. Text description suboptions that *follow* the TICK= suboption affect only the specified legend entry. For example, suppose you specify this option for a legend with three entries:

```
value=(color=red font=swiss tick=2 color=blue)
```

The text of all three entries would use the Swiss font; the first and third entries would be red and only the second entry would be blue.

Alias: T=n

Using Text Description Suboptions

Text description suboptions affect all the strings that follow them unless the suboption is changed or turned off. If the value of a suboption is changed, the new value affects all the text strings that follow it. Consider this example:

```
label=(font=albany amt height=4 "Weight"
       justify=right height=3 "(in tons)")
```

FONT=ALBANY applies to both **Weight** and **(in tons)**. HEIGHT=4 affects **Weight**, but is respecified as HEIGHT=3 for **(in tons)**. JUSTIFY=RIGHT affects only **(in tons)**.

Using the LEGEND Statement

LEGEND statements can be located anywhere in your SAS program. They are global and remain in effect until canceled or until you end your SAS session. LEGEND statements are not applied automatically, and must be explicitly assigned by an option in the procedure that uses them.

You can define up to 99 different LEGEND statements. If you define two LEGEND statements of the same number, the most recently defined statement replaces the previously defined statement of the same number. A LEGEND statement without a number is treated as a LEGEND1 statement.

Cancel individual LEGEND statements by defining a LEGEND statement of the same number without options (a null statement):

```
legend4;
```

Canceling one LEGEND statement does not affect any other LEGEND definitions. To cancel all current LEGEND statements, use RESET= in a GOPTIONS statement:

```
goptions reset=legend;
```

Specifying RESET=GLOBAL or RESET=ALL cancels all current LEGEND definitions as well as other settings.

To display a list of current LEGEND definitions in the LOG window, use the GOPTIONS procedure with the LEGEND option:

```
proc goptions legend nolist;
run;
```

Positioning the Legend

By default, the legend shares the procedure output area with the procedure output, such as a map or bar chart. (See "How Graphic Elements are Placed in the Graphics Output Area" on page 65.) However, several LEGEND statement options enable you to position a legend anywhere on the graphics output area and even to overlay the procedure output. This section describes these options and their effect on each other.

Positioning the Legend on the Graphics Output Area   There are two ways you can position the legend on the graphics output area:

- Describe the general location of the legend with the POSITION= option. If necessary, fine-tune the position with the OFFSET= option.
- Position the legend explicitly with the ORIGIN=option.

Using POSITION= and OFFSET=   The values of the POSITION= option affect the legend in two ways:

- OUTSIDE and INSIDE determine whether the legend is located outside or inside the axis area.
- BOTTOM or MIDDLE or TOP (vertical position) and LEFT or CENTER or RIGHT (horizontal position) determine where the legend is located in relation to its OUTSIDE or INSIDE position.

Figure 14.10 on page 237 shows the legend positions inside the axis area.

Figure 14.10   Legend Positions Inside the Axis Area

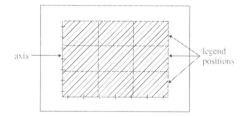

Figure 14.11 on page 237 shows legend positions outside the axis area.

Figure 14.11   Legend Positions Outside the Axis Area

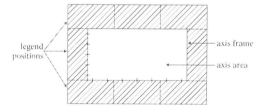

The default combination is POSITION=(BOTTOM CENTER OUTSIDE). The combination (OUTSIDE MIDDLE CENTER) is not valid.

Use OFFSET=(x,y) to adjust the position of the legend specified by the POSITION= option. The x value shifts the legend either left or right and the y value shifts the legend either up or down.

The offset values are always applied *after* the POSITION= request. For example, if POSITION=(TOP RIGHT OUTSIDE), the legend is located in the upper right corner of the graphics output area. If OFFSET=(0,0) is specified, the legend does not move. If OFFSET=(-5,-8)CM, the legend moves 5 centimeters to the left and 8 centimeters down.

Using ORIGIN= Use ORIGIN=(*x,y*) to specify the coordinates of the exact location of the lower left corner of the legend box. Because ORIGIN=(0,0) is the lower left corner of the graphics output area, the values of *x* and *y* must be positive. If you specify negative values, a warning is issued and the default value is used.

Relating Legends to Other Graphic Elements By default, the legend is inside the procedure output area and the space it occupies reduces the size of the graph itself. To control the way the legend relates to the other elements of the graph, use the MODE= option. These are values for the MODE= option:

- RESERVE reserve space for the legend outside the axis area and move the graph to make room for the legend. This is the default setting and is valid only when POSITION=OUTSIDE.
- PROTECT prevents the legend from being overwritten by the procedure output. PROTECT blanks out graphics elements, allowing only legend elements to be displayed in the legend's space.
- SHARE displays both graphics elements and legend elements in the same space. This setting is usually used when the legend is positioned inside the axis area. SHARE is useful when the graph has a space that the legend can fit into.

Interactions Between POSITION= and MODE= You cannot specify both POSITION=INSIDE and MODE=RESERVE because MODE=RESERVE assumes that the legend is *outside* the axis area, and POSITION=INSIDE positions the legend *inside* the axis area. Therefore, when you specify POSITION=INSIDE, change the value of the MODE= option to SHARE or PROTECT. Otherwise, SAS/GRAPH issues a warning and automatically changes the MODE= value to PROTECT.

Creating Drop Shadows and Block Effects

To produce a drop shadow or a three-dimensional block effect behind the legend use the CSHADOW= or CBLOCK= option in the LEGEND statement in conjunction with the graphics option OFFSHADOW=(*x,y*).

The value of *x* determines how far the shadow or block extends to the right (positive numbers) or to the left (negative numbers) of the legend. The value of *y* determines how far the shadow or block extends above (positive numbers) or below (negative numbers) the legend. If OFFSHADOW=(0,0) is specified, the shadow or block is not visible.

By default, OFFSHADOW=(0.0625, -0.0625) IN; that is, the shadow or block extends 1/16th of an inch to the right and 1/16th of an inch below the legend.

NOTE Statement

Writes lines of text in the output.

See: "TITLE, FOOTNOTE, and NOTE Statements" on page 279

Syntax

NOTE <*text-arguments(s)*>;

ODS HTML Statement

Opens or closes the HTML destination.

Used by: GANNO, GAREABAR, GBARLINE, GCHART, GCONTOUR, GIMPORT, GMAP, GPLOT, GRADAR, GREPLAY, GSLIDE, and G3D procedures

Requirements: On mainframes, either GPATH= or PATH= is required.

Syntax

ODS HTML <(<ID=>*identifier*)> <*action*>;

ODS HTML <(<ID=>*identifier*)> <*option(s)*>;

Description

This section describes the ODS HTML statement as it relates to SAS/GRAPH procedures. For complete information on the ODS HTML statement, see *SAS Output Delivery System: User's Guide*

The ODS HTML statement opens or closes the HTML destination. If the destination is open, the procedure produces output that is written in Hypertext Markup Language in the form of an HTML file. If no device is specified, SAS/GRAPH, by default, creates a PNG file containing the graph. The HTML file references the PNG file in order to display the graph in a Web page.

If DEVICE=JAVAMETA, graphics output is produced as metagraphics data. The browser passes the metacodes as a parameter to the Metaview applet. The Metaview applet renders the output defined by the metacodes, and displays the interactive graph in a Web page. For more information on DEVICE=JAVAMETA see "Developing Web Presentations for the Metaview Applet" on page 531.

You can also use the DEVICE=JAVA and DEVICE=ACTIVEX options to create interactive graphics presentations for the Web.

SAS/GRAPH adds datatip text to some graphs depending on the device specified. These datatips are generated by default using the values of fields in a SAS data set. You can specify the DESCRIPTION= option on the SAS/GRAPH procedure to change or remove the datatip text. For more information on using data tips see "Data Tips for Web Presentations" on page 598.

The FILE= option identifies the file that contains the HTML version of the procedure output. With SAS/GRAPH, the body file contains references to the graphs. If DEVICE=PNG, the graphs are stored in separate PNG files. When you view the body file in a browser, the graphs are automatically displayed. By default with ODS processing, the PNG files are stored in the current directory. To specify a destination for all the HTML and PNG files, use the PATH= option. To store thePNG files in a different location than the HTML files, use the GPATH= option to specify a location for thePNG files, and the PATH= option to specify the location of the HTML files. In both cases,the destination must be an aggregate storage location.

Anchors

ODS HTML automatically creates an *anchor* for every piece of output generated by the SAS procedures. An anchor specifies a particular location within an HTML file. In SAS/GRAPH, an anchor usually defines a link target such as a graph whose location is defined in an IMG element.

In order for the links from the contents, page, or frame file to work, each piece of output in the body files must have a unique anchor to link to. The anchor for the first piece of output in a body file acts as the anchor for that file. These anchors are used by the frame and contents files, if they are created, to identify the targets for the links that ODS HTML automatically generates. For more information about using anchors with the ODS HTML statement see *SAS Output Delivery System: User's Guide.* .

PATTERN Statement

Defines the characteristics of patterns used in graphs.

Used by: GCHART, GBARLINE, GCONTOUR, GMAP, GPLOT procedures; SYMBOL statement; Annotate facility.

Type: Global

Syntax

PATTERN<1...255> <COLOR=*pattern-color* | *_style_*>
    <REPEAT=*number-of-times*>
    <VALUE=*bar/block-pattern* | *map/plot-pattern* | *pie/star-pattern* >;

- *bar/block-pattern* can be one of these:

    EMPTY

    SOLID

    style <density>

- *map/plot-pattern* can be one of these:

    MEMPTY

    MSOLID

    M*density <style <angle>>*

- *pie/star-pattern* can be one of these:

    PEMPTY

    PSOLID

    P*density <style <angle>>*

Description

PATTERN statements create PATTERN definitions that define the color and type of area fill for patterns used in graphs. These are the procedures and the graphics areas that they create that use PATTERN definitions:

GCHART	color, fill pattern, or image for the bars in two-dimensional bar charts; color and fill pattern for the segments of three-dimensional bar charts, pie charts, and star charts.
GCONTOUR	contour levels in contour plots
GMAP	map areas in choropleth, block, and prism maps; blocks in block maps
GPLOT	areas beneath or between plotted lines

In addition, the SYMBOL statement and certain Annotate facility functions and macros can use pattern specifications. For details see the "SYMBOL Statement" on page 252 and Chapter 29, "Using Annotate Data Sets," on page 641.

You can use the PATTERN statement to control the fill and color of a pattern, and whether the pattern is repeated. There are three types of patterns:

- bar and block patterns
- map and plot patterns
- pie and star patterns

Pattern fills can be solid or empty, or composed of parallel or crosshatched lines. For two-dimensional bar charts, the PATTERN statement can specify images to fill horizontal or vertical bars. In addition, you can specify device-dependent hardware patterns for rectangle, polygon, and pie fills on devices that support hardware patterns.

If you do not create PATTERN definitions, SAS/GRAPH software generates them as needed and assigns them to your graphs by default. Generally, the default behavior is to rotate a solid pattern through the current color list. For details, see "About Default Patterns" on page 248.

Options

COLOR=*pattern-color* | *_style_*

specifies the color of the fill. *Pattern-color* is any SAS/GRAPH color name. The _STYLE_ value specifies the appropriate color based on the current style. See Chapter 12, "SAS/GRAPH Colors and Images," on page 167 the *SAS/GRAPH: Reference* for more information on specifying colors and images.

Note: ActiveX assigns colors in a different order from Java, so the same data can appear differently with those two drivers. △

Using the COLOR= option with a null value cancels the color specified in a previous PATTERN statement of the same number without affecting the values of other options.

The COLOR= option overrides the CPATTERN= graphics option.

The CFILL= option in the PIE and STAR statements overrides the COLOR= option. For details, see "Controlling Slice Patterns and Colors" on page 1053.

CAUTION:

Omitting the COLOR= option in a PATTERN statement can cause the PATTERN statement to generate multiple PATTERN definitions. △

If no color is specified for a PATTERN statement, that is, if neither the COLOR= nor the CPATTERN= option is used, the PATTERN statement rotates the specified fill through each color in the color list before the next PATTERN statement is used. .

Alias: C=*pattern-color*

See also: "Working with PATTERN Statements" on page 249

Featured in: "Example 7. Using BY-group Processing to Generate a Series of Charts" on page 309

Restriction: Partially supported by Java and ActiveX

IMAGE= *fileref* | *"external-file"*

specifies an image file that is used to fill one or more bars of a bar chart, as generated by the HBAR, HBAR3D, VBAR, and VBAR3D statements of the GCHART procedure. The format of the external file specification varies across operating environments. See also the IMAGESTYLE= option.

Note: When you specify an image file to fill a bar, the bar is not outlined. Also, the COLOR= and VALUE= options are ignored. △

Note: If an image is specified on a PATTERN statement that is used with another type of chart, then the PATTERN statement is ignored and default pattern rotation is affected. For example, if you submit a PIE statement when an image has been specified in a PATTERN statement, the default fill pattern is used for the pie slices, with each slice in the pie displaying the fill pattern in the same color.

For DEVICE=ACTIVEX and DEVICE=ACTXIMG, if you do not specify a pathname to the image, then the ActiveX control searches a predefined list of locations to try to find the image. If all else fails, the ActiveX control looks for the image on the Web. It is recommended that you specify the pathname to the image.

For DEVICE=JAVA and DEVICE=JAVAIMG, the IMAGE= option works only for the VBAR and HBAR statements. △

See also: For related information, see "Displaying Images on Data Elements" on page 185

Restriction: Partially supported by Java and ActiveX

IMAGESTYLE = TILE | FIT
   specifies how the image specified in the IMAGE= option is to be applied to fill a bar in a bar chart. The TILE value, which is the default, repeats the image as needed to fill the bar. The FIT value stretches a single instance of the image to fill the bar.

   Restriction: Partially supported by Java and ActiveX

REPEAT=*number-of-times*
   specifies the number of times that a PATTERN definition is applied before the next PATTERN definition is used. By default, REPEAT=1.

   The behavior of the REPEAT= option depends on the color specification:

   □ If you use both the COLOR= and the REPEAT= options in a PATTERN statement, the pattern is repeated the specified number of times in the specified color. The fill can be either the default solid or a fill specified with the VALUE= option.

   □ If you use the CPATTERN= option in a GOPTIONS statement to specify a single pattern color, and use the REPEAT= option either alone or with the VALUE= option in a PATTERN statement, the resulting hatch pattern is repeated the specified number of times.

   □ If you omit both the COLOR= and CPATTERN= options, and use the REPEAT= option either alone (generates default solids) or with the VALUE= option in a PATTERN statement, the resulting pattern is rotated through each color in the color list, and then the entire group generated by this cycle is repeated the number of times specified in the REPEAT= option. Thus, the total number of patterns produced depends on the number of colors in the current color list.

   Using REPEAT= with a null value cancels the repetition specified in a previous PATTERN statement of the same number without affecting the values of other options. Note that in most cases, it is preferable to use LEVELS=1 in the GMAP procedure rather than using this option in the PATTERN statement.

   Alias: R=*number-of-times*

   See also: "Understanding Pattern Sequences" on page 251

   Restriction: Partially supported by Java and ActiveX

VALUE=*bar*/*block-pattern*
   specifies patterns for:

      □ bar charts produced by the HBAR, HBAR3D, VBAR, and VBAR3D statements in the GCHART procedure including two-dimensional and three-dimensional bar shapes.

□ the front surface of blocks in block charts produced by the BLOCK statement in the GCHART procedure.

□ the blocks in block maps produced by the BLOCK statement in the GMAP procedure. (The map area from which the block rises takes a map pattern as described on the option VALUE= on page 244). See also "About Block Maps and Patterns" on page 1268.

Values for *bar/block-pattern* are as follows:

EMPTY
E
: an empty pattern. Neither the Java applet nor the ActiveX control supports EMPTY.

SOLID
S
: a solid pattern (the only valid value for three-dimensional charts).

style<density>
: a shaded pattern.

   Note: *style<density>* is not supported by the Java or ActiveX device drivers. △

   Style specifies the direction of the lines:

   L   left-slanting lines.

   R   right-slanting lines.

   X   crosshatched lines.

   Density specifies the density of the pattern's shading:

   1...5   1 produces the lightest shading and 5 produces the heaviest shading.

Figure 14.12 on page 243 shows all of the patterns available for bars and blocks.

Figure 14.12 Bar and Block Patterns

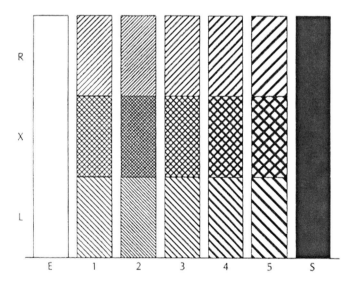

If no valid patterns are available, default bar and block fill patterns are selected in this order:

1 SOLID
2 X1–X5
3 L1–L5
4 R1–R5

Each fill is used once with every color in the color list unless a pattern color is specified. The entire sequence is repeated as many times as required to provide the necessary number of patterns.

Alias: V=*bar/block-pattern*

Restriction: Partially supported by Java and ActiveX

VALUE=*map/plot-pattern*

specifies patterns for the following:

- contour levels in contour plots produced by the GCONTOUR procedure
- map area surfaces in block, choropleth, and prism maps produced by the BLOCK, CHORO, AND PRISM statements in the GMAP procedure.
- areas under curves in plots produced by the AREAS= option in the PLOT statement in the GPLOT procedure.

Values for *map/plot-pattern* are as follows:

MEMPTY an empty pattern. EMPTY or E are also valid aliases, except
ME when used with the map areas in block maps created by the GMAP procedure.

MSOLID a solid pattern. SOLID or S are also valid aliases, except when
MS used with the map areas in block maps created by the GMAP procedure.

M*density*<*style*<*angle*>>shaded pattern.

> *Note:* M*density*<*style*<*angle*>> is not supported by the Java or ActiveX device drivers. △

Density specifies the density of the pattern's shading:

1...5 1 produces the lightest shading and 5 produces the heaviest shading.

Style specifies the type of the pattern lines:

N parallel lines (the default).

X crosshatched lines.

Angle specifies the angle of the pattern lines:

0...360 the degrees at which the parallel lines are drawn, measured from the horizontal. By default, *angle* is 0 (lines are horizontal).

Figure 14.13 on page 245 shows some typical map and plot patterns.

Figure 14.13 Map and Plot Patterns

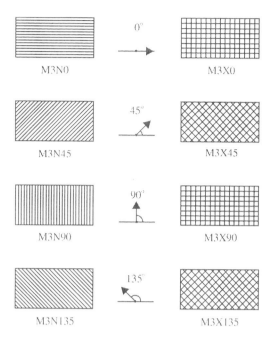

If no valid patterns are available, default map and plot fill patterns are selected in this order:

1 MSOLID
2 M2N0
3 M2N90
4 M2X45
5 M4N0
6 M4N90
7 M4X90

Each fill is used once with every color in the color list unless a pattern color is specified. The entire sequence is repeated as many times as required to provide the necessary number of patterns.

Alias: V=*map/plot-pattern*

Restriction: Partially supported by Java and ActiveX.

VALUE=*pie/star-pattern*
   specifies patterns for pie and star charts produced by the PIE and STAR statements in the GCHART procedure. Values for *pie/star-pattern* are

   PEMPTY   an empty pattern. EMPTY or E are also valid aliases.
   PE

   PSOLID   a solid pattern. SOLID or S are also valid aliases.
   PS

   P*density*<*style*<*angle*>> a shaded pattern.

   Note: P*density*<*style*<*angle*>> is not supported by the Java or ActiveX device drivers. △

   Density specifies the density of the pattern's shading:

1...5	1 produces the lightest shading and 5 produces the heaviest shading.

Style specifies the type of the pattern lines:

N	parallel lines (the default).
X	crosshatched lines.

Angle specifies the angle of the pattern lines:

0...360	the angle of the lines, measured in degrees from perpendicular to the radius of the slice. By default, *angle* is 0.

The FILL= option in the PIE and STAR statements in the GCHART procedure overrides VALUE=.

Figure 14.14 on page 246 shows some typical pie and star patterns.

Figure 14.14 Pie and Star Patterns

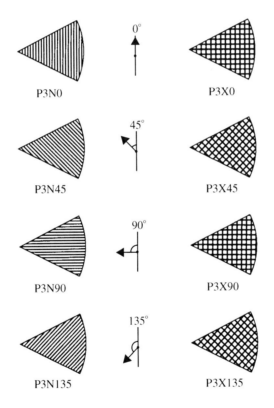

If no valid patterns are available, default pie and star fill patterns are selected in this order:

1 PSOLID
2 P2N0
3 P2N90
4 P2X45
5 P4N0
6 P4N90

7 P4X90

Each fill is used once with every color in the color list unless a pattern color is specified. The entire sequence is repeated as many times as required to provide the necessary number of patterns.

Note: If you use hatch patterns and request a legend instead of slice labels, the patterns in the slices are oriented to be visually equivalent to the legend. △

Alias: V=*pie/star-pattern*

Restriction: Partially supported by Java and ActiveX

Using the PATTERN Statement

PATTERN statements can be located anywhere in your SAS program. They are global and remain in effect until redefined, canceled, or until the end of your SAS session.

You can define up to 255 different PATTERN statements. A PATTERN statement without a number is treated as a PATTERN1 statement.

PATTERN statements generate one or more PATTERN definitions, depending on how the COLOR=, VALUE=, and IMAGE= options are used. For information on PATTERN definitions, see "Working with PATTERN Statements" on page 249, as well as the description of COLOR= on page 241, VALUE= on page 244, and IMAGE= on page 241 options.

PATTERN definitions are generated in the order in which the statements are numbered, regardless of gaps in the numbering or the statement's position in the program. Although it is common practice, you do not have to start with PATTERN1, and you do not have to use sequential statement numbers.

PATTERN definitions are applied automatically to all areas of the graphics output that require patterns. When assigning PATTERN definitions, SAS/GRAPH starts with the lowest-numbered definition with an appropriate fill specification or with no fill specification. It continues to use the specified patterns until all valid PATTERN definitions have been used. Then, if more patterns are required, SAS/GRAPH returns to the default pattern rotation, but continues to outline the areas in the same color as the fill.

Altering or Canceling PATTERN Statements   PATTERN statements are additive. If you define a PATTERN statement and later submit another PATTERN statement with the same number, the new PATTERN statement redefines or cancels only the options that are included in the new statement. Options not included in the new statement are not changed and remain in effect. For example, assume you define PATTERN4 as follows:

```
pattern4 value=x3 color=red repeat=2;
```

This statement cancels only REPEAT= without affecting the rest of the definition:

```
pattern4 repeat=;
```

Add or change options in the same way. This statement changes the color of the pattern from red to blue:

```
pattern4 color=blue;
```

After all these modifications, PATTERN4 has these characteristics:

```
pattern4 value=x3 color=blue;
```

Cancel individual PATTERN statements by defining a PATTERN statement of the same number without options (a null statement):

```
pattern4;
```

Canceling one PATTERN statement does not affect any other PATTERN definitions. To cancel all current PATTERN statements, use the RESET= option in a GOPTIONS statement:

```
goptions reset=pattern;
```

Specifying RESET=GLOBAL or RESET=ALL cancels all current PATTERN definitions as well as other settings.

To display a list of current PATTERN definitions in the LOG window, use the GOPTIONS procedure with the PATTERN option:

```
proc goptions pattern nolist;
run;
```

About Default Patterns

When a procedure produces a graph that needs one or more patterns, SAS/GRAPH either does one of the following:

- automatically generates the appropriate default patterns and outlines to fill the areas, or
- uses patterns, colors, and outlines that are defined by PATTERN statements, graphics options, and procedure options.

In order to understand how SAS/GRAPH generates and assigns patterns defined with PATTERN statements it is helpful to understand how it generates and assigns default patterns. The following sections describe the default pattern behavior for all procedures. See "Working with PATTERN Statements" on page 249 for details about defining patterns.

How Default Patterns and Outlines Are Generated In general, the default pattern that the SAS/GRAPH uses is a solid fill. The default colors are determined by the current style and the device.

SAS/GRAPH uses default patterns when no PATTERN statements are defined. The default colors are determined by the current style and the device.

Because the system option-GSTYLE-is in effect by default, the procedure uses the style's default bar fill colors, plot line colors, widths, symbols, patterns, and outline colors when producing output. Specifically, SAS/GRAPHuses the default values when you do not specify any of the following:

- any PATTERN statements
- the CPATTERN= graphics option
- the COLORS= graphics options (that is, you use the device's default color list and it has more than one color)
- the COUTLINE= option in the action statement

If all of these conditions are true, then SAS/GRAPH performs the following operations:

- selects the first default fill for the appropriate pattern, which is always solid, and rotates it once through the list of colors available in the current style, generating one solid pattern for each color. If you use the default style colors and the first color in the list is either black or white, the procedure does not create a pattern in that color. If you specify a color list with the COLORS= graphics option, then the procedure uses all the colors in the list to generate the patterns.

    Note: The one exception to the default solid pattern is the map area pattern in a block map produced by the GMAP procedure, which uses a hatch fill by default. By default the map areas and their outlines use the first color in the color list, regardless of whether the list is the default device list or one specified with COLORS= in the GOPTIONS statement. △

□ uses the style's outline color to outline every patterned area.

If a procedure needs additional patterns, SAS/GRAPH selects the next default pattern fill appropriate to the graph and rotates it through the color list, skipping the foreground color as before. SAS/GRAPH continues in this fashion until it has generated enough patterns for the chart.

Things That Affect Default Patterns   Changing any of these conditions can change or override the default behavior:

- If you specify a color list with the COLORS= option in a GOPTIONS statement and the list contains more than one color, SAS/GRAPH rotates the default fills, beginning with SOLID, through that list. In this case, it uses every color, even if the foreground color is black (or white). The default outline color remains the foreground color.
- If you specify either COLORS=(*one-color*) or the CPATTERN= graphics option, the default fill changes from SOLID to the appropriate list of hatch patterns. SAS/GRAPH uses the specified color to generate one pattern definition for each hatch pattern in the list.

For a description of these graphics options, see Chapter 15, "Graphics Options and Device Parameters Dictionary," on page 327.

Working with PATTERN Statements

With PATTERN statements, you can specify the following:

- the type of fill (VALUE=)
- the color of the fill (COLOR=)
- the images used to fill the bars in a 2D chart (IMAGE=)
- how many times to apply the statement before using the next one (REPEAT=.

See "Displaying Images on Data Elements" on page 185 for information on filling the bars of two–dimensional bar charts with images using the PATTERN statement.

You can also use procedure options to specify the pattern outline color and the CPATTERN= graphics option to specify a default color for all patterns.

Whether you use PATTERN statement options alone or with each other affects the number and kind of patterns your PATTERN statements generate. Depending on the options you use, you can explicitly specify every pattern used by your graphs or you can let the PATTERN statement generate a series of pattern definitions using either the color list or the list of default fills.

Explicitly Specifying Patterns   To explicitly specify all the patterns in your graph, you need to do one of the following for every pattern your graph requires:

- Provide a PATTERN statement that uses the COLOR= option to specify the pattern color, for example:

    ```
    pattern1 color=red;
    ```

    By default, the fill type SOLID.

- Provide a PATTERN statement that uses both the COLOR= option and the VALUE= option to specify the fill, for example:

    ```
    pattern1 color=blue value=r3;
    ```

Including the COLOR= option in the PATTERN statement is the simplest way to assure that you get exactly the patterns you want. When you use the COLOR= option, the PATTERN statement generates exactly one PATTERN definition for that statement. If you also use the REPEAT= option, the PATTERN definition is repeated the specified number of times.

Generating Multiple Pattern Definitions You can also use PATTERN statements to generate multiple PATTERN definitions. To do this use the VALUE= option to specify the type of fill you want but omit the COLOR= option – for example:

```
pattern1 value=r3;
```

In this case, the PATTERN statement rotates the R3 fill through all the colors in the color list. For more information on pattern rotation, see "Understanding Pattern Sequences" on page 251.

Selecting an Appropriate Pattern The type of fill you specify depends on the type of graph you are producing:

With this type of graph	Use this type of fill
bar and block charts (PROC GCHART), block maps (PROC GMAP)	VALUE= bar/block-pattern on page 242
contour plots (PROC GCONTOUR), map area surfaces (PROC GMAP)	VALUE=map/plot-pattern on page 244
pie and star charts (PROC GCHART)	VALUE=pie/star-pattern on page 245

Note: If you specify a fill that is inappropriate for the type of graph you are generating (for example, if you specify VALUE=L1 in a PATTERN statement for a choropleth map), SAS/GRAPH ignores the PATTERN statement and continues searching for a valid pattern. If it does not find a definition with a valid fill specification, it uses default patterns instead. △

Controlling Outline Colors Whenever you use PATTERN statements, the default outline color uses the style's outline color to outline every patterned area.

To change the outline color of any pattern, whether the pattern is default or user-defined, use the COUTLINE= option in the action statement that generates the chart.

The Effect of the CPATTERN= Graphics Option Although the CPATTERN= graphics option is used most often with default patterns, it does affect the PATTERN statement. With default patterns (no PATTERN statements specified) it does the following:

☐ specifies the color for all patterns

☐ causes default patterns to use hatched fills instead of the default SOLID.

In conjunction with the PATTERN statement it does the following:

☐ With a PATTERN statement that only specifies a fill (VALUE=), the CPATTERN= option determines the color of that fill. For example, these statements produce two green, hatched patterns:

```
goptions cpattern=green;
pattern1 value=x3;
pattern2 value=x1;
```

☐ With a PATTERN statement that only specifies a color (COLOR=), the COLOR= option overrides the CPATTERN= color, but CPATTERN= causes the fill to be hatched, not the default SOLID. For example, these statements produce one red, hatched pattern:

```
goptions cpattern=green;
pattern1 color=red;
```

See also the description of CPATTERN="CPATTERN" on page 343.

Understanding Pattern Sequences

Pattern sequences are sets of PATTERN definitions that SAS/GRAPH automatically generates when a PATTERN statement specifies a fill but not a color. In this case, the specified fill is used once with every color in the color list. If the REPEAT= option is also used, the resulting PATTERN definitions are repeated the specified number of times.

Generating Pattern Sequences SAS/GRAPH generates pattern sequences when a PATTERN statement uses VALUE= to specify a fill and all of the following conditions are also true:

- The COLOR= option is not used in the PATTERN statement.
- The CPATTERN= graphics option is not used.
- The color list, either default or user-specified, contains more than one color.

In this case, the PATTERN statement rotates the fill specified by the VALUE= option through every color in the color list, generating one PATTERN definition for every color in the list. After every color has been used once, SAS/GRAPH goes to the next PATTERN statement. For example, suppose you specified the following color list and PATTERN statements for bar/block patterns:

```
goptions colors=(blue red green) ctext=black;
pattern1 color=red    value=x3;
pattern2 value=r3;
pattern3 color=blue   value=l3;
```

Here, **PATTERN1** generates the first PATTERN definition. **PATTERN2** omits the COLOR= option, so the specified fill is rotated through all three colors in the color list before the PATTERN3 statement is used. This table shows the color and fill of the PATTERN definitions that would be generated if nine patterns were required:

Definition Number	Source	Characteristics: Color	Fill
1	PATTERN1	red	x3
2	PATTERN2	blue	r3
3	PATTERN2	red	r3
4	PATTERN2	green	r3
5	PATTERN3	blue	l3
6	first default	blue	solid
7	first default	red	solid
8	first default	green	solid
9	second default	blue	x1

Notice that after all the PATTERN statements are exhausted, the procedure begins using the default bar and block patterns, beginning with SOLID. Each fill from the default list is rotated through all three colors in the color list before the next default fill is used.

Repeating Pattern Sequences If you use the REPEAT= option but not the COLOR= option, the sequence generated by cycling the definition through the color list is repeated the number of times specified by the REPEAT= option. For example, these statements illustrate the effect of the REPEAT= option on PATTERN statements both with and without explicit color specifications:

```
goptions colors=(red blue green);
pattern1 color=gold repeat=2;
pattern2 value=x1 repeat=2;
```

Here, **PATTERN1** is used twice and **PATTERN2** cycles through the list of three colors and then repeats this cycle a second time:

Sequence Number	Source	Characteristics: Color	Fill
1	PATTERN1	gold	solid (first default)
2	PATTERN1	gold	solid (first default)
3	PATTERN2	red	x1
4	PATTERN2	blue	x1
5	PATTERN2	green	x1
6	PATTERN2	red	x1
7	PATTERN2	blue	x1
8	PATTERN2	green	x1

SYMBOL Statement

Defines the characteristics of symbols that display the data plotted by a PLOT statement used by PROC GBARLINE, PROC GCONTOUR, and PROC GPLOT.

Used by: GBARLINE, GCONTOUR, GPLOT procedures

Type Global

Syntax

SYMBOL<1...255> <COLOR=*symbol-color* | *_style_*>
    <MODE=EXCLUDE | INCLUDE> <REPEAT=*number-of-times*>
    <STEP=*distance*<*units*>> <*appearance-option(s)*>
    <*interpolation-option*> <SINGULAR=*n*>;
  appearance-options can be one or more of these:
  BWIDTH=*box-width*

CI=*line-color* | *_style_*
CO=*color*
CV=*value-color* | *_style_*
FONT=*font*
HEIGHT=*symbol-height*<*units*>
LINE=*line-type*
POINTLABEL<=(*label-description(s)*) | NONE>
VALUE=*special-symbol* | *text-string* | NONE
WIDTH=*thickness-factor*

interpolation-option can be one of these:

- general methods
   - INTERPOL=JOIN
   - INTERPOL=*map/plot-pattern*
   - INTERPOL=NEEDLE
   - INTERPOL=NONE
   - INTERPOL=STEP<*placement*><J><S>
- high-low interpolation methods
   - INTERPOL=BOX<*option(s)*><00...25>
   - INTERPOL=HILO<C><*option(s)*>
   - INTERPOL=STD<1 | 2 | 3><*variance*><*option(s)*>
- regression interpolation methods
   - INTERPOL=R<*type*><0><CLM | CLI<50...99>>
- spline interpolation methods
   - INTERPOL=L<*degree*><P><S>
   - INTERPOL=SM<*nn*><P><S>
   - INTERPOL=SPLINE<P><S>

Description

SYMBOL statements create SYMBOL definitions, which are used by the GPLOT, GBARLINE and GCONTOUR procedures.

For the GPLOT and GBARLINE procedure, SYMBOL definitions control the following:

- the appearance of plot symbols and plot lines, including bars, boxes, confidence limit lines, and area fills
- interpolation methods
- how plots handle data out of range

For the GCONTOUR procedure, SYMBOL definitions control the following:

- the appearance and text of contour labels
- the appearance of contour lines

If you create SYMBOL definitions, they are automatically applied to a graph by the procedure. If you do not create SYMBOL definitions, these procedures generate default definitions and apply them as needed to your plots.

Options

When the syntax of an option includes *units*, use one of these:

CELLS	character cells
CM	centimeters
IN	inches
PCT	percentage of the graphics output area
PT	points.

If you omit *units*, a unit specification is searched for in this order:

1 the GUNIT= option in a GOPTIONS statement
2 the default unit, CELLS.

BWIDTH=*box-width*
: specifies the width of the box generated by either the INTERPOL=BOX or INTERPOL=HILOB option. *Box-width* can be any number greater than 0. By default, the value of *box-width* is the same as the value of the WIDTH= option, whose default value is 1. Therefore, if you specify a WIDTH= value for and omit the BWIDTH= option, the width of the box changes accordingly.

  Featured in: "Example 4. Creating and Modifying Box Plots" on page 302.

CI=*line-color* | *_style_*
: specifies a color for an interpolation line (GPLOT and GBARLINE) or a contour line (GCONTOUR). The _STYLE_ value specifies the appropriate color based on the current style. If you omit the CI= option but specify the CV= option, the CI= option assumes the value of the CV= option. In this case, the CI= and CV= options specify the same color, which is the same as specifying the COLOR= option alone.

  If you omit the CI= option, the color specification is searched for in this order:

  1 the COLOR= option
  2 the CV= option
  3 the CSYMBOL= option in a GOPTIONS statement
  4 each color in the color list sequentially before the next SYMBOL definition is used.

  See also: "Using Color" on page 275

  Featured in: "Example 1. Ordering Axis Tick Marks with SAS Date Values" on page 294

CO=*color*
: specifies a color for the following:

  - outlines of filled areas generated by the INTERPOL=*map/plot-pattern* option
  - confidence limit lines generated by the INTERPOL=R *series* option
  - staffs, boxes, and bars generated by the high-low interpolation methods: INTERPOL=HILO, INTERPOL=BOX, and INTERPOL=STD

  If you omit the CO= option, the search order for a color specification depends on the interpolation method being used.

  See also: "Using Color" on page 275

  Featured in: "Example 5. Filling the Area between Plot Lines" on page 304 and "Example 4. Creating and Modifying Box Plots" on page 302

COLOR=*symbol-color* | *_style_*
: specifies a color for the entire definition, unless it is followed by a more explicit specification. For the GPLOT and GBARLINE procedures, this includes plot symbols, the plot line, confidence limit lines, and outlines. For the GCONTOUR

procedure, this includes contour lines and labels. The _STYLE_ value specifies the appropriate color from the current style.

Using the COLOR= option is exactly the same as specifying the same color for both the CI= and CV= options.

If COLOR= precedes the CI= or CV= option in the same statement, the CI= or CV= option is used instead.

If you do not use the COLOR=, CI=, CV=, or CO= option, the color specification is searched for in this order:

1 the CSYMBOL= option in a GOPTIONS statement

2 each color in the color list sequentially before the next SYMBOL definition is used.

If you do not use a SYMBOL statement to specify a color for each symbol, but you do specify a color list in a GOPTIONS statement, then Java and ActiveX assign colors to symbols differently than other devices. To ensure consistency on all devices, you should specify the desired color of each symbol. If you do not specify a symbol color, SAS/GRAPH uses the first default color and the first symbol. It uses each color in the list of default colors until the list is exhausted. SAS/GRAPH then selects the next symbol and begins again with the first default color. It rotates the new symbol through the list of default colors before selecting another symbol. It continues selecting new symbols and colors until no more symbols are needed.

Note: Neither the Java applet nor the ActiveX control supports using COLOR= with PROC GCONTOUR. △

Style Reference: Color attribute of the GraphLabelText style element.

Alias: C=*symbol-color*

See also: "Using Color" on page 275

Restriction: Partially supported by Java and ActiveX

CV=*value-color* | _style_

specifies a color for the following:

- plot symbols in the GPLOT procedure
- the filled areas generated by the INTERPOL=*map/plot-pattern* option
- contour labels in the GCONTOUR procedure

The _STYLE_ value specifies the appropriate color based on the current style. If you omit the CV= option but specify the CI=, the CV= option assumes the value of the CI= option. In this case, the CV= and CI= options specify the same color, which is the same as specifying the COLOR= option alone.

If you omit the CV= option, the color specification is searched for in this order:

1 the COLOR= option

2 the CI= option

3 the CSYMBOL= option in a GOPTIONS statement

4 each color in the color list sequentially before the next SYMBOL definition is used.

Note: Neither the Java applet nor the ActiveX control supports using the CV= option with PROC GCONTOUR. △

See also: "Using Color" on page 275

Featured in: "Example 1. Ordering Axis Tick Marks with SAS Date Values" on page 294, "Example 5. Filling the Area between Plot Lines" on page 304, and "Example 4. Creating and Modifying Box Plots" on page 302

Restriction: Partially supported by Java and ActiveX

FONT="*font*"

specifies the font for the plot symbol (GPLOT, GBARLINE) or contour labels (GCONTOUR) specified by the VALUE= option. The *font* specification must be enclosed in quotes and can include the **/bold** and **/italic** font modifiers.

By default, the symbol specified by the VALUE= option is taken from the special symbol table shown in Figure 14.21 on page 271. To use symbols from the special symbol table, you must omit the FONT= option.

To use a symbol that is not in that special symbol table, specify the font containing the symbol and the character code or hexadecimal code of the symbol that you want to use. You can also specify text instead of special symbols. For example:

```
symbol font="Albany AMT" value="80"x;    /* hexadecimal code for the Euro symbol
symbol font="Monotype Sorts" value="s";  /* character code for a filled triangl
symbol font="Cumberland AMT/bo" value="F";   /* prints the letter F in bold */
```

To cancel a font specification and return to the default special symbol table, enter a null font specification:

```
symbol font= value=dot;
```

Alias: F=*font*

See also: the VALUE= option on page 269, "Specifying Plot Symbols" on page 274, and "Specifying Special Characters Using Character and Hexadecimal Codes" on page 160.

Featured in: Example 2 on page 1116

Restriction: Not supported by Java and ActiveX

HEIGHT=*symbol-height<units>*

specifies the height in number of units of plot symbols (GPLOT, GBARLINE) or contour labels (GCONTOUR).

Note: The HEIGHT= option affects only the height of the symbols and labels on the plot; it does not affect the height of any symbols that might appear in a legend.

The HEIGHT option overrides the MarkerSize attribute in graph styles. For more information on graph styles, see *SAS Output Delivery System: User's Guide*. △

Note: With the Java device driver, the minimum height is two pixels; with ActiveX a symbol can be so small as to be invisible.

Neither the Java applet nor the ActiveX control supports HEIGHT= with PROC GCONTOUR. △

Alias: H=*symbol-height<units>*

See also: the option SHAPE= on page 231 in the LEGEND statement

Featured in: "Example 4. Creating and Modifying Box Plots" on page 302 and "Example 3. Rotating Plot Symbols Through the Color List" on page 299

Restriction: Partially supported by Java and ActiveX

INTERPOL=BOX*<option(s)><00...25>*

produces box and whisker plots. The bottom and top edges of the box are located at the sample 25th and 75th percentiles. The center horizontal line is drawn at the 50th percentile (median). By default, INTERPOL=BOX. In this case the vertical lines, or whiskers, are drawn from the box to the most extreme point less than or equal to 1.5 interquartile ranges. (An interquartile range is the distance between the 25th and the 75th sample percentiles.) Any value more extreme than this is marked with a plot symbol.

Values for *option(s)* are one or more of these:

F   fills the box with the color specified by CV= and outlines the box with the color specified by CO=

J   joins the median points of the boxes with a line

T   draws tops and bottoms on the whiskers.

In addition, you can specify a percentile to control the length of the whiskers within the range 00 through 25. These are examples of percentile specifications and their effect:

00   high/low extremes. INTERPOL=BOX00 is *not* the same as the default, INTERPOL=BOX.

01   1st percentile low, 99th high

05   5th percentile low, 95th high

10   10th percentile low, 90th high

25   25th percentile low, 75th high; since the box extends from the 25th to the 75th percentile, no whiskers are produced.

Figure 14.15 on page 257 shows the type of plot INTERPOL=BOX produces.

Figure 14.15   Box Plot

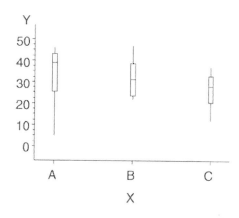

Note: If you use the HAXIS= or VAXIS= options in the PLOT statement or the ORDER= option in an AXIS definition to restrict the range of axis values, by default any observations that fall outside the axis range are excluded from the interpolation calculation. See the MODE= option on page 266 △

You cannot use the GPLOT procedure PLOT statement option AREAS= with INTERPOL=BOX.

To increase the thickness of all box plot lines, including the box, whiskers, join line, and top and bottom ticks, use the WIDTH= option.

To increase the width of the box itself, use the BWIDTH= option. By default the value of the BWIDTH= option is the same as the value of the WIDTH= option. Therefore, if you specify a value for the WIDTH= option and omit BWIDTH=, the width of the box changes.

For a scatter effect with the box, use a multiple plot request, as in this example:

```
symbol1 i=none v=star color=green;
symbol2 i=box v=none color=blue;
proc gplot data=test;
   plot (y y)*x / overlay;
```

Note: When using DEVICE=JAVA and DEVICE=JAVAIMG with overlaid plots, different interpolations are supported per overlay unless any of the interpolations is BOX, HILO or STD. When any of these interpolations are encountered, the first interpolation specified becomes the only interpolation that is used for all overlays. All other interpolations are ignored. △

Alias: I=BOX<*option(s)*><00...25>

Featured in: "Example 4. Creating and Modifying Box Plots" on page 302

INTERPOL=HILO<C><*option*>

specifies that a solid vertical line connect the minimum and maximum Y values for each X value. The data should have at least two values of Y for every value of X; otherwise, the single value is displayed without the vertical line.

By default, for each X value, the mean Y value is marked with a tick. This is shown in Figure 14.16 on page 259.

To specify high, low, close stock market data, include this option:

C
: draws tick marks at the close value instead of at the mean value. Specifying C assumes that there are three values of Y (HIGH, LOW, and CLOSE) for every value of X. If more or fewer than three Y values are specified, the mean is ticked. The Y values can be in any order in the input data set.

In addition, you can specify one of these values for *option*:

B
: connects the minimum and maximum Y values with bars instead of lines. Use the BWIDTH= option to increase the width of the bars.

J
: joins the mean values or the close values (if HILOC is specified) with a line. This point is not marked with a tick mark. You cannot use the PLOT statement option AREAS= with INTERPOL=HILOJ.

T
: adds tops and bottoms to each line.

BJ
: connects maximum and minimum values with a bar and joins the mean or close values.

TJ
: adds tops and bottoms to the lines and joins the mean or close values.

Figure 14.16 on page 259 shows the type of plot INTERPOL=HILO produces. Plot symbols in the form of dots have been added to this figure.

Figure 14.16 High-Low Plot

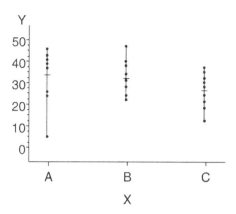

To increase the thickness of all lines generated by the INTERPOL=HILO option, use the WIDTH= option.

Note: If you use the HAXIS= or VAXIS= options in the PLOT statement or the ORDER= option in an AXIS definition to restrict the range of axis values, by default any observations that fall outside the axis range are excluded from the interpolation calculation. See the option MODE= on page 266. △

When using DEVICE=JAVA and DEVICE=JAVAIMG with overlaid plots, different interpolations are supported per overlay unless any of the interpolations is BOX, HILO or STD. When any of these interpolations are encountered, the first interpolation specified becomes the only interpolation that is used for all overlays. All other interpolations are ignored.

Alias: I=HILO<C><*option*>

Featured in: "Example 1. Ordering Axis Tick Marks with SAS Date Values" on page 294

Restriction: Partially supported by Java

INTERPOL=JOIN

connects data points with straight lines. Points are connected in the order they occur in the input data set. Therefore, the data should be sorted by the independent (horizontal axis) variable.

If the data contain missing values, the observations are omitted. However, the plot line is not broken at missing values unless the SKIPMISS option is used.

Alias: I=JOIN

See also: the SKIPMISS on page 1358 option and "Missing Values" on page 1331

INTERPOL=L<*degree*><P><S>

specifies a Lagrange interpolation to smooth the plot line. Specify one of these values for *degree*:

1 | 3 | 5   specifies the degree of the Lagrange interpolation polynomial. By default, *degree* is 1.

In addition, you can specify one or both of these:

P   specifies a parametric interpolation

S   sorts a data set by the independent variable before plotting its data.

The Lagrange methods are useful chiefly when data consist of tabulated, precise values. A polynomial of the specified degree (1, 3, or 5) is fitted through the nearest 2, 4, or 6 points. In general, the first derivative is not continuous. If the values of the horizontal variable are not strictly increasing, the corresponding parametric method (L1P, L3P, or L5P) is used.

Specifying INTERPOL=L1P, INTERPOL=L3P, or INTERPOL=L5P results in a parametric Lagrange interpolation of degree 1, 3, or 5, respectively. Both the horizontal and vertical variables are processed with the Lagrange method and a parametric interpolation of degree 1, 3, or 5, using the distance between points as a parameter.

INTERPOL=*map/plot-pattern*
I=*map/plot-pattern*

specifies that a pattern fill the polygon that has been defined by the data points. Values for *map/plot-pattern* are as follows:

MEMPTY
ME

an empty pattern. EMPTY and E are valid aliases.
The Java applet does not support this option.

MSOLID
MS

a solid pattern. SOLID and S are valid aliases

M*density<style<angle>>*

a shaded pattern. (The Java applet does not support this option.)
Density specifies the density of the pattern's shading:

1...5   1 produces the lightest shading and 5 produces the heaviest.

Style specifies the direction of pattern lines:

N   parallel lines (the default)

X   crosshatched lines.

Angle specifies the starting angle for parallel or crosshatched lines:

0...360   the degree at which the parallel lines are drawn. By default, *angle* is 0 (lines are parallel to the horizontal axis).

The INTERPOL=*map/plot-pattern* option only works if the data are structured so that the data points and, consequently, the plot lines form an enclosed area. The plot lines should not cross each other.

Alias: I=L*<degree><P><S>*

See also: the "PATTERN Statement" on page 240

Featured in: "Example 5. Filling the Area between Plot Lines" on page 304

Restriction: Partially supported by Java

INTERPOL=NEEDLE

draws a vertical line from each data point to a horizontal line at the 0 value on the vertical axis or the minimum value on the vertical axis. The horizontal line is drawn automatically.

Figure 14.17 on page 261 shows the type of plot INTERPOL=NEEDLE produces. Plot symbols are not displayed in this figure.

Figure 14.17 Needle Plot

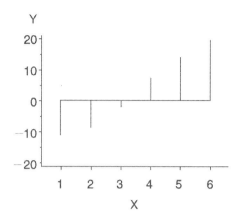

You cannot use the PLOT statement option AREAS= with INTERPOL=NEEDLE.

Alias: I=NEEDLE

INTERPOL=NONE
I=NONE
> suppresses any interpolation and, if the VALUE= option is not specified, also suppresses plot points. If no interpolation method is specified in a SYMBOL statement and if the graphics option INTERPOL= is not used, INTERPOL=NONE is the default.
>
> You cannot use the PLOT statement option AREAS= with INTERPOL=NONE.

INTERPOL=R<*type*><0><CLM | CLI<50...99>>
> specifies that a plot is a regression analysis. By default, regression lines are not forced through plot origins and confidence limits are not displayed.
>
> *Type* specifies the type of regression. Specify one of these values for *type*:
>
> L requests linear regression representing the regression equation
>
> $$Y = \beta_0 + \beta_1 X$$
>
> Q requests quadratic regression representing the regression equation
>
> $$Y = \beta_0 + \beta_1 X + \beta_2 X^2$$
>
> C requests cubic regression representing the regression equation
>
> $$Y = \beta_0 + \beta_1 X + \beta_2 X^2 + \beta_3 X^3$$
>
> > *Note:* When least-square solutions for the parameters are not unique, the SAS/GRAPH uses a quadratic equation by default for the interpolation whereas the Java and ActiveX device drivers might pick a cubic solution to use. △
>
> By default, *type* is L. The regression line is drawn in the line type specified in the LINE= option. By default, the type of the regression line is 1.
>
> *Note:* You must specify *type* if you use either 0, or CLI, or CLM. △

To force the regression line through a (0,0) origin, specify:

0   eliminates the β_0 parameter, or intercept, from the regression equation. If the origin is at (0,0), also forces the regression line through the origin. For example, if you specify 0 for a linear regression, the plot line represents the equation

$$Y = \beta_1 X$$

Note: To force the regression line through the origin (0,0) when the data ranges do not place the origin at (0,0), use the GPLOT procedure options HZERO and VZERO (ignored if the data contain negative values), or use the HAXIS= and VAXIS= options to specify axes ranges from 0 to maximum data value. If the data ranges contain negative values and the HAXIS= and VAXIS= options specify ranges starting at 0, only values within the displayed range are used in the interpolation calculations. △

To display confidence limits, specify one of these:

CLM   displays confidence limits for mean predicted values

CLI   displays confidence limits for individual predicted values.

You can specify confidence levels from 50% to 99%. By default, the confidence level is 95%. Include a confidence level specification only if you use CLM or CLI.

The line type used for the confidence limit lines is determined by adding 1 to the values of LINE=. By default, the line type of confidence limit lines is 2.

Figure 14.18 on page 262 shows the type of plot INTERPOL=RCCLM95 produces (cubic regression analysis with 95% confidence limits).

Figure 14.18   Plot of Regression Analysis and Confidence Limits

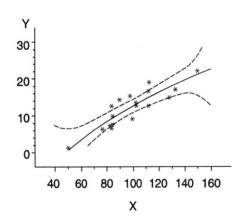

Alias: I=R*<type>*<0><CLM | CLI<50...99>>

Featured in: Example 4 on page 1372

Restriction: Partially supported by Java

INTERPOL=SM<*nn*><P><S>

specifies that a smooth line is fit to data using a spline routine. INTERPOL=SM is a method for smoothing noisy data. The points on the plot do not necessarily fall on the line.

The relative importance of plot values versus smoothness is controlled by *nn*. Values for *nn* are as follows:

0...99   produces a cubic spline that minimizes a linear combination of the sum of squares of the residuals of fit and the integral of the square of the second derivative (Reinsch 1967)*. The greater the *nn* value, the smoother the fitted curve. By default, the value of *nn* is 0.

In addition, specify one or both of these:

P   specifies a parametric cubic spline

S   sorts data by the independent variable before plotting.

Restriction: Not supported by Java

INTERPOL=SPLINE<P><S>

specifies that the interpolation for the plot line use a spline routine. INTERPOL=SPLINE produces the smoothest line and is the most efficient of the nontrivial spline interpolation methods.

Spline interpolation smoothes a plot line using a cubic spline method with continuous second derivatives (Pizer 1975)**This method uses a piecewise third-degree polynomial for each set of two adjacent points. The polynomial passes through the plotted points and matches the first and second derivatives of neighboring segments at the points.

Specify one or both of these:

P   specifies a parametric spline interpolation method. This interpolation uses a parametric spline method with continuous second derivatives. Using the method described earlier for the spline interpolation, a parametric spline is fitted to both the horizontal and vertical values. The parameter used is the distance between points

$$t = \sqrt{(x^2 + y^2)}$$

If two points are so close together that the computations overflow, the second point is not used.

S   sorts a data set by the independent variable before plotting its data.

Note: When points on the graph are out of range of the axis values, the curve is clipped. If an end point is out of range, no curve is drawn. Out-of-range conditions can be caused by restricting the range of axis values with the HAXIS= or VAXIS= option in the PLOT statement or the ORDER= option in an AXIS definition.

Note: When points on the graph are close together and a spline interpolation is used, the Java applet is unable to draw some line types correctly. △

△

Alias: I=SPLINE<P><S>

---

*   Reinsch, C.H. (1967), "Smoothing by Spline Functions," *Numerische Mathematik*, 10, 177–183.
** Pizer, Stephen M. (1975), *Numerical Computing and Mathematical Analysis*, Chicago: Science Research Associates, Inc., Chapter 4.

INTERPOL=STD<1 | 2 | 3><*variance*><*option(s)*>
specifies that a solid line connect the mean Y value with ± 1, 2, or 3 standard deviations for each X.

Note: By default, two standard deviations are used. △

The sample variance is computed about each mean, and from it, the standard deviation s_y is computed. *Variance* can be one or both of these:

M computes $s_{\bar{y}}$,

P computes sample variances using a pooled estimate, as in a one-way ANOVA model.

In addition, specify one of these values for *option(s)*:

B connects the minimum and maximum Y values with bars instead of lines.

J connects the means from bar to bar with a line.

T adds tops and bottoms to each line.

BJ connects maximum and minimum values with a bar and joins the mean values.

TJ adds tops and bottoms to the lines and joins the mean values.

Figure 14.19 on page 264 shows the type of plot INTERPOL=STD produces. A horizontal tick is drawn at the mean. Plot symbols in the form of dots have been added to this figure.

Figure 14.19 Plot of Standard Deviations

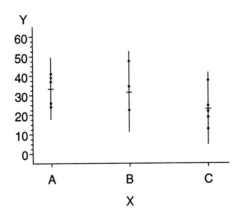

Note: By default, the vertical axis ranges from the minimum to the maximum Y value in the data. If the requested number of standard deviations from the mean covers a range of values that exceeds the maximum or is less than the minimum, the STD lines are cut off at the minimum and maximum Y values. When this cutoff occurs, rescale the axis using VAXIS= in the PLOT statement or ORDER= in an AXIS definition so that the STD lines are shown. △

If you restrict the range of axis values by using the HAXIS= or VAXIS= option in a PLOT statement or the ORDER= option in an AXIS definition, by default any observations that fall outside the axis range are excluded from the interpolation calculation. See the MODE= option on page 266 option.

To increase the thickness of all lines generated by the INTERPOL=STD option, use the WIDTH= option.

You cannot use the PLOT statement option AREAS= with INTERPOL=STD.

When using DEVICE=JAVA and DEVICE=JAVAIMG with overlaid plots, different interpolations are supported per overlay unless any of the interpolations is BOX, HILO or STD. When any of these interpolations are encountered, the first interpolation specified becomes the only interpolation that is used for all overlays. All other interpolations are ignored.

Alias: I=STD<1 | 2 | 3><variance><option(s)>

Restriction: Partially supported by Java

INTERPOL=STEP<placement><J><S>

specifies that the data are plotted with a step function. By default, the data point is on the left of the step, the steps are not joined with a vertical line, and the data are not sorted before processing.

Specify one of these values for *placement*:

L	displays the data point on the left of the step.
R	displays the data point on the right of the step.
C	displays the data point in the center of the step.

Note: When a step is retraced in order to locate its center point, the GIF, JPEG, PNG, ACTXIMG, Java, and JAVAIMG devices treat this as effectively not drawing that part of the step at all. ActiveX, however, draws each part of the step—resulting in a somewhat different graph. △

In addition, specify one or both of these:

J	produces steps joined with a vertical line.
S	sorts unordered data by the independent variable before plotting.

Figure 14.20 on page 265 shows the type of plot INTERPOL=STEPJR produces. Plot symbols in the form of dots have been added to this figure.

Figure 14.20 Step Plot

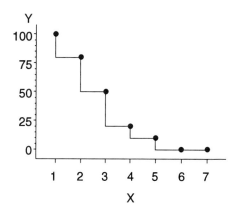

Alias: I=STEP<placement><J><S>

LINE=*line-type*
L=*line-type*

specifies the line type of the plot line in the GPLOT procedure, or the contour line in the GCONTOUR procedure:

1   a solid line.

2...46   a dashed line.
Line types are shown in Figure 14.22 on page 277. By default, LINE=1.

Note: This option overrides the LineStyle attribute in graph styles.
Neither the Java applet nor ActiveX control supports GCONTOUR. △

Restriction: Partially supported by Java and ActiveX

MODE=EXCLUDE | INCLUDE
specifies that any interpolation method exclude or include data values that are outside the range of plot axes. By default, MODE=EXCLUDE prevents values outside the axis range from being displayed.

If you control the range of values displayed on an axis by using HAXIS= and VAXIS= in the GPLOT procedure, or ORDER= in an AXIS definition, any data points that lie outside the range of the axes are discarded before interpolation is applied to the data. Using these options to control value ranges has a particularly noticeable effect on the high-low interpolation methods, which include INTERPOL=HILO, INTERPOL=BOX, and INTERPOL=STD. Regression analysis also represents only part of the original data.

Restriction: Not supported by Java and partially supported by ActiveX

See also: "Values Out of Range" on page 1331

POINTLABEL<=(*label-description(s)*) | NONE>
labels plot points. The labels always use the format that is assigned to the variable or variables whose values are used for the labels. POINTLABEL without any specified descriptions labels points with the Y value. NONE suppresses the point labels. *Label-description(s)* can be used to change the variable whose values are used to label points, and to change features of the label text, such as the color, font, or size of the text.

Note: If you do not specify a color on a SYMBOL statement, the symbol definition is rotated through the color list before the next SYMBOL statement is used. Thus, if your plot contains multiple plot lines and you want to limit your POINTLABEL specification to a single line, you must specify a color in the SYMBOL statement that contains the POINTLABEL description. △

Label-description(s) can be one or more of these:

COLOR=*text-color*
    specifies the color of the label text. The default is the first color from the color list.

    Alias: C=*text-color*

DROPCOLLISIONS | NODROPCOLLISIONS

    specify DROPCOLLISIONS to drop new labels if they collide with a label already in use. Specify NODROPCOLLISIONS to retain all labels. The default is DROPCOLLISIONS.

    The algorithm for the placement of markers tries to avoid placing labels such that they collide. If the algorithm is unable to avoid a collision, then the default DROPCOLLISIONS is to drop the new label, whereas NODROPCOLLISIONS retains even colliding labels.

FONT=*font* | NONE
    specifies the font for the text. See Chapter 11, "Specifying Fonts in SAS/ GRAPH Programs," on page 155 for details on specifying *font*. If you omit FONT=, a font specification is searched for in this order:

    1 the FTEXT= option in a GOPTIONS statement
    2 the default hardware font, NONE.

Alias: F=*font* | NONE

HEIGHT=*text-height* <*units* >
 specifies the height of the text characters in number of units. By default, HEIGHT=1 CELL. If you omit HEIGHT=, a text height specification is searched for in this order:
 1 the HTEXT= option in a GOPTIONS statement
 2 the default value, 1.

 Alias: H=*text-height* <*units* >

JUSTIFY=CENTER | LEFT | RIGHT
 specifies the horizontal alignment of the label text. The default is CENTER. The location of the point label is relative to the location of the corresponding data point.

POSITION=TOP | MIDDLE | BOTTOM
 specifies the vertical placement of the label text. The default is TOP. The location of the point label is relative to the location of the corresponding data point.

 Alias: J=C | L | R

"#*var*" | "#*x*:#*y* <$*char*>" | "#*y*:#*x* <$*char*>"
 specifies the variable or variables whose values label the plot points. The variable specification must be enclosed in either single or double quotation marks. The first specified variable must be prefixed with a pound sign (#). If a second variable is specified, it must be prefixed with a colon and a pound sign (:#). When you specify both the X and Y variables, you can also specify the character to display as the delimiter between variable values in the plot label.
  By default if the POINTLABEL= option is specified without naming a label variable, the Y values label the plot points. You can change the default by using "#*var*" to specify a different variable whose values should label the points. For example, you might specify the name of the X variable. The following option specifies the variable SALES as the variable whose values label plot points:

  POINTLABEL=("#sales")

 Alternatively, you can label the plot points with the values of the X and Y variables, in either order. The order that you specify X and Y in the variable specification determines the order that the values are displayed in the label. The following option specifies variables HEIGHT and WEIGHT; in the label, the value for HEIGHT is displayed, followed by the value for WEIGHT:

  POINTLABEL=("#height:#weight")

By default when you specify both the X and Y variables, a colon (:) displays in the label to separate the values in each label. To change the character that displays as the delimiter, use the $ syntax to specify an alternative character. The following option specifies a vertical bar (|) as the delimiter in the label:

  POINTLABEL=("#height:#weight $|")

The $ syntax must be within the same quotation marks as the variable specification. The $ specification can precede or follow the variable specification, but it must be separated from the variable specification by at least one space.

 Note: Specifying a delimiting character with the $ only changes the character that displays in the label. It does not change the syntax of the variable specification, which requires a colon and pound sign (:#) to precede the second variable. △

Note: There is a sixteen character length limit for each variable. A maximum character length limit of thirty-three characters is possible. This can be composed of X and Y variables, any other valid data set variable, and a separator as required. △

When creating output using the ActiveX or Java devices, the variables that you specify in the POINTLABEL= option must be for the plot's X and Y variables. Specifying any other variables causes unexpected labeling.

Specify as many label-description suboptions as you want. Enclose them all within a single set of parentheses, and separate each suboption from the others by at least one space.

Restriction: Partially supported by Java and ActiveX

REPEAT=*number-of-times*

specifies the number of times that a SYMBOL definition is applied before the next SYMBOL definition is used. By default, REPEAT=1.

The behavior of REPEAT= depends on whether any of the SYMBOL color options (CI=, CV=, CO=, and COLOR=) or the CSYMBOL= graphics option also is used:

- If any SYMBOL color option also is used in the SYMBOL definition, that SYMBOL definition is repeated the specified number of times in the specified color.

- If no SYMBOL color option is used but the CSYMBOL= graphics option is currently in effect, the SYMBOL definition is repeated the specified number of times in the specified color.

- If no SYMBOL statement color options are used and the CSYMBOL= graphics option is not used, the SYMBOL definition is cycled through each color in the color list, and then the entire group generated by this cycle repeats the number of times specified by the REPEAT= option. Thus, the total number of iterations of the SYMBOL definition depends on the number of colors in the current color list.

Neither the Java applet nor ActiveX control supports GCONTOUR.

Alias: R=*number-of-times*

See also: "Using the SYMBOL Statement" on page 272

Restriction: Partially supported by Java and ActiveX

SINGULAR=*n*

tunes the algorithm used to check for singularities. The default value is machine dependent but is approximately 1E-7 on most machines. This option is rarely needed.

STEP=*distance<units>*

specifies the minimum distance between labels on contour lines. The value of *distance* must be greater than zero. By default, STEP=65PCT.

Note: If you specify units of PCT or CELLS, the STEP= option calculates the distance between the labels based on the width of the graphics output area, not the height. For example, if you specify STEP=50PCT and if the graphics output area is 9 inches wide, the distance specified is 4.5 inches. A value less than 10 percent is ignored and 10 percent is used instead. △

When you use the STEP= option, specify the minimum distance that you want between labels. The option then calculates how many labels it can fit on the contour line, taking into account the length of the labels and the minimum distance you specified. Once it has calculated how many labels it can fit while retaining the minimum distance between them, it places the labels, evenly spaced, along the line. Consequently, the space between labels can be greater than what you specify, although it will never be less.

In general, to increase the number of labels from the default, reduce the value of *distance*.

If the procedure cannot write the label at a particular location on the contour, for example because the contour line makes a sharp turn, the label might be placed farther along the line or omitted. If labels are omitted, a note appears in the log. Specifying a low value for the GCONTOUR procedure's TOLANGLE= option can also cause labels to be omitted, since this forces the procedure to select smoother labeling locations, which might not be available on some contours.

Featured in: Example 2 on page 1116

Restriction: Not supported by Java and ActiveX

VALUE=*special-symbol* | *text-string* | NONE

- specifies a plot symbol for the data points (GPLOT and GBARLINE). If you omit the SYMBOL statement, plot points are generated using the default plot symbol. The default symbol is a square if you use the ActiveX or Java devices and a PLUS sign for other devices. If you specify a SYMBOL statement, but do not specify the VALUE= option, plot symbols are suppressed.

  Note: For ActiveX output, the VALUE= option is not supported when INTERPOL=HILO or INTERPOL=STD. You can use the OVERLAY option with GPLOT to get symbols to appear on the data points. △

- specifies contour-label text in a contour plot (GCONTOUR). By default with the AUTOLABEL option, GCONTOUR labels contour lines with the contour variable's value at that contour level.

- VALUE=NONE suppresses plot symbols at the data points, or labels on the contour lines. You can set the VALUE=NONE option independent of the INTERPOL= option.

Values for *special-symbol* are the names and characters shown in Figure 14.21 on page 271. The special symbol table can be used only if the FONT= option is not used or a null value is specified:

```
font=,
```

To specify a single quotation mark, you must enclose it in double quotation marks

```
value="'"
```

To specify a double quotation mark, you must enclose it in single quotation marks:

```
value='"'
```

In some operating environments, punctuation characters might require single quotes.

If you use VALUE=*text-string* to specify a plot symbol, you must also use the FONT= option to specify a symbol font or a text font. If you specify a symbol font, the characters in the string are character codes for the symbols in the font. If you specify a text font, the characters in the string are displayed. If you specify a text string containing quotes or blanks, enclose the string in single quotes.

For example, if you specify this statement, the plot symbol is the word "plus" instead of the symbol +:

```
symbol font=swiss value=plus;
```

Java and ActiveX support the following characters from the marker font for *special-symbol*:

Table 14.2 Marker-font symbols supported by Java and ActiveX

Character	Aliases
Marker	Cone, Pyramid, Default
Square	Cube
Star	
Circle	Sphere, Dot, Balloon
Plus	Cross
Flag	Y
X	
Prism	Z
Spade	"
Heart	#
Diamond	$
Club	%
Hexagon	Paw
Cylinder	Hash

Note: If you do not use a SYMBOL statement to specify a color for each symbol, but you do specify a color list in a GOPTIONS statement, then Java and ActiveX assign colors to symbols differently than do the other device drivers. To ensure consistency on all devices, you should specify the desired color of each symbol. If you do not specify a symbol color, SAS/GRAPH uses the first default color and the first symbol. It uses each color in the list of default colors until the list is exhausted. SAS/GRAPH then selects the next symbol and begins again with the first default color. It rotates the new symbol through the list of default colors before selecting another symbol. It continues selecting new symbols and colors until no more symbols are needed. △

Note: The VALUE option overrides the MarkerSymbol attribute in graph styles. △

See also: the option FONT= on page 256 and "Specifying Plot Symbols" on page 274.

Alias: V=*special-symbol* | *text-string* | NONE

Featured in: "Example 3. Rotating Plot Symbols Through the Color List" on page 299, "Example 4. Creating and Modifying Box Plots" on page 302, and Example 2 on page 1116

Restriction: Partially supported by Java and ActiveX

WIDTH=*thickness-factor*
specifies the thickness of interpolated lines (GPLOT) or contour lines (GCONTOUR), where *thickness-factor* is a number. The thickness of the line increases directly with *thickness-factor*. By default, WIDTH=1.

WIDTH= also affects all the lines in box plots (INTERPOL=BOX), high-low plots with bars (INTERPOL=HILOB), and standard deviation plots

(INTERPOL=STD). It also affects the outlines of the area generated by the AREAS= option in the PLOT statement of the GPLOT procedure.

Note: By default, the value specified by WIDTH= is used as the default value for the BWIDTH= option. For example, specifying WIDTH=6 also sets BWIDTH= to 6 unless you explicitly assign a value to BWIDTH=.

Java and ActiveX do not provide the same measure of control for width as SAS/GRAPH device drivers. Measurements are translated to pixels rather than a percentage. For DEVICE=JAVA and DEVICE=ACTIVEX the maximum width is 6. △

Style Reference: LineThickness attribute of the GraphAxisLines element

Alias: W=*thickness-factor*

Featured in: "Example 1. Ordering Axis Tick Marks with SAS Date Values" on page 294 and "Example 4. Creating and Modifying Box Plots" on page 302

Restriction: Partially supported by Java and ActiveX

Figure 14.21   Special Symbols for Plotting Data Points

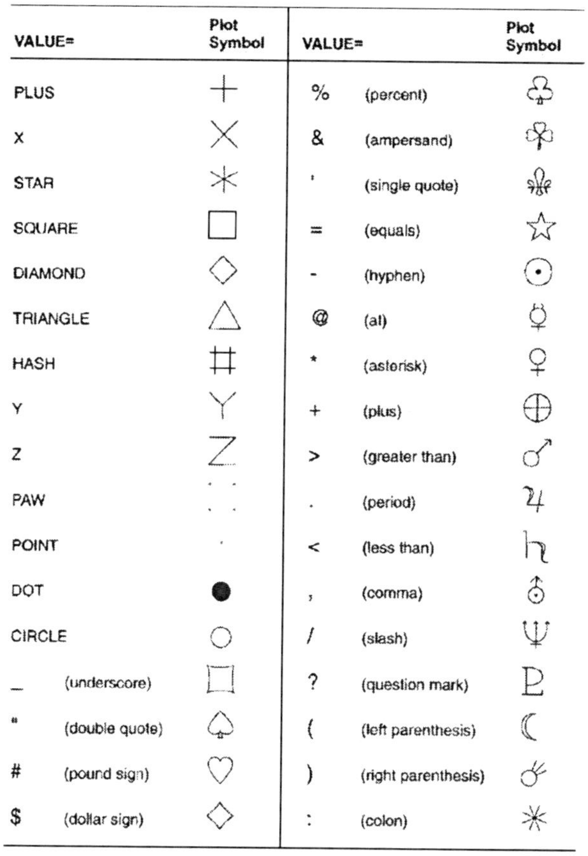

VALUE=	Plot Symbol	VALUE=		Plot Symbol
PLUS	+	%	(percent)	♣
X	×	&	(ampersand)	♧
STAR	✷	'	(single quote)	☥
SQUARE	□	=	(equals)	☆
DIAMOND	◇	-	(hyphen)	⊙
TRIANGLE	△	@	(at)	☿
HASH	#	*	(asterisk)	♀
Y	Y	+	(plus)	⊕
Z	Z	>	(greater than)	♂
PAW	⋰	.	(period)	♃
POINT	·	<	(less than)	♄
DOT	●	,	(comma)	☊
CIRCLE	○	/	(slash)	♆
_	(underscore)	□	?	(question mark)	♇
"	(double quote)	♤	(	(left parenthesis)	☾
#	(pound sign)	♡	)	(right parenthesis)	☋
$	(dollar sign)	◇	:	(colon)	✸

Note: The words or special characters in the VALUE= column are entered exactly as shown. △

Using the SYMBOL Statement

A SYMBOL statement specifies one or more options that indicate the color and other attributes used by the GPLOT, GBARLINE, and GCONTOUR procedures. For GPLOT and GBARLINE, the main attributes include the plot symbol, interpolation method, and type of plot line. For GCONTOUR, the main attributes include the type of contour lines used and the text used to label those lines.

Note: SYMBOL statements can be applied only to contour plots when the AUTOLABEL option is specified on GCONTOUR. △

You can define up to 255 different SYMBOL statements. A SYMBOL statement without a number is treated as a SYMBOL1 statement.

SYMBOL definitions can be defined anywhere in your SAS program. They are global and remain in effect until canceled or until you end your SAS session. Once defined, SYMBOL definitions can be used as follows:

- assigned by default by GPLOT or explicitly selected with the plot request
- used by GCONTOUR to control the labels and attributes of contour lines

SYMBOL statements generate one or more symbol definitions, depending on how color is used and whether a plot symbol or type of contour line is specified. For more information, see "Controlling Consecutive SYMBOL Statements" on page 273 and "Using Generated Symbol Sequences" on page 277.

Although it is common practice, you do not have to start with SYMBOL1, and you do not have to use sequential statement numbers. When assigning SYMBOL definitions, SAS/GRAPH software starts with the lowest-numbered definition and works upward, ignoring gaps in the numbering.

Altering or Canceling SYMBOL Statements SYMBOL statements are additive. If you define a SYMBOL statement and later submit another SYMBOL statement with the same number, the new SYMBOL statement defines or cancels only the options that are included in the new statement. Options that are not included in the new statement are not changed and remain in effect.

Note: An exception to this rule is presented by POINTLABEL= suboptions which are not carried over to subsequent SYMBOL statements. △

Assume you define SYMBOL4 as follows:

```
symbol4 value=star cv=red height=4;
```

The following statement cancels only HEIGHT= without affecting the rest of the definition:

```
symbol4 height=;
```

Add or change options in the same way. This statement adds an interpolation method to SYMBOL4:

```
symbol4 interpol=join;
```

This statement changes the color of the plot symbol from red to blue:

```
symbol4 cv=blue;
```

After all these modifications, SYMBOL4 has these characteristics:

```
symbol4 value=star cv=blue interpol=join;
```

Cancel individual SYMBOL statements by defining a SYMBOL statement of the same number without options (a null statement):

```
symbol4;
```

Canceling one SYMBOL statement does not affect any other SYMBOL definitions. To cancel all current SYMBOL statements, use the RESET= option in a GOPTIONS statement:

```
goptions reset=symbol;
```

Specifying RESET=GLOBAL or RESET=ALL cancels all current SYMBOL definitions as well as other settings.

To display current SYMBOL definitions in the Log window, use the GOPTIONS procedure with the SYMBOL option:

```
proc goptions symbol nolist;
run;
```

Controlling Consecutive SYMBOL Statements

If you specify consecutively numbered SYMBOL statements and you want SAS/GRAPH to use each definition only once, use color specifications to ensure that each SYMBOL statement generates only one symbol definition. You can do the following actions:

- specify colors on each SYMBOL statement, using the COLOR=, CI=, CV=, or CO= options. This method lets you explicitly assign colors for each definition. For example, these statements generate two definitions:

  ```
  symbol1 value=star color=green;
  symbol2 value=square color=yellow;
  ```

- specify a default color for all SYMBOL statements using the CSYMBOL= option in the GOPTIONS statement. This method makes it easy to specify the same color for each definition when you do not need more explicit color specifications.

- limit the color list to a single color using the COLORS= option in the GOPTIONS statement. This method makes it easy to specify the same color for each definition when you want the color to apply to other definitions also, such as PATTERN definitions.

For more information on specifying colors for symbol definitions, see "Using Color" on page 275.

If you do not use color to limit a SYMBOL statement to a single symbol definition, SAS/GRAPH generates multiple symbol definitions from that statement by rotating the current definition through the color list (for more details, see "Using Generated Symbol Sequences" on page 277). Because SAS/GRAPH uses symbol definitions in the order they are generated, this means that the nth symbol definition applied to a graph does not necessarily correspond to the SYMBOLn statement.

For example, assuming that no color is specified on the CSYMBOL= graphics option, these statements generate four definitions:

```
goptions colors=(red blue green);
symbol1 value=star;
symbol2 value=square color=yellow;
```

Because no color is specified on SYMBOL1, SAS/GRAPH rotates the symbol definition through the color list, which has three colors. Thus, SYMBOL1 defines the first three applied symbol definitions, and SYMBOL2 defines the 4th:

Sequence Number	Source	Characteristics: Color	Symbol
1	SYMBOL1	red	star
2	SYMBOL1	blue	star

Sequence Number	Source	Characteristics: Color	Symbol
3	SYMBOL1	green	star
4	SYMBOL2	yellow	square

In this case, if a graph needs only three symbols, the SYMBOL2 definition is not used.

To make the *n*th applied symbol definition correspond to the SYMBOL*n* statement, limit each SYMBOL statement to a single color, using one of the techniques listed at the beginning of this section.

Setting Definitions for PROC GPLOT and PROC GBARLINE

The following topics apply only for SYMBOL statements used with PROC GPLOT and PROC GBARLINE:

- specifying plot symbols
- specifying default interpolation methods
- sorting data with spline interpolation

Specifying Plot Symbols The VALUE= option specifies the plot symbols that PROC GPLOT and PROC GBARLINE uses to mark the data points on a plot. Plot symbols can be in the following forms:

- special symbols as shown in Figure 14.21 on page 271
- characters from symbol fonts
- text strings

By default, the plot symbol is the + symbol. To specify a special symbol, use the VALUE= option to specify a name or a character from Figure 14.21 on page 271:

```
symbol1 value=hash color=green;
symbol2 value=) color=blue;
```

This example uses color to ensure that each SYMBOL statement generates only one definition. You can omit color specifications to let SAS/GRAPH rotate symbol definitions through the color list. For details, see "Using Generated Symbol Sequences" on page 277.

To use plot symbols other than those in Figure 14.21 on page 271, use the FONT= option to specify a font for the plot symbol. If the font is a symbol font, such as Marker, the string specified with the VALUE= option is the character code for the symbol to be displayed. If the font is a text font, the string specified with the VALUE= option is displayed as the plot symbol. (See VALUE= on page 269 and FONT= on page 256.)

This table illustrates some of the ways you can define a plot symbol:

Definition	Plot Symbol
symbol1 value=plus;	✚
symbol2 value=+;	⊕
symbol3 font=swiss value=plus;	plus

Definition	Plot Symbol
symbol4 font=marker value=U;	■
symbol5value="";	⚜

Specifying a Default Interpolation Method The INTERPOL= option in a GOPTIONS statement specifies a default interpolation method to be used with all SYMBOL definitions. This default interpolation method is in effect unless you specify a different interpolation in a SYMBOL statement. If the GOPTIONS statement does not specify an interpolation method, the default for each SYMBOL statement is NONE.

Sorting Data with Spline Interpolation If you want the GPLOT procedure to sort by the horizontal axis variable before plotting, add the letter S to the end of any of the spline interpolation methods (INTERPOL=L, INTERPOL=SM, and INTERPOL=SPLINE). For example, suppose you want to overlay three plots (Y1*X1, Y2*X2, and Y3*X3) and for each plot, you want the X variable sorted in ascending order. Use these statements:

```
symbol1 i=splines c=red;
symbol2 i=splines c=blue;
symbol3 i=splines c=green;

proc gplot;
   plot y1*x1 y2*x2 y3*x3 / overlay;
run;
```

Using Color

Generally, there are two ways to explicitly specify color for SYMBOL statements:
- specify colors on the SYMBOL statements
- specify a color on the CSYMBOL= graphics option

You can also let SAS/GRAPH rotate symbol definitions through the color list. For details, see "Using Generated Symbol Sequences" on page 277.

Specifying Colors with SYMBOL Statements The SYMBOL statement has these options for specifying color:
- The CV= option specifies color for plot symbols in GPLOT and GBARLINE, or for contour labels in GCONTOUR.
- The CO= option specifies color for confidence limit lines and area outlines in GPLOT and GBARLINE.
- The CI= option specifies color for plot lines in GPLOT and GBARLINE, or contour lines in GCONTOUR.
- The COLOR= option specifies color for the entire symbol. For GPLOT and GBARLINE, this includes plot symbols, plot lines, and outlines. For GCONTOUR, this includes contour lines and labels.

The CV= and CI= options have the same effect as using the COLOR= option when they are used in these ways:
- Only CV= or CI= option is used. (The option that is not used is assigned the value of the option used.)
- Both the CV= and CI= options specify the same color.

In general, the CI=, CV=, and CO= options color specific areas of the symbol. Use these options to produce symbols and plot lines of different colors without having to overlay multiple plot pairs. For example, if you request regression analysis with confidence limits, use this statement to assign red to the plot symbol, blue to the regression lines, and green to the confidence limit lines:

```
symbol cv=red ci=blue co=green;
```

The COLOR= option colors the entire symbol or those portions of it not colored by one of the other color options. If the COLOR= option precedes the CI= or CV= options, the CI= or CV= specification is used instead. If none of the SYMBOL color options is used, color specifications are searched for in this order:

1 the CSYMBOL= option in a GOPTIONS statement

2 each color in the color list sequentially before the next SYMBOL definition is used

CAUTION:
  If no color options are used, the SYMBOL definition cycles through each color in the color list. △

If the SYMBOL color options and the CSYMBOL= graphics option are not used, the SYMBOL definition cycles through each color in the color list before the next definition is used. For details, see "Using Generated Symbol Sequences" on page 277.

Specifying Color with CSYMBOL=   The CSYMBOL= option in the GOPTIONS statement specifies the default color to be used by all SYMBOL definitions:

```
goptions csymbol=green;
symbol1 value=star;
symbol2 value=square;
```

In this example, both SYMBOL statements use green.

CSYMBOL= is overridden by any of the SYMBOL statement color options. See "Using Color" on page 275 for details.

If more SYMBOL definitions are needed, SAS/GRAPH returns to generating default symbol sequences.

Specifying Line Types

To specify the type of line for plot or contour lines, use the LINE= option to specify a number from 1 through 46. Figure 14.22 on page 277 shows the line types represented by these numbers. By default, the line type is 1 for plot and contour lines, and 2 for confidence limit lines.

Figure 14.22 Line Types

Note: These line types are also used by other statements and procedures. Some options accept a line type of 0, which produces no line. △

Using Generated Symbol Sequences

Symbol sequences are sets of SYMBOL definitions that are automatically generated by SAS/GRAPH software if any of these conditions is true:

- no valid SYMBOL definition is available. In this case, default symbol sequences are generated by rotating symbol definitions through the color specified in the GOPTIONS statement's CSYMBOL= option. If a CSYMBOL= color is not in effect, the definitions are rotated through the color list.

- a SYMBOL statement specifies color but not a plot symbol for the GPLOT procedure, or a line type for the GCONTOUR procedure (assuming that GCONTOUR does not specify the needed line types). In this case, a default plot symbol or line type is used with the specified color and only one definition is generated.

- a SYMBOL statement specifies a plot symbol for GPLOT or a line type for GCONTOUR, but no color options. In this case, the specified plot symbol or line type is used once with the color specified by the CSYMBOL= graphics option. If a

CSYMBOL= color is not in effect, the specified plot symbol or line type is rotated through the color list.

If the REPEAT= option is also used, the resulting SYMBOL definition is repeated the specified number of times.

Default Symbol Sequences   Default symbol sequences are generated by rotating symbol definitions through the current color list.

- Definitions used for GPLOT rotate plot symbols through the color list; the first default plot symbol is a plus sign (+).
- Definitions used for GCONTOUR rotate line types; the first default line type is a solid line (line type 1).

Each time a default definition is required, SAS/GRAPH takes the first default plot symbol or line type and uses it with the first color in the color list. If more than one definition is required, it uses the same plot symbol or line type with the next color in the color list and continues until all the colors have been used once. If more definitions are needed, SAS/GRAPH selects the second default plot symbol or line type and rotates it through the color list. It continues in this fashion, selecting default plot symbols or line types and cycling them through the color list until all the required definitions are generated.

If a color has been specified with the CSYMBOL= option in the GOPTIONS statement, each default plot symbol or line type is used once with the specified color, and the colors in the color list are ignored.

Symbol Sequences Generated from SYMBOL Statements   If a SYMBOL statement does not specify color, and if the CSYMBOL= graphics option is not used, the symbol definition is rotated through every color in the color list before the next SYMBOL definition is used:

```
goptions colors=(blue red green);
symbol1 cv=red i=join;
symbol2 i=spline v=dot;
symbol3 cv=green v=star;
```

Here, the SYMBOL1 statement generates the first SYMBOL definition. The SYMBOL2 statement does not include color, so the first default plot symbol is rotated through all colors in the color list before the SYMBOL3 statement is used. This table shows the colors and symbols that would be used if nine symbol definitions were required for PROC GPLOT:

Sequence Number	Source	Characteristics: Color	Symbol	Interpolation
1	SYMBOL1	cv=red	first default	join
2	SYMBOL2	color=blue	dot	spline
3	SYMBOL2	color=red	dot	spline
4	SYMBOL2	color=green	dot	spline
5	SYMBOL3	cv=green	star	NONE
6	first default	color=blue	first default	default
7	first default	color=red	first default	default

Sequence Number	Source	Characteristics: Color	Symbol	Interpolation
8	first default	color=green	first default	default
9	second default	color=blue	second default	default

Notice that after the SYMBOL statements are exhausted, the procedure begins using the default definitions (sequences 6 through 9). Each plot symbol from the default list is rotated through all colors in the color list before the next plot symbol is used. Also, SYMBOL1 does not specify a plot symbol, so the default sequencing provides the first default symbol (a + sign). When sequencing resumes in sequence number 6, it starts at the beginning again, selecting the first default plot symbol and rotating it through the color list.

If you use the REPEAT= option but no color, the sequence generated by cycling the definition through the color list is repeated the number of times specified by the REPEAT= option. For example, these statements define a color list and illustrate the effect of the REPEAT= option on SYMBOL statements both with and without explicit color specifications:

```
goptions colors=(blue red green);
symbol1 color=gold repeat=2;
symbol2 value=star color=cyan;
symbol3 value=square repeat=2;
```

Here, SYMBOL1 is used twice, SYMBOL2 is used once, and SYMBOL3 rotates through the list of three colors and then repeats this cycle a second time:

Sequence Number	Source	Characteristics: Color	Symbol	Interpolation
1	SYMBOL1	gold	first default	default
2	SYMBOL1	gold	first default	default
3	SYMBOL2	cyan	star	default
4	SYMBOL3	blue	square	default
5	SYMBOL3	red	square	default
6	SYMBOL3	green	square	default
7	SYMBOL3	blue	square	default
8	SYMBOL3	red	square	default
9	SYMBOL3	green	square	default

TITLE, FOOTNOTE, and NOTE Statements

Control the content, appearance, and placement of text.

Used by: GANNO, GAREABAR, GBARLINE, GCHART, GCONTOUR, GFONT, GIMPORT, GMAP, GPLOT, GRADAR, GREPLAY, GSLIDE, G3D, and G3GRID
Global: TITLE and FOOTNOTE
Local: NOTE

Syntax

TITLE<1...10> <*text-argument(s)*>;

FOOTNOTE<1...10> <*text-argument(s)*>;

NOTE <*text-arguments(s)*>;
   text-argument(s) can be one or more of these:
  "*text-string*"
  text-options (text options must precede text-string.)

text-options can be one or more of the following, in any order:
- appearance options
  - COLOR=*color*
  - FONT=*font*
  - HEIGHT=*text-height*<*units*>
- placement and spacing options
  - JUSTIFY=LEFT | CENTER | RIGHT
  - LSPACE=*line-space*<*units*>
  - MOVE=(*x,y*)<*units*>
  - WRAP
- baseline angling and character rotation options
  - ANGLE=*degrees*
  - LANGLE=*degrees*
  - ROTATE=*degrees*
- boxing, underlining, and line drawing options
  - BCOLOR=*background-color*
  - BLANK=YES
  - BOX=1...4
  - BSPACE=*box-space*<*units*>
  - DRAW=(*x,y...,x-n,y-n*)<*units*>
  - UNDERLIN=0...3
- linking option
  - LINK= "*URL*"

Options

When the syntax of an option includes *units*, use one of these:

CELLS	character cells
CM	centimeters
IN	inches
PT	points
PCT	percentage of the graphics output area

If you omit *units*, a unit specification is searched for in this order:

1 the GUNIT= option in a GOPTIONS statement
2 the default unit, CELLS.

ANGLE=*degrees*
A=*degrees*

specifies the angle of the *baseline* of the entire text string with respect to the horizontal. A positive *degrees* value angles the baseline counterclockwise; a negative value angles it clockwise. By default, ANGLE=0 (horizontal).

Angled titles or footnotes might require more vertical space and, consequently, might increase the size of the title area or the footnote area, thereby reducing the vertical space in the procedure output area.

Using the BOX= option with angled text does not produce angled boxes; the box is sized to accommodate the angled note.

Using the ANGLE= option after one text string and before another can reset some options to their default values. See "Using Options That Can Reset Other Options" on page 293.

The ANGLE= option has the same effect on the text as LANGLE=, except when you specify an angle of 90 degrees or –90 degrees. In these angle specifications, the procedure output area is shrunk from the left or right to accommodate the angled title or footnote. The result depends on the statement in which you use the option:

□ *With the TITLE statement:*

Figure 14.23 on page 281 shows how ANGLE=90 degrees or ANGLE=–90 degrees positions and rotates title text.

ANGLE=90
positions the title at the left edge of the graphics output area, angled 90 degrees (counterclockwise) and centered vertically.

ANGLE=–90
positions the title at the right edge of the graphics output area, angled –90 degrees (clockwise) and centered vertically.

Figure 14.23 Positioning Titles with the ANGLE= Option

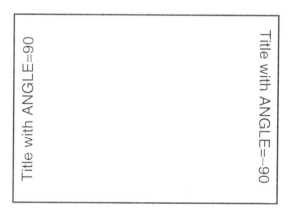

□ *With the FOOTNOTE statement:*

Figure 14.24 on page 282 shows how ANGLE=90 degrees or ANGLE=–90 degrees positions and rotates footnote text.

ANGLE=90

    positions the footnote at the right edge of the graphics output area, angled 90 degrees (counterclockwise) and centered vertically.

ANGLE=–90

    positions the footnote at the left edge of the graphics output area, angled –90 (clockwise) and centered vertically.

Figure 14.24 Positioning Footnotes with the ANGLE=Option

☐ *With the NOTE statement:*

Figure 14.25 on page 282 shows how ANGLE= 90 degrees or -90 degrees positions and rotates note text.

ANGLE=90

    positions the note at the bottom of the left edge of the graphics output area, angled 90 degrees (counterclockwise) and reading from bottom to top.

ANGLE=–90

    positions the note at the top of the right edge of the graphics output area, angled –90 (clockwise) and reading from top to bottom.

Figure 14.25 Positioning Notes with the ANGLE= Option

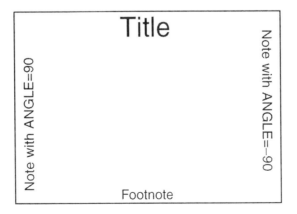

See also: the options LANGLE= on page 287 and ROTATE= on page 290

Featured in: "Example 6. Enhancing Titles" on page 307

Restriction: Not supported by Java and ActiveX

BCOLOR=*background-color*

specifies the background color of a box produced by the BOX= option. If you omit BOX=, BCOLOR= is ignored. By default, the background color of the box is the same as the background color for the entire graph. The color of the frame of the box is determined by the color specification used in BOX=.

Note: The BCOLOR= option can be reset by the ANGLE= or JUSTIFY= options, or by the MOVE= optionwith absolute coordinates. See "Using Options That Can Reset Other Options" on page 293 for details. △

Alias: BC=*background-color*

See also: the option BOX= on page 283

Featured in: "Example 6. Enhancing Titles" on page 307.

BLANK=YES

protects the box and its contents from being overwritten by any subsequent graphics elements by blanking out the area where the box is displayed. The BLANK= option enables you to overlay graphics elements with boxed text. It is ignored if you omit the BOX= option. Because titles and footnotes are written from the highest numbered to the lowest numbered, the BLANK= option only blanks out titles and footnotes of a lower number.

Note: The BLANK= option can be reset by the ANGLE= or JUSTIFY= options, or by the MOVE= option with absolute coordinates. See "Using Options That Can Reset Other Options" on page 293 for details. △

Alias: BL=YES

See also: the option BOX= on page 283

Featured in: "Example 6. Enhancing Titles" on page 307

Restriction: Not supported by Java and ActiveX

BOX=1...4

draws a box around one line of text. A value of 1 produces the thinnest box lines; 4 produces the thickest. Boxing angled text does not produce an angled box; the box is sized to include the angled text.

The color of the box is either:

- the color specified by the COLOR= option in the statement
- the default text color.

The COLOR= option affects only the frame of the box. To color the background of the box, use the BCOLOR= option.

You can include more than one text string in the box as long as no text break occurs between the strings; that is, you cannot use the JUSTIFY= option to create multiple lines of text within a box.

To draw a box around multiple lines of text, you can either

- Use the MOVE= option with relative coordinates to position the lines of text where you want them and enclose them with the BOX= option. For example, this statement produces the boxed note shown in Figure 14.26 on page 284:

```
note font=swiss justify=center box=3
     "Office Hours"      move=(40pct,-12pct) "9-5";
```

- Use the DRAW= option to draw the box and do not use the BOX= option.

Figure 14.26 Using the BOX= Option and the MOVE= Option to Box Multiple Lines of Text

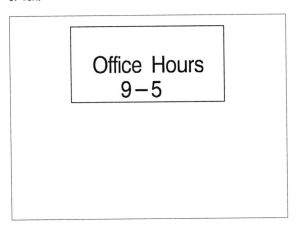

Note: The BOX= option can be reset by the ANGLE= or JUSTIFY= options, or by the MOVE= options with absolute coordinates. See "Using Options That Can Reset Other Options" on page 293 for details. △

Alias: BO=1...4

See also: the options BCOLOR= on page 283, BLANK= on page 283, and BSPACE= on page 284.

Featured in: "Example 6. Enhancing Titles" on page 307

Restriction: Not supported by Java and ActiveX

BSPACE=*box-space<units>*
 specifies the amount of space between the boxed text and the box. The space above the text is measured from the font maximum, and the space below the text is measured from the font minimum. By default, BSPACE=1. If the BOX= option is not used, the BSPACE= option is ignored.

The spacing is uniform around the box. For example, BSPACE=.5IN leaves one-half inch of space between the text and the top, bottom, and sides of the box.

Note: The BSPACE= option can be reset by the ANGLE= or JUSTIFY= options, or by the MOVE= option with absolute coordinates. See "Using Options That Can Reset Other Options" on page 293 for details. △

Alias: BS=*box-space<units>*

See also: the option BOX= on page 283.

Restriction: Not supported by Java and ActiveX.

COLOR=*color*
 specifies the color for the following text, box, or line. The COLOR= option affects all text, lines, and boxes that follow it and stays in effect until another COLOR= specification is encountered.

Change colors as often as you like. For example, this statement produces a title with red text in a box with a blue frame and a cream background:

```
title color=red "Total Sales" color=blue
   box=3 bcolor=cream;
```

Although the BCOLOR= option controls the background color of the box, the frame color is controlled with the COLOR= option that precedes the BOX= option.

If you omit the COLOR= option, a color specification is searched for in this order:

1 the CTITLE= option in a GOPTIONS statement
2 the CTEXT= option in a GOPTIONS statement
3 the default, the first color in the color list.

Alias: ~~ C=*color*

See also: the option BCOLOR= on page 283, and "Controlling Titles and Footnotes with Java and ActiveX Devices in HTML Output" on page 194

DRAW=(*x,y...,x-n,y-n*)<*units*>
draws lines anywhere on the graphics output area using *x* and *y* as absolute or relative coordinates. The following table shows the specifications for absolute and relative coordinates:

Absolute Coordinates	Relative Coordinates
x<*units*>	±*x*<*units*>
y<*units*>	±*y*<*units*>

The coordinate position (0,0) is the lower-left corner of the graphics output area. Specify at least two coordinate pairs. Commas between coordinates are optional; blanks can be used instead. The DRAW= option does not affect the positioning of text.

The starting point for lines specified with relative coordinates begins at the end of the most recently drawn text or line in the current statement. If no text or line has been drawn in the current statement, a warning is issued and the relative draw is measured from where a zero-length text string would have ended, given the normal placement for the statement.

You can mix relative and absolute coordinates. For example, DRAW=(+0,+0,+0,1IN) draws a vertical line from the end of the text to one inch from the bottom of the graphics output area.

Alias: D=(*x,y...,x-n, y-n*)<*units*>

Restriction: Not supported by Java and ActiveX

FONT=*font*
specifies the font for the subsequent text. See Chapter 11, "Specifying Fonts in SAS/GRAPH Programs," on page 155 for details on specifying SAS/GRAPH fonts. If you omit this option, a font specification is searched for in this order:

□ for a TITLE1 statement

1 the FTITLE= option in a GOPTIONS statement
2 the FTEXT= option in a GOPTIONS statement
3 the default font, SWISS (COMPLEX in Release 6.06 and earlier).

□ for all other TITLE statements and the FOOTNOTE and NOTE statements:

1 the FTEXT= option in a GOPTIONS statement
2 the default hardware font, NONE.

Note: Font names greater than eight characters in length must be enclosed in quotation marks. △

Note: If the TITLE or FOOTNOTE is being output through an ODS markup destination and the corresponding NOGTITLE or NOGFOOTNOTE option is specified, then the *bold* and *italic* FONT attributes are on by default. However, if you specify different attributes with the FONT= option, the *bold* and *italic* attributes are turned off. △

Alias: F=*font*

See also: "Controlling Titles and Footnotes with Java and ActiveX Devices in HTML Output" on page 194

Featured in: "Example 6. Enhancing Titles" on page 307

HEIGHT=*text-height<units>*
> specifies the height of text characters in number of units. By default, HEIGHT=1. Height is measured from the font minimum to the capline. Ascenders can extend above the capline, depending on the font.
>
> If your text line is too long to be displayed in the height specified in the HEIGHT= option, the height specification is reduced so that the text can be displayed. A note in the SAS log tells you what percentage of the specified size was used.
>
> If you omit the HEIGHT= option, a text height specification is searched for in this order:
>
> □ *for a TITLE1 statement:*
>   1 the HTITLE= option in a GOPTIONS statement
>   2 the HTEXT= option in a GOPTIONS statement
>   3 the default value, 2.
>
>   By default, a TITLE1 title is twice the height of all other titles.
>
> □ *for all other TITLE statements and the FOOTNOTE and NOTE statements:*
>   1 the HTEXT= option in a GOPTIONS statement
>   2 the default value, 1.
>
> *Note:* The Java applet and ActiveX control allow you to control the relative height of text with the HEIGHT= option, but not the absolute height in terms of specific units. △

Alias: H=*text-height<units>*

See also: "Controlling Titles and Footnotes with Java and ActiveX Devices in HTML Output" on page 194

Featured in: "Example 1. Ordering Axis Tick Marks with SAS Date Values" on page 294 and "Example 6. Enhancing Titles" on page 307

Restriction: Partially supported by Java and ActiveX

JUSTIFY=LEFT | CENTER | RIGHT
> specifies the alignment of the text string. The default depends on the statement with which you use the JUSTIFY= option:
>
> □ *for a FOOTNOTE statement* the default is CENTER
> □ *for a NOTE statement* the default is LEFT
> □ *for a TITLE statement* the default is CENTER.
>
> All the text strings following JUSTIFY= are treated as a single string and are displayed as one line that is left-, right-, or center-aligned.
>
> You can change the justification within a single line of text. For example, this NOTE statement displays a date on the left side of the output and the page number on the same line on the right:
>
> ```
> note "June 28, 1997" justify=right "Page 3";
> ```
>
> In addition, you can use the JUSTIFY= option to produce multiple lines of text by repeating the JUSTIFY= option with the same value before the text string for each line. Multiple lines of text with the same justification are blocked together. For example, this TITLE statement produces a three-line title with each line right-justified:

```
title justify=right "First Line"
      justify=right "Second Line"
      justify=right "Third Line";
```

You can get the same effect with three TITLE statements, each specifying JUSTIFY=RIGHT. If you produce a block of text by specifying the same justification for multiple text strings, and then change the justification for an additional text string, that text is placed on the same line as the first string specified in the statement.

Note: Using the JUSTIFY= option after one text string and before another can reset some options to their default values. See "Using Options That Can Reset Other Options" on page 293 for details. △

Alias: J=L | C | R

Featured in: "Example 3. Rotating Plot Symbols Through the Color List" on page 299

LANGLE=*degrees*

specifies the angle of the *baseline* of the entire text string(s) with respect to the horizontal. A positive value for *degrees* moves the baseline counterclockwise; a negative value moves it clockwise. By default, LANGLE=0 (horizontal).

Angled titles or footnotes might require more vertical space and consequently can increase the size of the title area or the footnote area, thereby reducing the vertical space in the procedure output area.

Using the BOX= option with angled text does not produce an angled box; the box is sized to accommodate the angled note.

Unlike the ANGLE= option, the LANGLE= option does not reset any other options. Therefore, the LANGLE= option is easier to use because you do not need to repeat options after a text break.

The LANGLE= option has the same effect on the text as the ANGLE= option, except when an angle of 90 degrees or –90 degrees is specified. The result depends on the statement in which you use the option:

□ *With the TITLE statement:*

Figure 14.27 on page 287 shows how LANGLE=90 degrees and LANGLE=–90 degrees positions and rotates titles.

LANGLE=90

angles the title 90 degrees (counterclockwise) so that it reads from bottom to top. The title is centered horizontally and positioned at the top of the picture.

LANGLE=-90

angles the title –90 degrees (clockwise) so that it reads from top to bottom. The title is centered horizontally and positioned at the top of the picture.

Figure 14.27 Positioning Titles with the LANGLE= Option

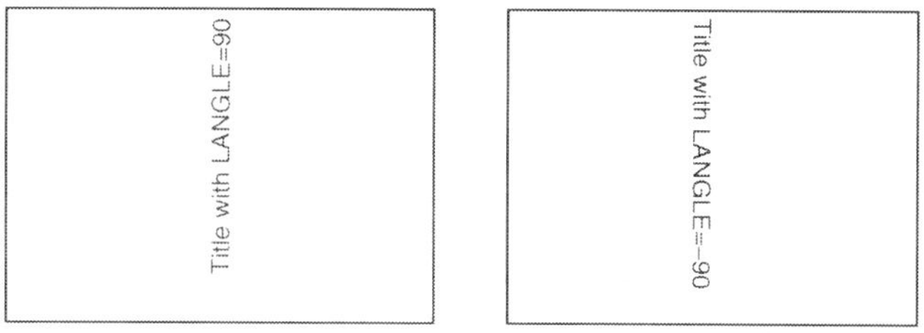

□ *With the FOOTNOTE statement:*

Figure 14.28 on page 288 shows how LANGLE=90 degrees and LANGLE=–90 degrees positions and rotates footnotes.

LANGLE=90
angles the footnote 90 degrees (counterclockwise) so that it reads from bottom to top. The footnote is centered horizontally and positioned as the bottom of the picture.

LANGLE=–90
angles the footnote –90 degrees (clockwise) so that it reads from top to bottom. The footnote is centered horizontally and positioned at the bottom of the picture.

Figure 14.28 Positioning Footnotes with the LANGLE= Option

□ *With the NOTE statement:*

Figure 14.29 on page 288 shows how LANGLE=90 degrees and LANGLE=–90 degrees positions and rotates notes.

LANGLE=90
positions the note at the top of the left edge of the procedure output area, angled 90 degrees (counterclockwise) so that it reads from bottom to top.

LANGLE=–90
positions the note at the top of the left edge of the procedure output area, angled –90 degrees (clockwise) so that it reads from top to bottom.

Figure 14.29 Positioning Notes with the LANGLE= Option

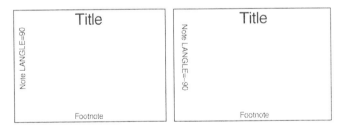

Alias: LA=*degrees*

See also: the option ANGLE= on page 281

Restriction: Not supported by Java and ActiveX

LINK= *"URL"*
specifies a uniform resource locator (URL) that a title or footnote links to.

The text-string that you use to specify the URL can contain occurrences of the variables #BYVAL, #BYVAR, and #BYLINE, as described in text-string on page 290.

Note: If the title or footnote is being output through an ODS markup destination (such as HTML) and the corresponding ODS option NOGTITLE or NOGFOOTNOTE is specified, then the title or footnote is rendered in the body of the HTML file rather than in the graphic itself. Specifying the NOGTITLE or NOGFOOTNOTE options results in increasing the amount of space allowed for the procedure output area, which can result in increasing the size of the graph. Space that would have been used for the title or footnote is devoted instead to the graph. You might need to be aware of this possible difference if you are using annotate or map coordinates. △

See also: "Controlling Where Titles and Footnotes are Rendered" on page 194

LSPACE=*line-space* *<units>*

specifies the amount of spacing *above* lines of note and title text and the amount of spacing *below* lines of footnote text. For notes and titles, the spacing is measured from the capline of the current line to the font minimum of the line above. For footnotes, the spacing is measured from the font minimum of the current line to the capline of the line below. By default, LSPACE=1.

Note: The LSPACE= option can be reset by the ANGLE= or JUSTIFY= option, or by the MOVE= option with absolute coordinates. See "Using Options That Can Reset Other Options" on page 293 for details. △

Alias: LS=*line-space* *<units>*

Restriction: Not supported by Java and ActiveX

MOVE=(*x,y*) *<units>*

positions subsequent text or lines anywhere on the graphics output area using *x* and *y* as absolute or relative coordinates. The following table shows the specifications for absolute and relative coordinates:

Absolute Coordinates	Relative Coordinates
x<units>	±*x<units>*
y<units>	±*y<units>*

Commas between coordinates are optional; you can use blanks instead.

The starting point for lines specified with relative coordinates begins with the end of the most recently drawn text or line in the current statement. If no text or line has been drawn in the current statement, a warning is issued and the relative move is measured from where a zero-length text string would have ended, given the normal placement for the statement. You can mix relative and absolute coordinates.

The MOVE= option overrides a JUSTIFY= option specified for the same text string.

If a NOTE, FOOTNOTE, or TITLE statement uses the MOVE= option to position the text so that the statement does not use its default position, the text of the next NOTE, FOOTNOTE, or TITLE statement occupies the unused position and no blank lines are displayed.

Note: If you specify the MOVE= option with at least one absolute coordinate and if the option follows one text string and precedes another, some options can be reset to their default values. If you specify the GUNIT graphics option, then that unit is the default unit. If you do not specify the GUNIT= graphics option, then the default unit is CELLS. See "Using Options That Can Reset Other Options" on page 293 for details △

Alias: M=(*x,y*) <*units*>

Featured in: "Example 2. Specifying Logarithmic Axes" on page 297 and "Example 6. Enhancing Titles" on page 307

Restriction: Not supported by Java and ActiveX

ROTATE=*degrees*

specifies the angle at which *each character* of text is rotated with respect to the baseline of the text string. The angle is measured from the current text baseline angle, which is specified by the ANGLE= or LANGLE= options. By default, the baseline is horizontal. A positive value for *degrees* rotates the character counterclockwise; a negative value rotates it clockwise. By default, ROTATE=0 (parallel to the baseline).

Figure 14.30 on page 290 shows how characters are positioned when ROTATE=90 is used with the default (horizontal) baseline.

Figure 14.30 Tilting Characters with the ROTATE= Option

Alias: R=*degrees*

See also: the option ANGLE= on page 281

Featured in: "Example 6. Enhancing Titles" on page 307

Restriction: Not supported by Java and ActiveX

text-string(s)

is one or more strings up to 200 characters. You must enclose text strings in single or double quotation marks. The text appears exactly as you type it in the statement, including uppercase and lowercase characters and blanks.

To use single quotation marks or apostrophes within the title, you can either

☐ use a pair of single quotation marks together:

```
footnote 'All''s Well That Ends Well';
```

☐ enclose the text in double quotation marks:

```
footnote "All's Well That Ends Well";
```

Because FOOTNOTE, NOTE, and TITLE statements concatenate all text strings, the strings must contain the correct spacing. With a series of strings, add blanks at the beginning of a text string rather than at the end, as in this example:

```
note color=red "Sales:" color=blue " 2000";
```

With some fonts, you produce certain characters by specifying a hexadecimal value. A trailing **x** identifies a string as a hexadecimal value. For example, this statement* produces the title **Profits Increase £ 3,000**:

---

* This statement assumes you are using a U.S. key map.

```
title font=swiss "Profits Increase " "18'x "3,000";
```

For more information see "Specifying Special Characters Using Character and Hexadecimal Codes" on page 160.

In addition, you can embed one or more of the following in the string:

#BYLINE
: substitutes the entire BY line without leading or trailing blanks for #BYLINE in the text string, and displays the BY line in the footnote, note, or title produced by the statement.

#BYVAL*n* | #BYVAL(*BY-variable-name*)
: substitutes the current value of the specified BY variable for #BYVAL in the text string and displays the value in the footnote, note, or title produced by the statement. Specify the variable with one of these:

  n
  : specifies which variable in the BY statement #BYVAL should use. The value of *n* indicates the position of the variable in the BY statement. For example, #BYVAL2 specifies the second variable in the BY statement.

  BY-variable-name
  : names the BY variable. For example, #BYVAL(YEAR) specifies the BY variable, YEAR. *Variable-name* is not case sensitive.

  Featured in: "Example 7. Using BY-group Processing to Generate a Series of Charts" on page 309 and "Example 9. Combining Graphs and Reports in a Web Page" on page 315

#BYVAR*n* | #BYVAR(*BY-variable-name*)
: substitutes the name of the BY-variable or label associated with the variable (whatever the BY line would normally display) for #BYVAR in the text string and displays the name or label in the footnote, note, or title produced by the statement. Specify the variable with one of these:

  n
  : specifies which variable in the BY statement #BYVAR should use. The value of *n* indicates the position of the variable in the BY statement. For example, #BYVAR2 specifies the second variable in the BY statement.

  BY-variable-name
  : names the BY variable. For example, #BYVAR(SITES) specifies the BY variable, SITES. *Variable-name* is not case sensitive.

  A BY variable name displayed in a title, note, or footnote is always in uppercase. If a label is used, it appears as specified in the LABEL statement.

  For more information , see "Substituting BY Line Values in a Text String" on page 294

UNDERLIN=0...3
: underlines subsequent text. Values of 1, 2 and 3 underline with an increasingly thicker line. UNDERLIN=0 halts underlining for subsequent text.

  Underlines follow the text baseline. If you use an LANGLE= or ANGLE= option for the line of text, the underline is drawn at the same angle as the text. Underlines do not break up to follow rotated characters. See the option ROTATE= on page 290.

  To make the text and the underline the same color, specify a COLOR= option *before* the UNDERLIN= option that precedes the text string. To make the text a different color, specify the COLOR= option *after* the UNDERLIN= option.

  Note: The UNDERLIN= option can be reset by the ANGLE= or JUSTIFY= option, or by the MOVE= option with absolute coordinates. See "Using Options That Can Reset Other Options" on page 293 for details.

Note: The Java applet and ActiveX control underline text when the UNDERLIN= option is specified, but they do not vary the thickness of the line. △

△

Alias: U=

Featured in: "Example 6. Enhancing Titles" on page 307

Restriction: Partially supported by Java and ActiveX

WRAP

wraps the text to a second line if the text does not fit on one line. If the WRAP option is omitted, the text font-size is reduced until the text fits on one line. Wrapping occurs at the last blank before the text meets the end of the window. If there are no blanks in the text string, then there is no wrapping.

Restriction: The WRAP option does not work with the BOX, BLANK, UNDERLINE, and MOVE options.

Using TITLE and FOOTNOTE Statements

You can define TITLE and FOOTNOTE statements anywhere in your SAS program. They are global and remain in effect until you cancel them or until you end your SAS session. All currently defined FOOTNOTE and TITLE statements are automatically displayed.

You can define up to ten TITLE statements and ten FOOTNOTE statements in your SAS session. A TITLE or FOOTNOTE statement without a number is treated as a TITLE1 or FOOTNOTE1 statement. You do not have to start with TITLE1 and you do not have to use sequential statement numbers. Skipping a number in the sequence leaves a blank line.

You can use as many text strings and options as you want, but place the options before the text strings they modify. See "Using Multiple Options" on page 293.

The most recently specified TITLE or FOOTNOTE statement of any number completely replaces any other TITLE or FOOTNOTE statement of that number. In addition, it cancels all TITLE or FOOTNOTE statements of a higher number. For example, if you define TITLE1, TITLE2, and TITLE3, resubmitting the TITLE2 statement cancels TITLE3.

To cancel individual TITLE or FOOTNOTE statements, define a TITLE or FOOTNOTE statement of the same number without options (a null statement):

```
title4;
```

But remember that this cancels all other existing statements of a higher number.

To cancel all current TITLE or FOOTNOTE statements, use the RESET= graphics option in a GOPTIONS statement:

```
goptions reset=footnote;
```

Specifying RESET=GLOBAL or RESET=ALL also cancels all current TITLE and FOOTNOTE statements as well as other settings.

Using the NOTE Statement

NOTE statements are local, not global, and they must be defined within a procedure or RUN-group with which they are used. They remain in effect for the duration of the procedure that includes NOTE statements in any of its RUN-groups or until you end your SAS session. All notes defined in the current RUN group, as well as those defined in previous RUN-groups, are displayed in the output as long as the procedure remains active.

You can use as many text strings and options as you want, but place the options before the text strings they modify. See "Using Multiple Options" on page 293.

Using Multiple Options

In each statement you can use as many text strings and options as you want, but you must place the options before the text strings they modify. Most options affect all text strings that follow them in the same statement, unless the option is explicitly reset to another value. In general, TITLE, FOOTNOTE, and NOTE statement options stay in effect until one of these events occurs:

- The end of the statement is reached.
- A new specification is made for that option.

For example, this statement specifies that one part of the note is red and another part is blue, but the height for all of the text is 4:

```
note height=4 color=red "Red Tide"
    color=blue " Effects on Coastal Fishing";
```

Setting Defaults

You can set default characteristics for titles (including TITLE1 definitions), footnotes, and notes by using the following graphics options in a GOPTIONS statement:

CTITLE=*color*
: sets the default color for all titles, footnotes, and notes; overridden by the COLOR= option in a TITLE, FOOTNOTE, or NOTE statement.

CTEXT=*text-color*
: sets the default color for all text; overridden by the CTITLE= option for titles, footnotes, and notes.

FTITLE=*title-font*
: sets the default font for TITLE1 definitions; overridden by the FONT= option in the TITLE1 statement.

FTEXT=*text-font*
: sets the default font for all text, including the TITLE1 statement if the FTITLE= option is not used; overridden by the FONT= option a TITLE, FOOTNOTE, or NOTE statement.

HTITLE=*height<units>*
: sets the default height for TITLE1 definitions; overridden by the HEIGHT= option in the TITLE1 statement.

HTEXT=*n<units>*
: sets the default height for all text, including the TITLE1 statement if the HTITLE= option is not used; overridden by the HEIGHT= option a TITLE, FOOTNOTE, or NOTE statement.

See Chapter 15, "Graphics Options and Device Parameters Dictionary," on page 327 for a complete description of each option.

Using Options That Can Reset Other Options

The ANGLE=, MOVE=, and JUSTIFY= options affect the position of the text and cause *text breaks*. (To cause a text break, the MOVE= option must have at least one absolute coordinate.) When a statement contains multiple text strings, the resulting text break can cause the following options to reset to their default values:

- BCOLOR=
- BLANK=
- BOX=
- BSPACE=
- LSPACE=
- UNDERLIN=.

Note: The LANGLE= option does not cause a text break. △

If in a TITLE, FOOTNOTE, or NOTE statement, before the first text string, you use an option that can be reset (such as the UNDERLIN= option) and before the second string you use an option that resets it (such as the JUSTIFY= option), the first option does not affect the second string. In order for the first option to affect the second string, repeat the option and position it *after* the resetting option and *before* the text string.

For example, this statement produces a two-line title in which only the first line is underlined:

```
title underlin=2 "Line 1" justify=left "Line 2";
```

To underline Line 2, repeat the UNDERLIN= option *before* the second text string and *after* the JUSTIFY= option:

```
title underlin=2 "Line 1" justify=left
      underlin=2 "Line 2";
```

Substituting BY Line Values in a Text String

To use the #BYVAR and #BYVAL options, insert the option in the text string at the position you want the substitution text to appear. Both #BYVAR and #BYVAL specifications must be followed by a delimiting character, either a space or other nonalphanumeric character, such as the quotation mark that ends the text string. If not, the specification is completely ignored and its text remains intact and is displayed with the rest of the string. To allow a #BYVAR or #BYVAL substitution to be followed immediately by other text, with no delimiter, use a trailing dot (as with macro variables). The trailing dot is not displayed in the resolved text. If you want a period to be displayed as the last character in the resolved text, use two dots after the #BYVAR or #BYVAL substitution.

If you use a #BYVAR or #BYVAL specification for a variable that is not named in the BY statement (such as #BYVAL2 when there is only one BY-variable or #BYVAL(ABC) when ABC is not a BY-variable or does not exist), or if there is no BY statement at all, the substitution for #BYVAR or #BYVAL does not occur. No error or warning message is issued and the option specification is displayed with the rest of the string. The graph continues to display a BY line at the top of the page unless you suppress it by using the NOBYLINE option in an OPTION statement.

For more information, see "BY Statement" on page 216.

Note: This feature is not available in the DATA Step Graphics Interface or in the Annotate facility since BY lines are not created in a DATA step. △

Example 1. Ordering Axis Tick Marks with SAS Date Values

Features:
AXIS statement options:
    LABEL=
    OFFSET=
    ORDER=

FOOTNOTE statement option:
   JUSTIFY=
SYMBOL statement options:
   INTERPOL=
   WIDTH=
GOPTIONS statement options:
   BORDER

Sample library member: GAXTMDV1

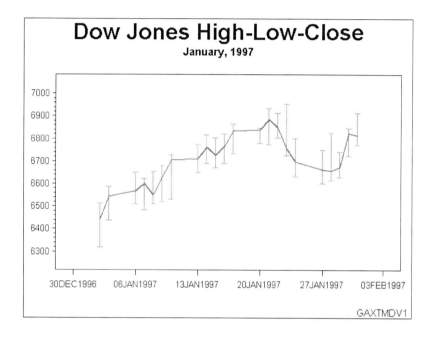

This example uses SAS datetime values with an AXIS statement's ORDER= option to set the major tick marks on the horizontal axis. It adjusts the position of the first and last major tick marks.

The example also uses HILOCTJ interpolation in a SYMBOL statement to join minimum and maximum values.

Set the graphics environment.. BORDER draws a border around the graph.

```
goptions reset=all border;
```

Create the data set. DOWHLC contains the high, low, and close values of the Dow Jones Industrial index for each business day for a month.

```
data dowhlc;
    input date date9. high low close;
    format date date9.;
    datalines;
02JAN1997    6511.38    6318.96    6442.49
03JAN1997    6586.42    6437.10    6544.09
06JAN1997    6647.22    6508.30    6567.18
07JAN1997    6621.82    6481.75    6600.66
08JAN1997    6650.30    6509.84    6549.48
09JAN1997    6677.24    6520.23    6625.67
10JAN1997    6725.35    6530.62    6703.79
13JAN1997    6773.45    6647.99    6709.18
```

```
14JAN1997    6816.17   6689.94   6762.29
15JAN1997    6800.77   6669.93   6726.88
16JAN1997    6818.47   6688.40   6765.37
17JAN1997    6863.88   6732.66   6833.10
20JAN1997    6839.13   6777.30   6843.87
21JAN1997    6934.69   6771.14   6883.90
22JAN1997    6913.14   6801.16   6850.03
23JAN1997    6953.55   6724.19   6755.75
24JAN1997    6798.08   6629.91   6696.48
27JAN1997    6748.82   6598.73   6660.69
28JAN1997    6823.48   6612.20   6656.08
29JAN1997    6673.39   6627.98   6740.74
30JAN1997    6845.03   6719.96   6823.86
31JAN1997    6912.37   6769.99   6813.09
;
```

Prepare the data for a high-low plot. DOWHLC2 generates three records for each date, storing each date's high, low, and close values in variable DOW.

```
data dowhlc2;
   set dowhlc;
   drop high low close;
   dow=high; output;
   dow=low; output;
   dow=close; output;
run;
```

Define titles and footnote. JUSTIFY=RIGHT in the FOOTNOTE statement causes the footnote to be displayed in the bottom right.

```
title1 "Dow Jones High-Low-Close";
title2 "January, 1997";
footnote justify=right "GAXTMDV1 ";
```

Define symbol characteristics. INTERPOL=HILOCTJ specifies that the minimum and maximum values of DOW are joined by a vertical line with a horizontal tick mark at each end. The close values are joined by straight lines. The CV= option controls the color of the symbol. The CI= and WIDTH= options control the color and the thickness of the line that joins the close points.

```
symbol interpol=hiloctj
       cv=red
       ci=blue
       width=2;
```

Define characteristics of the horizontal axis. The ORDER= option uses a SAS date value to set the major tick marks. The OFFSET= option moves the first and last tick marks to make room for the tick mark value.

```
axis1 order=("30DEC1996"d to "03FEB1997"d by week)
      offset=(3,3)
      label=none ;
```

Define characteristics of the vertical axis. LABEL=NONE suppresses the AXIS label.

```
axis2
      label=none
      offset=(2,2);
```

Generate the plot and assign AXIS definitions. The HAXIS= option assigns AXIS1 to the horizontal axis, and the VAXIS= option assigns AXIS2 to the vertical axis.

```
proc gplot data=dowhlc2;
   plot dow*date / haxis=axis1
                   vaxis=axis2;
run;
quit;
```

Example 2. Specifying Logarithmic Axes

Features:

AXIS statement options:
- LABEL=
- LENGTH=
- LOGBASE=
- LOGSTYLE=
- MAJOR=
- MINOR=
- VALUE=

TITLE statement option:
- MOVE=

GOPTIONS statement options:
- GUNIT

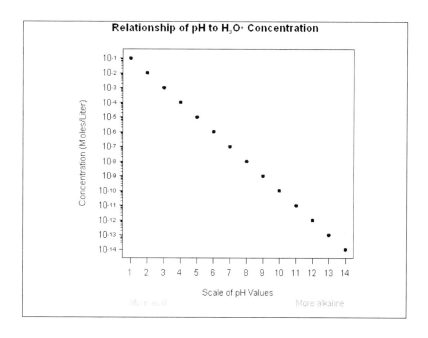

This example illustrates the AXIS statement options LOGBASE= and LOGSTYLE=. The horizontal axis represents pH level. The vertical axis, which represents the concentration of the hydroxide ion expressed as moles per liter, is scaled logarithmically.

In addition, this example shows how the TICK= parameter of the VALUE= option modifies individual tick marks.

The example uses the MOVE= option in a TITLE statement to position the title's subscript and superscript text.

Set the graphics environment. The GUNIT option specifies the default unit of measure to use with height specifications.

```
goptions reset=all  gunit=pct;
```

Create the data set. The CONCENTR option contains the pH values and the concentration amount.

```
data concentr;
   input ph conc;
   datalines;
1  1E-1
2  1E-2
3  1E-3
4  1E-4
5  1E-5
6  1E-6
7  1E-7
8  1E-8
9  1E-9
10 1E-10
11 1E-11
12 1E-12
13 1E-13
14 1E-14
;
run;
```

Define title and footnote. The MOVE= option positions subscript 3 and superscript +. Each new position is relative to the last position specified by the MOVE= option.

```
title1 h=3.7 "Relationship of pH to H"
       move=(-0,-.75) h=2 "3"
       move=(+0,+.75) h=2 "O"
       move=(+0,+.75) h=2 "+"
       move=(-0,-.75) h=2 " Concentration";
```

Define symbol characteristics.

```
symbol value=dot color=black height=2;
```

Define characteristics for horizontal axis. The LABEL= option uses the JUSTIFY= suboption to create a descriptive two-line label that replaces the variable name PH. MINOR=NONE removes all minor tick marks. The LENGTH= option controls the length of the horizontal axis. The OFFSET= option specifies the distance from the first and last major tick marks to the ends of the axis line.

```
axis1 label=(h=3 "Scale of pH Values"
             justify=left color=red h=2 "More acid"
             justify=right color=blue "More alkaline")
      minor=none
      length=60
      offset=(2,2);
```

Define characteristics for vertical axis. LOGBASE=10 scales the vertical axis logarithmically, using a base of 10. Each major tick mark represents a power of 10. LOGSTYLE=EXPAND displays minor tick marks in logarithmic progression. The LABEL= option uses the ANGLE= suboption to place the label parallel to the vertical axis. The VALUE= option displays the major tick mark values as 10 plus an exponent. The HEIGHT= suboption for each TICK= specification affects only the text following it.

```
axis2 logbase=10
     logstyle=expand
      label=(angle=90 h=2 color=black
            "Concentration (Moles/Liter)" )
     value=(tick=1 "10" height=1.2 "-14"
            tick=2 "10" height=1.2 "-13"
            tick=2 "10" height=1.2 "-13"
            tick=3 "10" height=1.2 "-12"
            tick=4 "10" height=1.2 "-11"
            tick=5 "10" height=1.2 "-10"
            tick=6 "10" height=1.2 "-9"
            tick=7 "10" height=1.2 "-8"
            tick=8 "10" height=1.2 "-7"
            tick=9 "10" height=1.2 "-6"
            tick=10 "10" height=1.2 "-5"
            tick=11 "10" height=1.2 "-4"
            tick=12 "10" height=1.2 "-3"
            tick=13 "10" height=1.2 "-2"
            tick=14 "10" height=1.2 "-1")
            offset=(3,3);
```

Generate the plot and assign AXIS definitions. AXIS1 modifies the horizontal axis and AXIS2 modifies the vertical axis. The AUTOHREF and AUTOVREF options draw reference lines at all major tick marks on both axes. The CHREF and CVREF options specify the color for these reference lines.

```
proc gplot data= concentr;
   plot conc*ph / haxis=axis1
                  vaxis=axis2
                  autohref chref=graydd
                  autovref cvref=graydd;
run;
quit;
```

Example 3. Rotating Plot Symbols Through the Color List

Features:

GOPTIONS statement options:

    COLORS=

LEGEND statement options:

    LABEL=

SYMBOL statement options:

    VALUE=

TITLE statement option:

    JUSTIFY=

HEIGHT=

Sample library member: GSYRPSC1

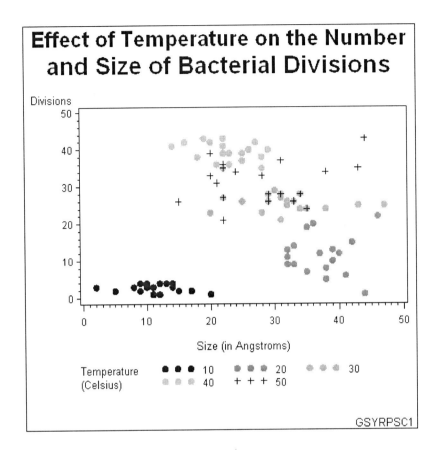

This example specifies a plot symbol on a SYMBOL statement and rotates the symbol through the specified color list. Temperature values in the data are represented by the same plot symbol in a different color. The example also shows how default symbol sequencing provides a default plot symbol if a plot needs more plot symbols than are defined.

It also uses a LEGEND statement to specify a two-line legend label, and to align the label with the legend values.

Set the graphics environment. The COLORS= option specifies the color list. This list is used by the SYMBOL statement.

```
goptions reset=all  border
         colors=(black blue green red)
         ;
```

Create the data set. BACTERIA contains information about the number and size of bacterial divisions at various temperatures.

```
data bacteria;
    input temp div mass life @@;
    datalines;
10  3 10  1   20 22 46  0   30 23 20  9   40 42 16 16   50 33 20  6
10  1 11  2   20 01 44  2   30 21 31 10   40 41 14 12   50 31 21  7
10  4 14  3   20 13 32  4   30 24 34  9   40 43 22 14   50 34 24  2
10  2 09  2   20 12 40  6   30 26 29  8   40 42 20 16   50 26 29  4
10  3 08  3   20 09 33  8   30 24 38 11   40 39 23 18   50 34 38  2
```

```
10 2 09 1   20 08 38 1   30 25 47 14  40 38 18 12  50 43 44 1
10 4 10 3   20 15 42 3   30 29 30 14  40 35 22 14  50 39 20 8
10 3 11 2   20 20 36 5   30 28 31 9   40 40 26 15  50 28 31 0
10 2 15 3   20 19 35 7   30 26 25 11  40 39 25 17  50 26 15 4
10 4 12 3   20 14 33 2   30 27 22 8   40 36 23 12  50 27 22 3
10 4 13 3   20 12 37 4   30 26 33 9   40 42 27 14  50 26 33 5
10 2 17 1   20 10 39 6   30 25 43 13  40 40 29 16  50 35 43 7
10 3 14 1   20 08 38 4   30 28 34 8   40 38 28 14  50 28 34 4
10 1 12 1   20 06 41 2   30 26 32 14  40 36 21 12  50 21 22 2
10 1 11 4   20 09 32 2   30 27 31 8   40 39 22 12  50 37 31 2
10 1 20 2   20 11 32 5   30 25 32 16  40 41 22 15  50 35 22 5
10 4 09 2   20 13 39 1   30 28 29 12  40 43 19 15  50 28 29 1
10 3 02 2   20 09 32 5   30 26 32 9   40 39 22 15  50 36 22 5
10 2 05 3   20 07 35 4   30 24 35 15  40 37 25 14  50 24 35 4
10 3 08 1   20 05 38 6   30 23 28 9   40 35 28 16  50 33 28 6
;
proc sort data=bacteria;
   by temp;
run;
```

Define title and footnote. J= breaks the title into two lines. H= specifies the size of the title.

```
title1 "Effect of Temperature on the Number"
       j=c  h=2 "and Size of Bacterial Divisions";
footnote1  j=r "GSYRPSC1";
```

Define symbol shape. The VALUE= option specifies a dot for the plot symbol. Because no color is specified, the symbol is rotated through the color list. Because the plot needs a fifth symbol, the default plus sign is rotated into the color list to provide that symbol.

```
symbol1   value=dot;
```

Define axis characteristics.

```
axis1 label=("Size (in Angstroms)") ;
axis2 label=("Divisions");
```

Define legend characteristics. The LABEL= option specifies text for the legend label. J=L specifies a new line and left-justifies the second string under the first. The POSITION= option aligns the top label line with the first (and in this case only) value row.

```
legend1 label=(position=(top left)
              "Temperature" j=l "(Celsius)")
          ;
```

Generate the plot.

```
proc gplot data= bacteria;
    plot div*mass=temp / haxis=axis1
                         vaxis=axis2
                         legend=legend1;
run;
quit;
```

Example 4. Creating and Modifying Box Plots

Features:

SYMBOL statement options:

> BWIDTH=
>
> CO=
>
> CV=
>
> HEIGHT=
>
> INTERPOL=
>
> VALUE=

Sample library member: GSYCMBP1

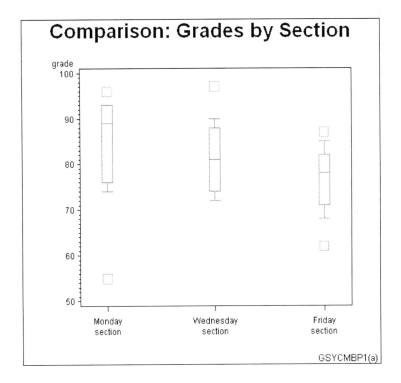

This example shows how to create box plots and how to specify SYMBOL definitions so data outside the box-plot range can be represented with data points. It also shows how to change a box plot's percentile range to see whether the new range encompasses the data.

The first plot in the example uses a SYMBOL definition with INTERPOL=BOXT20 to specify a box plot with whisker tops at the 80th percentile and whisker bottoms at the 20th percentile. Data points that are outside this percentile range are represented with squares.

As illustrated in the following output, the example then changes the SYMBOL definition to INTERPOL=BOXT10, which expands the whisker range to the 90th percentile for tops and the 10th percentile for bottoms. There are no data points outside the new percentile range.

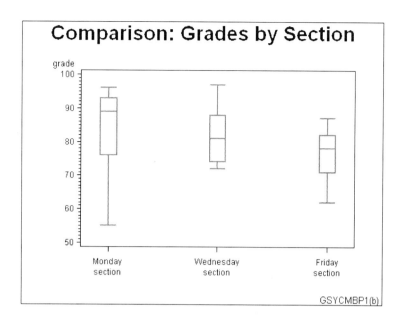

Set the graphics environment.

```
goptions reset=all border;
```

Create the data set. GRADES contains codes to identify each class section, and the grades scored by students in each section.

```
data grades;
    input section $ grade @@;
    datalines;
A 74 A 89 A 91 A 76 A 87 A 93 A 93 A 96 A 55
B 72 B 72 B 84 B 81 B 97 B 78 B 88 B 90 B 74
C 62 C 74 C 71 C 87 C 68 C 78 C 80 C 85 C 82
;
```

Define title and footnote.

```
title1 "Comparison: Grades by Section";
footnote1 j=r  "GSYCMBP1(a) ";
```

Define symbol characteristics. INTERPOL=BOXT20 specifies a box plot with tops and bottoms on its whiskers, and the high and low bounds at the 80th and 20th percentiles. The CO= option colors the boxes and whiskers. The BWIDTH= option affects the width of the boxes. The VALUE= option specifies the plot symbol that marks the data points outside the range of the box plot. The CV= option colors the plot symbols. The HEIGHT= option specifies a symbol size.

```
symbol interpol=boxt20 /* box plot              */
       co=blue         /* box and whisker color */
       bwidth=4        /* box width             */
       value=square    /* plot symbol           */
       cv=red          /* plot symbol color     */
       height=2;       /* symbol height         */
```

Define axis characteristics.

```
axis1 label=none
      value=(t=1 "Monday" j=c "section"
             t=2 "Wednesday" j=c "section"
```

```
                    t=3 "Friday" j=c "section")
         offset=(5,5)
         length=50;
```

Generate the first plot.

```
proc gplot data= grades;
   plot grade*section / haxis=axis1
                       vaxis=50 to 100 by 10;
run;
```

Define the footnote for the second plot.

```
footnote j=r ''GSYCMBP1(b)'';
```

Change symbol characteristics. INTERPOL=BOXT10 changes the high and low bounds to the 90th percentile at the top and the 10th percentile on the bottom. All other symbol characteristics remain unchanged.

```
symbol interpol=boxt10 width=2;
```

Generate the second plot.

```
plot grade*section / haxis=axis1
                    vaxis=50 to 100 by 10;
run;
quit;
```

Example 5. Filling the Area between Plot Lines

Features:

AXIS statement option:
    ORDER=

SYMBOL statement options:
    CO=
    CV=
    INTERPOL=

Sample library member: GSYFAPL1

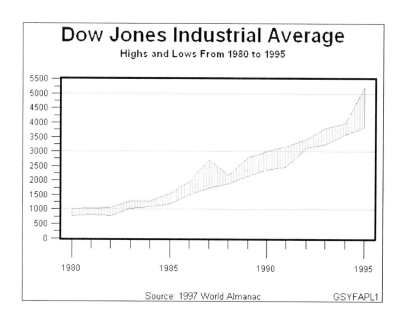

This example shows how to fill the area between two plot lines by concatenating two data sets into one to form a polygon with the data points. It uses a SYMBOL statement to specify a pattern to fill the polygon and to determine the color of the area fill and the outline around the area.

The example plots yearly highs and lows for the Dow Jones Industrial Average. It separates the dependent variables HIGH and LOW to produce an upper plot line and a lower plot line. The dependent variable is named VALUE and the independent variable is named YEAR. When concatenated into one data set, AREA, the data sets form the polygon.

Set the graphics environment.

```
goptions reset=all border;
```

Create the data set. STOCKS contains yearly highs and lows for the Dow Jones Industrial Average, and the dates of the high and low values each year.

```
data stocks;
   input year @7  hdate date9. @17 high
              @26 ldate date9. @36 low;
   format hdate ldate date9.;
   datalines;

1980   20NOV1980 1000.17   21APR1980  759.13
1981   27APR1981 1024.05   25SEP1981  824.01
1982   27DEC1982 1070.55   12AUG1982  776.92
1983   29NOV1983 1287.20   03JAN1983 1027.04
1984   06JAN1984 1286.64   24JUL1984 1086.57
1985   16DEC1985 1553.10   04JAN1985 1184.96
1986   02DEC1986 1955.57   22JAN1986 1502.29
1987   25AUG1987 2722.42   19OCT1987 1738.74
1988   21OCT1988 2183.50   20JAN1988 1879.14
1989   09OCT1989 2791.41   03JAN1989 2144.64
1990   16JUL1990 2999.75   11OCT1990 2365.10
1991   31DEC1991 3168.83   09JAN1991 2470.30
1992   01JUN1992 3413.21   09OCT1992 3136.58
1993   29DEC1993 3794.33   20JAN1993 3241.95
1994   31JAN1994 3978.36   04APR1994 3593.35
```

```
1995   13DEC1995  5216.47   30JAN1995  3832.08
;
```

Restructure the data so that it defines a closed area. Create the temporary data sets HIGH and LOW.

```
data high(keep=year value)
     low(keep=year value);
   set stocks;
   value=high; output high;
   value=low; output low;
run;
```

Reverse order of the observations in LOW.

```
proc sort data=low;
   by descending year;
run;
```

Concatenate HIGH and LOW to create data set AREA.

```
data area;
   set high low;
run;
```

Define titles and footnote.

```
title1 "Dow Jones Industrial Average";
title2  "Highs and Lows From 1980 to 1995";
footnote " Source: 1997 World Almanac"
         j=r "GSYFAPL1 ";
```

Define symbol characteristics. The INTERPOL= option specifies a map/plot pattern to fill the polygon formed by the data points. The pattern consists of medium-density parallel lines at 90 degrees. The CV= option colors the pattern fill. The CO= option colors the outline of the area. (If the CO= option is not used, the outline is the color of the area.)

```
symbol interpol=m3n90
       cv=red
       co=blue;
```

Define axis characteristics. The ORDER= option places the major tick marks at 5-year intervals.

```
axis1 order=(1980 to 1995 by 5)
      label=none
      major=(height=2)
      minor=(number=4 height=1)
      offset=(2,2)
      width=3;
axis2 order=(0 to 5500 by 500)
      label=none
      major=(height=1.5)  offset=(0,0)
      minor=(number=1 height=1);
```

Generate the plot using data set AREA.

```
proc gplot data=area;
   plot value*year / haxis=axis1
                    vaxis=axis2
```

```
                        vref=(1000 3000 5000);
run;
quit;
```

Example 6. Enhancing Titles

Features:

GOPTIONS statement options:
    BORDER
  TITLE statement options:
    BCOLOR=
    BLANK=
    BOX=
    COLOR=
    FONT=
    HEIGHT=
    MOVE=
    ROTATE=
    UNDERLIN=

Sample library member: GTIENTI1

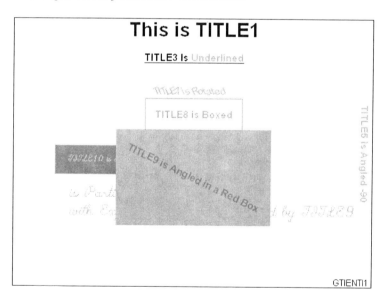

This example illustrates some ways you can format title text. The same options can be used to format footnotes.

Set the graphics environment. BORDER draws a border around the graph.

```
goptions reset=all border;
```

Define title1. TITLE1 uses the default font and height defined in the default style. The HEIGHT= option sets the height of the text.

```
title1 "This is TITLE1" height=4;
```

Define TITLE3. The UNDERLIN= option underlines both text strings.

```
title3 underlin=1
      "TITLE3 Is"
      color=red
      " Underlined";
```

Define TITLE5. The ANGLE= option tilts the line of text clockwise 90 degrees and places it at the right edge of the output.

```
title5 color=red
      angle=-90
      "TITLE5 is Angled -90";
```

Define TITLE7. The ROTATE= option rotates each character in the text string at the specified angle. The HEIGHT= option sets the height of the text.

```
title7 height=4
      color=red
      rotate=25
      "TITLE7 is Rotated";
```

Define TITLE8. The BOX= option draws a green box around the text.

```
title8 color=green
      box=1
      "TITLE8 is Boxed";
```

Define TITLE9. The BLANK= option prevents the boxed title from being overwritten by TITLE10. The first COLOR= option specifies the color of the box border, and the BCOLOR= option specifies the color of the box background. The second COLOR= option specifies the text color.

```
title9 color=red
      box=3
      blank=yes
      bcolor=red
      color=blue
      move=(70,20)
      angle=-25
      "TITLE9 is Angled in a Red Box";
```

Define TITLE10. In this statement, the BOX= option draws a box around the first text string. The BOX= option is turned off by the MOVE= option that uses absolute coordinates and causes a text break.

```
title10 color=red
       box=1
       bcolor=blue
       move=(60,20)
       font=script
       "TITLE10 is in Script and "
       move=(60,15)
       height=2
       "is Partially Boxed, Positioned"
       move=(60,10)
            height=2
       "with Explicit Moves, and Overlaid by TITLE9"
            ;
```

Define footnote.

```
footnote justify=right "GTIENTI1   ";
```

Display titles and footnote. All existing titles and footnotes are automatically displayed by the procedure.

```
proc gslide;
run;
quit;
```

Example 7. Using BY-group Processing to Generate a Series of Charts

Features:

AXIS statement options:
- LABEL=
- MAJOR=
- MINOR=
- NOPLANE
- ORDER=
- STYLE=
- VALUE=

BY statement

OPTIONS statement option:
- NOBYLINE

PATTERN statement option:
- COLOR=

TITLE statement:
- #BYVAL

Sample library member: GBYGMSC1

This example uses a BY statement with the GCHART procedure to produce a separate three-dimensional vertical bar chart for each value of the BY variable TYPE. The three charts, which are shown in Display 14.1 on page 312, Display 14.2 on page 312, and Display 14.3 on page 313 following the code, show leading grain producers for 1995 and 1996.

The program suppresses the default BY lines and instead uses #BYVAL in the TITLE statement text string to include the BY variable value in the title for each chart.

The AXIS1 statement that is assigned to the vertical (response) axis is automatically applied to all three graphs generated by the BY statement. This AXIS statement removes all the elements of the response axis except the label. The same AXIS statement also includes an ORDER= option. Because this option is applied to all the graphs, it ensures that they all use the same scale of response values.

Because no subgroups are specified and the PATTERNID= option is omitted, the color specified in the single PATTERN statement is used by all the bars.

Set the graphics environment.

```
goptions reset=all border;
```

Create the data set GRAINLDR. GRAINLDR contains data about grain production in five countries for 1995 and 1996. The quantities in AMOUNT are in thousands of metric tons. MEGTONS converts these quantities to millions of metric tons.

```
data grainldr;
   length country $ 3 type $ 5;
   input year country $ type $ amount;
   megtons=amount/1000;
   datalines;
1995 BRZ  Wheat    1516
1995 BRZ  Rice     11236
1995 BRZ  Corn     36276
1995 CHN  Wheat    102207
1995 CHN  Rice     185226
1995 CHN  Corn     112331
1995 INS  Wheat    .
1995 INS  Rice     49860
1995 INS  Corn     8223
1995 USA  Wheat    59494
1995 USA  Rice     7888
1995 USA  Corn     187300
1996 BRZ  Wheat    3302
1996 BRZ  Rice     10035
1996 BRZ  Corn     31975
1996 IND  Wheat    62620
1996 IND  Rice     120012
1996 IND  Corn     8660
1996 USA  Wheat    62099
1996 USA  Rice     7771
;
```

Create a format for the values of COUNTRY.

```
proc format;
   value $country "BRZ" = "Brazil"
                  "CHN" = "China"
                  "IND" = "India"
                  "INS" = "Indonesia"
                  "USA" = "United States";
run;
```

Suppress the default BY line and define a title that includes the BY-value. #BYVAL inserts the value of the BY variable COUNTRY into the title of each report.

```
options nobyline;
title1 "Leading #byval(type) Producers"
       j=c "1995 and 1996";
footnote1 j=r  "GBYGMSC1 ";
```

Specify a color for the bars.

```
pattern1 color=green;
```

Define the axis characteristics for the response axes. The ORDER= option specifies the range of values for the response axes. ANGLE=90 in the LABEL= option rotates the label 90 degrees. All the other options remove axis elements. The MAJOR=, MINOR=, and VALUE= options remove the tick marks and values. STYLE=0 removes the line. The NOPLANE option removes the three-dimensional plane.

```
axis1 order=(0 to 550 by 100)
      label=(angle=90 "Millions of Metric Tons")
      major=none
```

```
        minor=none
        value=none
        style=0
        noplane;
```

Define midpoint axis characteristics. The SPLIT= option defines the character that causes an automatic line break in the axis values.

```
axis2 label=none
      split=" ";
```

Sort data according to values of BY variable. The data must be sorted before running PROC GCHART with the BY statement.

```
proc sort data=grainldr out=temp;
    by type;
run;
```

Generate the vertical bar charts using a BY statement. The BY statement produces a chart for each value of SITE. The FORMAT statement assigns the $COUNTRY. format to the chart variable. Assigning AXIS1 to the RAXIS= option causes all three charts to have the same response axis.

```
proc gchart data=temp (where=(megtons gt 31));
   by type;
   format country $country.;
   vbar3d country / sumvar=megtons
                    outside=sum
                    descending
                    shape=hexagon
                    width=8
                    coutline=black
                    cframe=grayaa
                    maxis=axis2
                    raxis=axis1 name="GBYGMSC1";
run;
quit;
```

Display 14.1 Output for BY Value Corn

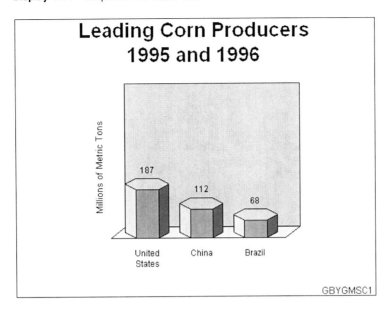

Display 14.2 Output for BY Value Rice

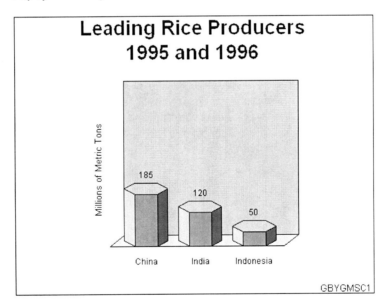

Display 14.3 Output for BY Value Wheat

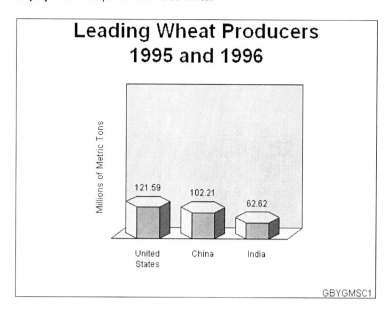

Example 8. Creating a Simple Web Page with the ODS HTML Statement

Features:
ODS HTML statement options:
  BODY=
  CLOSE
GOPTIONS statement options:
  RESET=
LEGEND statement options:
  ACROSS=
  LABEL=

Sample library member: GONCSWB1

Display 14.4 Displaying a Map in a Web Page

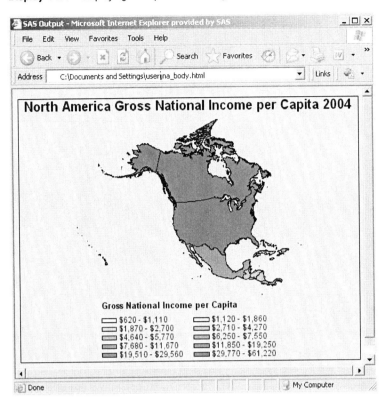

This example illustrates the simplest way to use the ODS HTML statement to create an HTML file and a GIF file that you can display in a Web browser. It generates one body file that displays one piece of SAS/GRAPH output—a map of average per capita income.

This example also illustrates default pattern behavior with maps and explicit placement of the legend on the graph. It shows how the default solid map pattern uses different shades of the default style color to differentiate between countries.

And it shows how to use a LEGEND statement to arrange and position a legend so it fits well with the graph's layout.

Close the ODS Listing destination for procedure output, and set the graphics environment. To conserve system resources, ODS LISTING CLOSE closes the Listing destination for procedure output. Thus, the graphics output is not displayed in the GRAPH window, although it is written to the graphics catalog and to the GIF files.

```
ods listing close;
goptions reset=all;
```

Open the ODS HTML destination. The BODY= option names the file for storing HTML output.

```
ods html body="na_body.html"
     ;
```

Define title for the map. By default, any defined title is included in the graphics output (GIF file).

```
title "North America Gross National Income per Capita 2004";
```

Define legend characteristics. The ACROSS= option defines the number of columns in the legend. The LABEL= option specifies a legend label and left-justifies it above the legend values.

```
legend across=2
       origin=(8,5)
       mode=share
       label=(position=top
              justify=left
              "Gross National Income per Capita")
;
```

Generate the prism map. Because the NAME= option is omitted, SAS/GRAPH assigns the default name GMAP to the GRSEG entry in the graphics catalog. This is the name that is assigned to the GIF file created by the ODS HTML statement.

```
proc gmap map=maps.namerica data=sashelp.demographics;
     id cont id;
   format gni dollar10.0;
     choro gni / levels=10 legend=legend1;
run;
quit;
```

Close the ODS HTML destination, and open the ODS Listing destination. You must close the HTML destination before you can view the output with a browser. ODS LISTING opens the Listing destination so that the destination is again available for displaying output during this SAS session.

```
ods html close;
ods listing;
```

Example 9. Combining Graphs and Reports in a Web Page

Features:

AXIS statement options:
    LENGTH=
    VALUE=

 BY statement

GOPTIONS statement options:
    BORDER
    DEVICE=
    TRANSPARENCY

ODS HTML statement options:
    BODY=
    CONTENTS=
    FRAME=
    PATH=
    NOGTITLE

OPTIONS statement option:
    NOBYLINE

TITLE statement option:

#BYVAL

Sample library member: GONCGRW1

This example generates several graphs of sales data that can be accessed from a single Web page. The graphs are two bar charts of summary sales data and three pie charts that break the data down by site. Each bar chart and an accompanying report is stored in a separate body file.

The three pie charts are generated with BY-group processing and are stored in one body file. The program suppresses the default BY lines and instead includes the BY variable value in the title for each chart. The SAS/GRAPH titles are displayed in the HTML output instead of in the graphics output.

The Web page contains two frames, one that displays a Table of Contents for all the graphs, and one that serves as the display area. Links to each piece of output appear in the table of contents, which is displayed in the left frame. Initially the frame file displays the first body file, which contains a bar chart and a report, as shown in the following figure.

Display 14.5 Browser View of Bar Chart and Quarterly Sales Report

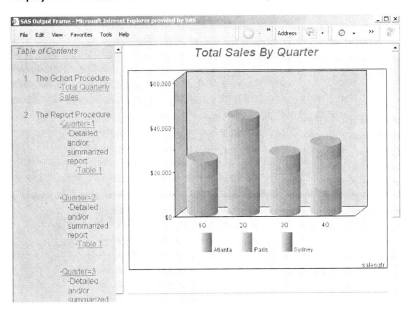

Notice that the chart title is displayed outside the graph as part of the HTML file.

Select the link to *Total Department Sales* to display the second bar chart, as shown in the following figure.

Display 14.6 Browser View of Bar Chart and Department Sales Report

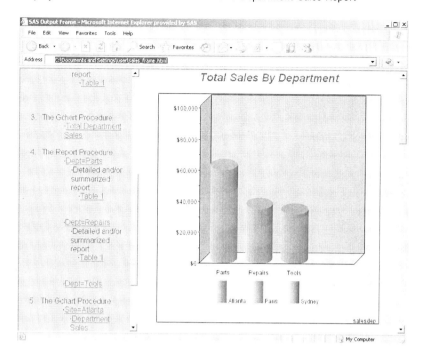

Selecting any link for *Department Sales* displays the corresponding pie chart as shown in the following figure.

Display 14.7 Browser View of Pie Charts of Site Sales

Because the pie charts are stored in one file, you can easily see all three by scrolling through the file.

Additional features include AXIS statements that specify the same length for both midpoint axes, so that the bar charts are the same width even though they have a different number of bars.

Close the ODS Listing destination for procedure output, and set the graphics environment. To conserve system resources, ODS LISTING CLOSE closes the Listing destination for procedure output. DEVICE=GIF causes the ODS HTML statement to generate the graphics output as GIF files. The TRANSPARENCY option causes the graphics output to use the Web-page background as the background of the graph. The

BORDER option is used so that the border around the graphics output area is compatible with the borders that are created for nongraphics output.

```
ods listing close;

goptions reset=all border   ;
```

Create the data set TOTALS. The data set contains quarterly sales data for three manufacturing sites for one year.

```
data totals;
   length Dept $ 7 Site $ 8;
   input Dept Site Quarter Sales;
   datalines;
Repairs Sydney  1  5592.82
Repairs Atlanta 1  9210.21
Tools   Sydney  1  1775.74
Tools   Atlanta 1  2424.19
Tools   Paris   1  5914.25
Parts   Atlanta 2  11595.07
Parts   Paris   2  9558.29
Repairs Sydney  2  5505.31
Repairs Paris   2  7538.56
Tools   Atlanta 2  1903.99
Tools   Paris   2  7868.34
Parts   Sydney  3  8437.96
Parts   Paris   3  6789.85
Tools   Atlanta 3  3048.52
Tools   Paris   3  9017.96
Parts   Sydney  4  6065.57
Parts   Atlanta 4  9388.51
Parts   Paris   4  8509.08
Repairs Atlanta 4  2088.30
Repairs Paris   4  5530.37
;
```

Open the ODS HTML destination. The FRAME= option names the HTML file that integrates the contents and body files. The CONTENTS= option names the HTML file that contains the table of contents to the HTML procedure output. The BODY= option names the file for storing the HTML output. The contents file links to each of the body files written to the HTML destination. The NOGTITLE option suppresses the graphics titles from the SAS/GRAPH output and displays them through the HTML page.

```
ods html frame="sales_frame.html"
   contents="sales_contents.html"
    body="sales_body1.html"
     nogtitle;
```

Define title and footnote.

```
title1 "Total Sales By Quarter";
footnote j=r   "salesqtr ";
```

Define axis characteristics for the first bar chart. In AXIS2, the LENGTH= option specifies the length of the midpoint axis.

```
axis1 order=(0 to 60000 by 20000)
      minor=(number=1)
```

```
          label=none;
   axis2 label=none length=70pct
         value=("1Q" "2Q" "3Q" "4Q");
```

Suppress the legend label and define the size of the legend values.

```
   legend1 label=none shape=bar(4,4);
```

Generate the vertical bar chart of quarterly sales. The NAME= option specifies the name of the catalog entry.

```
   proc gchart data=totals;
      format sales dollar8.;
      vbar3d quarter / discrete
                       sumvar=sales
                       shape=cylinder
                       subgroup=site
                       cframe=grayaa
                       caxis=black
                       width=12
                       space=4
                       legend=legend1
                       maxis=axis2
                       raxis=axis1
                       des="Total Quarterly Sales"
                       name="salesqtr";
   run;
   quit;
```

Sort the data set for the report of quarterly sales. The data must be sorted in order of the BY variable before running PROC REPORT with BY-group processing.

```
   proc sort data=totals out=qtrsort;
      by quarter site;
   run;
```

Reset the footnote and suppress the BY line. We suppress the BY line because otherwise #BYVAL inserts the value of the BY variable into the title of each report.

```
   footnote1;
   options nobyline;
```

Generate a report of quarterly sales. Because the HTML body file that references the GCHART procedure output is still open, the report is stored in that file. The chart and report are shown in Display 14.5 on page 316.

```
   title1 "Sales for Quarter #byval(quarter)";
   proc report data=qtrsort nowindows;
     by quarter;
     column quarter site dept sales;
     define quarter / noprint group;
     define site    / display group;
     define dept    / display group;
     define sales   / display sum format=dollar8.;
     compute after quarter;
             site="Total";
```

```
         endcomp;
      break after site    / summarize style=rowheader;
      break after quarter / summarize style=rowheader;
   run;
```

Open a new body file for the second bar chart and report. Assigning a new body file closes SALES_BODY1.HTML. The contents and frame files, which remain open, contains links to all body files.

```
   ods html body="sales_body2.html";
```

Define title and footnote for second bar chart.

```
   title1 "Total Sales By Department";
   footnote1 j=r  "salesdep ";
```

Define axis characteristics. These AXIS statements replace the ones defined earlier. As before, the LENGTH= option defines the length of the midpoint axis.

```
   axis1 label=none
         minor=(number=1);
         order=(0 to 100000 by 20000)
   axis2 label=none length=70pct;
```

Generate the vertical bar chart of departmental sales.

```
   proc gchart data=totals;
      format sales dollar8.;
      vbar3d dept / shape=cylinder
                    subgroup=site
                    cframe=grayaa
                    width=12
                    space=4
                    sumvar=sales
                    legend=legend1
                    maxis=axis2
                    raxis=axis1
                    caxis=black
                    des="Total Department Sales"
                    name="salesdep";
   run;
   quit;
```

Sort the data set for the report of department sales. The data must be sorted in order of the BY variable before running PROC REPORT with BY-group processing.

```
   proc sort data=totals out=deptsort;
      by dept site;
   run;
```

Reset the footnote, define a report title, and generate the report of department sales. #BYVAL inserts the value of the BY variable into the title of each report. The chart and report are shown in Display 14.5 on page 316.

```
   footnote1;
   title1 "Sales for #byval(dept)";
   proc report data=deptsort nowindows;
      by dept;
      column dept site quarter sales;
      define dept    / noprint group;
```

```
       define site    / display group;
       define quarter / display group;
       define sales   / display sum format=dollar8.;
       compute after dept;
              site="Total";
       endcomp;
       break after site / summarize style=rowheader;
       break after dept / summarize style=rowheader;
   run;
```

Open a new body file for the pie charts. Assigning a new file as the body file closes SALES_BODY2.HTML. The contents and frame files remain open. GTITLE displays the titles in the graph.

```
   ods html body="sales_body3.html"  gtitle;
```

Sort data set in order of the BY variable before running the GCHART procedure with BY-group processing.

```
   proc sort data=totals out=sitesort;
      by site;
   run;
```

Define title and footnote. #BYVAL inserts the value of the BY variable SITE into the title for each output.

```
   title "Departmental Sales for #byval(site)";
   footnote j=r "salespie ";
```

Generate a pie chart for each site. All the procedure output is stored in one body file. Because BY-group processing generates multiple graphs from one PIE3D statement, the name assigned by the NAME= option is incremented to provide a unique name for each piece of output.

```
   proc gchart data=sitesort;
        format sales dollar8.;
        by site;
        pie3d dept / noheading
                     coutline=black
                     sumvar=sales
                     des="Department Sales"
                     name="salespie";
   run;
   quit;
```

Close the ODS HTML destination, and open the ODS Listing destination.

```
   ods html close;
   ods listing;
```

Example 10. Creating a Bar Chart with Drill-Down Functionality for the Web

Features:
GOPTIONS statement option:
    RESET=
    TRANSPARENCY=

DEVICE=

ODS HTML statement options:

BODY=

NOGTITLE

PATH=

Sample library member: GONDDCW1

This example shows you how to create a drill-down graph in which the user can select an area of the graph in order to display additional information about the data. The program creates one vertical bar chart of total sales for each site and three reports that break down the sales figures for each site by department and quarter. The following figure shows the bar chart of sales.

Display 14.8 Vertical Bar Chart of Total Sales

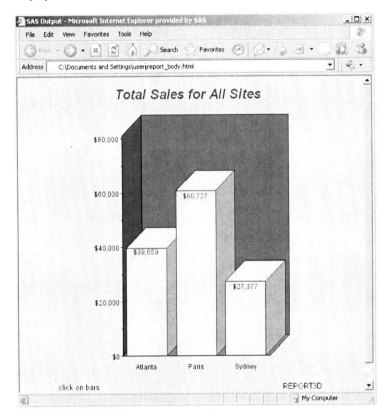

Display 14.9 on page 323 shows the PROC REPORT output that appears when you click on the bar for Atlanta.

Display 14.9 PROC REPORT Output Displayed in a Web Browser

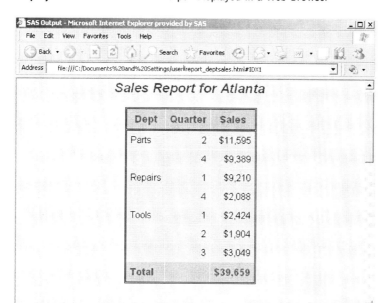

For additional information about this program, see "Details" on page 325.

Close the ODS Listing destination for procedure output, and set the graphics environment. To conserve system resources, ODS LISTING CLOSE closes the Listing destination for procedure output. In the GOPTIONS statement, DEVICE=GIF causes the ODS HTML statement to generate the graphics output as GIF files. The TRANSPARENCY option causes the graphics output to use the Web-page background as the background of the graph.

```
ods listing close;
goptions reset=all device=gif transparency noborder;
```

Add the HTML variable to TOTALS and create the NEWTOTAL data set. The HTML variable SITEDRILL contains the targets for the values of the variable SITE. Each HREF value specifies the HTML body file and the name of the anchor within the body file that identifies the target graph.

```
data newtotal;
   set totals;
   length sitedrill $40;
if site="Atlanta" then
   sitedrill="HREF='report_deptsales.html#IDX1'";

else if site="Paris" then
   sitedrill="HREF='report_deptsales.html#IDX2'";

if site="Sydney" then
   sitedrill="HREF='report_deptsales.html#IDX3'";
run;
```

Open the ODS HTML destination. The BODY= option names the file for storing HTML output. The NOGTITLE option suppresses the graph titles from the SAS/GRAPH output and displays them in the HTML.

```
ods html
    body="report_body.html"
    nogtitle;
```

Define title and footnote.

```
title1 "Total Sales for All Sites";
footnote1  j=l "click on bars" j=r "REPORT3D ";
```

Assign a pattern color for the bars. Each bar in the graph uses the same PATTERN definition.

```
pattern color=cyan;
```

Define axis characteristics. The VBAR3D statement assigns AXIS1 to the response axis and AXIS2 to the midpoint axis.

```
axis1 order=(0 to 80000 by 20000)
      minor=(number=1)
      label=none;
axis2 label=none offset=(9,9);
```

Generate the vertical bar chart of total sales for each site. The HTML= option specifies SITEDRILL as the variable that contains the name of the target. Specifying the HTML= option causes SAS/GRAPH to add an image map to the HTML body file. The NAME= option specifies the name of the catalog entry.

```
proc gchart data=newtotal;
    format sales dollar8.;
    vbar3d site / discrete
                  width=15
                  sumvar=sales
                  inside=sum
                  html=sitedrill
                  coutline=black
                  cframe=blue
                  maxis=axis2
                  raxis=axis1
                  name="report3d ";
run;
quit;
```

Open the file for the PROC REPORT output. Assigning a new body file closes REPORT_BODY.HTML.

```
ods html body="report_deptsales.html" ;
```

Sort the data set NEWTOTAL. The data must be sorted in order of the BY variable before running PROC REPORT with BY-group processing.

```
proc sort data=newtotal;
    by site dept quarter;
run;
quit;
```

Clear the footnote.

```
goptions reset=footnote1;
```

Suppress the default BY line and define a title that includes the BY-value. #BYVAL inserts the value of the BY variable SITE into the title of each report.

```
options nobyline;
title1 "Sales Report for #byval(site)";
```

Print a report of departmental sales for each site.

```
proc report data=newtotal nowindows;
   by site;
   column site dept quarter sales;
   define site    / noprint group;
   define dept    / display group;
   define quarter / display group;
   define sales   / display sum format=dollar8.;
   compute after site;
         dept="Total";
   endcomp;
   break after site / summarize style=rowheader page;
run;
quit;
```

Close the ODS HTML destination, and open the ODS Listing destination.

```
ods html close;
ods listing;
```

Details

This section provides additional information about the pieces of this program and how they work together to generate SAS/GRAPH output with drill-down functionality. It describes

- how an HREF value is built

- how the HTML= option creates an image map in the HTML file

- how the HTML file references the SAS/GRAPH output.

Building an HREF value

In the DATA step, the variable SITEDRILL is assigned a string that defines the link target for a data value. For example,

```
if site="Atlanta" then
   sitedrill="HREF='report_deptsales.html#IDX1'";
```

The link target is specified by the HTML HREF attribute. The HREF value tells the Web page where to link to when a user selects the region associated with the value **Atlanta**.

For example, clicking on the first bar in the chart links to the target defined by **report_deptsales.html#IDX1**. This target consists of a filename and an anchor. The file, **report_deptsales.html**, is generated by the PROC REPORT step. **IDX1** is the anchor that identifies the section of the file that contains the report for the first BY group, **Atlanta**.

Because anchor names increment, in order to assign them accurately you must know how many pieces of output your program generates and in what order. For example, this table lists in order the pieces of output generated by this example and their default anchor names:

Procedure	Output	Anchor name
GCHART	report3d.gif	IDX
REPORT	Atlanta report	IDX1
REPORT	Paris report	IDX2
REPORT	Sydney report	IDX3

Creating an image map

The HTML= option in the GCHART procedure is assigned the variable with the target information – in this case, SITEDRILL.

```
html=sitedrill
```

This option causes SAS/GRAPH to generate in the HTML body file the MAP and AREA elements that compose the image map. It loads the HREF attribute value from SITEDRILL into the AREA element. This image map, which is named **gqcke00k_map**, is stored in **report_body.html** (ODS generates unique map names each time you run the program, so the next time this program runs, the map name will be different):

```
<MAP NAME="gqcke00k_map">
   <AREA SHAPE="POLY"
      HREF="report_deptsales.html#IDX3"
      COORDS="423,409,423,242,510,242,510,409" >
   <AREA SHAPE="POLY"
      HREF="report_deptsales.html#IDX2"
      COORDS="314,409,314,139,401,139,401,409" >
   <AREA SHAPE="POLY"
      HREF="report_deptsales.html#IDX1"
      COORDS="205,409,205,199,292,199,292,409" >
<
/MAP>
```

The AREA element defines the regions within the graph that you can select to link to other locations. It includes attributes that define the shape of the region (the SHAPE= option) and position of the region (the COORDS= option) as well as the link target (the HREF= option).

The value assigned to the HREF= attribute is contained in the variable assigned to the HTML= option, in this case SITEDRILL.

Referencing SAS/GRAPH Output

In the GOPTIONS statement, DEVICE=GIF causes SAS/GRAPH to create GIF files from the SAS/GRAPH output. It also adds to the open body file an IMG element that points to the GIF file. In this case, SAS/GRAPH adds the following IMG element to **report_body.html**:

```
<IMG SRC="report3d.gif" USEMAP="#gqcke00k_map">
```

The IMG element tells the Web page to get the image from the file **report3d.gif**. It also tells the Web page to use the image map **#report3d_map** to define the hotspots of the bar chart.

CHAPTER

15

Graphics Options and Device Parameters Dictionary

Introduction **327**
Specifying Graphics Options and Device Parameters **327**
    Specifying Units of Measurement **328**
Dictionary of Graphics Options and Device Parameters **328**

Introduction

This chapter provides a detailed description of all of the graphics options and device parameters used with SAS/GRAPH software. These include
- all graphics options used by the GOPTIONS statement
- all device parameters that can be specified as options in the ADD and MODIFY statements in the GDEVICE procedure
- all device parameters that appear as fields in the GDEVICE windows.

The descriptions provide the syntax, defaults, and required information for each option and parameter.

The graphics options and device parameters are intermixed and listed alphabetically. When the graphics option and device parameter have the same name, they are discussed in the same dictionary entry and the description uses only that name and does not distinguish between the option and the parameter except where the distinction is necessary.

For a list of all the graphics options, see "GOPTIONS Statement" on page 220. For a list of all the device parameters, see "ADD Statement" on page 1129.

If the syntax for the graphics option and the device parameter is different, both forms are shown. If the syntax is the same, one form is shown.

Specifying Graphics Options and Device Parameters

Use a GOPTIONS statement to specify the graphics options. Some graphics options can also be specified in an OPTIONS statement. Use the GDEVICE procedure to specify the device parameters. (See "GOPTIONS Statement" on page 220 and Chapter 38, "The GDEVICE Procedure," on page 1125 for details.)

Note: The syntax for device parameters is the syntax for specifying parameters when using the GDEVICE procedure statements. With the GDEVICE windows, simply enter values into fields in the windows. △

Note: The values that you specify for any option or parameter must be valid for the device. If you specify a value that exceeds the device's capabilities, SAS/GRAPH software reverts to values that can be used with the device. △

Specifying Units of Measurement

When the syntax of an option includes *units*, use one of these unless the syntax specifies otherwise:

CELLS	character cells
CM	centimeters
IN	inches
PCT	percentage of the graphics output area
PT	points (there are approximately 72 points in an inch).

If you omit *units*, a unit specification is searched for in this order:
1. the value of GUNIT= in a GOPTIONS statement
2. the default unit, CELLS.

Dictionary of Graphics Options and Device Parameters

ACCESSIBLE

Generates descriptive text and summary statistics representing your graphics output.

Used in: GOPTIONS statement

Default: NOACCESSIBLE

Restriction: Only supported by JAVA and ActiveX when used with the ODS HTML output destination.

Syntax

ACCESSIBLE | NOACCESSIBLE

ACCESSIBLE
    enables you to comply with section 508 of the Rehabilitation Acts and meet usability requirements for disabled users. Specifying the ACCESSIBLE option, when used with the ODS HTML statement, generates descriptive text and data for your graphs. SAS/GRAPH writes accessibility information to the graph's output HTML file, and creates a left-justified footnote that provides a link to the information.

    The information and the link are not visible in the output HTML, however both are detected by accessibility aids, such as screen readers. You can also access the information by pressing the tab key and enter. The information will be displayed once you press enter on the link in the footnote. The information will also display if you move your mouse over the location of the left-justified footnote, and click the link when the mouse pointer shape changes.

Figure 15.1 Accessible

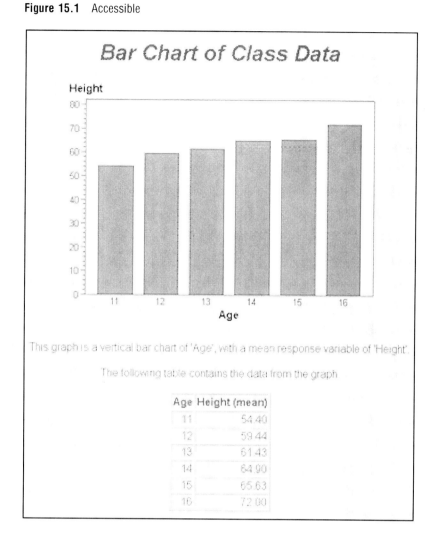

NOACCESSIBLE
    toggles off the ACCESSIBLE option.

Figure 15.2 Noaccessible

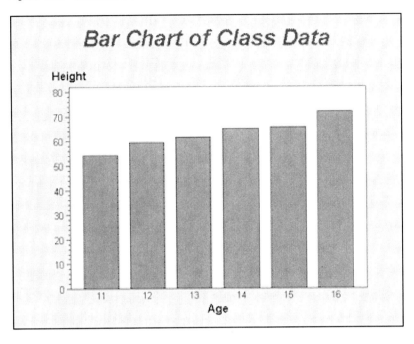

ADMGDF

Specifies whether to write an ADMGDF or GDF file when the GSFNAME= and GSFMODE= graphics options are used with a GDDM device driver.

Used in: GOPTIONS statement
Default: NOADMGDF
Restriction: GDDM device drivers on IBM mainframe systems only

Syntax

ADMGDF | NOADMGDF

ADMGDF
  instructs the GDDM device driver to write out an ADMGDF file.

NOADMGDF
  instructs the GDDM device driver to write out a GDF file.

ALTDESC

Specifies whether to write the DESCRIPTION= statement text to the ALT= text in an HTML file.

Used in: GOPTIONS statement

Default: ALTDESC

Restriction: Only supported when used with the HTML output destination and the DESCRIPTION= option.

Syntax

ALTDESC | NOALTDESC

ALTDESC
With ODS HTML output, by default the entire output has an HTML ALT tag that specifies which procedure was used, and which variables were plotted. Or, if you have specified text using the DESCRIPTION= option, then that value is used for the HTML ALT tag rather than the default ALT tag (many users add a textual description of the graph using this technique, to help the vision-impaired, and to help meet 508-compliance).

If you prefer not to have an ALT tag for the entire graph, you can suppress it by specifying DESCRIPTION=" " (which might be more convenient on a graph by graph basis) or by using GOPTIONS NOALTDESC (which might be more convenient for turning them off for all graphs, such as putting this in your AUTOEXEC.BAT).

NOALTDESC
toggles off the ALTDESC option.

ASPECT

Sets the aspect ratio for graphics elements.

Used in: GOPTIONS statement GDEVICE procedure GDEVICE Detail window
Default: device–dependent
Restriction: not supported by Java or ActiveX

Syntax

ASPECT=*scaling-factor*

scaling-factor
is a non-negative integer or real number that determines the ratio of width to height for graphics elements. If you specify ASPECT=1, each graphics element has equal horizontal and vertical scaling factors; ASPECT=2 scales the graphics element twice as wide as its height; and so on. If ASPECT= is not specified or is set to 0 or null, SAS/GRAPH uses the aspect ratio of the hardware device.

Details

The aspect ratio affects many graphics characteristics, such as the shape of software characters and the roundness of pie charts. Some graphics drivers do not produce correct output if the aspect ratio is anything other than the default. When you use a device that uses local scaling (that is, the device itself can scale the output, for example, some plotters), use ASPECT= to tell SAS/GRAPH the scaling factor.

Note: You can get more reliable results if you use the default aspect ratio and use the HSIZE= and VSIZE= graphics options to set the dimensions. △

AUTOCOPY

Specifies whether to generate hard copy automatically.

Used in: GOPTIONS statement; GDEVICE procedure; GDEVICE Parameters window
Defaults: GOPTIONS: NOAUTOCOPY; GDEVICE: AUTOCOPY=N
Restrictions: device-dependent; not supported by Java or ActiveX

Syntax

GOPTIONS: AUTOCOPY | NOAUTOCOPY

GDEVICE: AUTOCOPY=Y | N

AUTOCOPY
AUTOCOPY=Y
   prints a copy of the graph automatically.

NOAUTOCOPY
AUTOCOPY=N
   suppresses printing a copy of the graph. A blank `Autocopy` field in the Parameters window is the same as AUTOCOPY=N.

Details

AUTOCOPY is used only for older terminals that have printers attached directly to the device.

AUTOFEED

Specifies whether devices with continuous paper or automatic paper feed should roll or feed the paper automatically for the next graph.

Used in: GOPTIONS statement; GDEVICE procedure; GDEVICE Parameters window
Defaults: GOPTIONS: AUTOFEED (if a device is specified): GDEVICE: AUTOFEED=Y
Restrictions: device-dependent; not supported by Java or ActiveX
See also: PPDFILE

Syntax

GOPTIONS: AUTOFEED | NOAUTOFEED

GDEVICE: AUTOFEED=Y | N

AUTOFEED
AUTOFEED=Y
   causes the device to feed new paper automatically for the next graph. A blank `Autofeed` field in the Parameters window is the same as AUTOFEED=Y.

NOAUTOFEED
AUTOFEED=N
   suppresses the automatic paper feed.

Details

For PostScript devices, if AUTOFEED is unaltered, the PostScript file is unchanged. If you specify NOAUTOFEED and do not select a PPD file with the PPDFILE option, a PostScript Level 1 MANUALFEED command is added to the driver output. If you specify NOAUTOFEED and select a PPD that contains a MANUALFEED option, the procedure code for that MANUALFEED option is sent. If there is no MANUALFEED option in the PPD, no MANUALFEED code is sent. See "PPDFILE" on page 406.

AUTOSIZE

Controls whether to change the size of the character cells in order to preserve the number of rows and columns specified in the device entry.

Used in: GOPTIONS statement
Default: device-dependent
Restriction: not supported by Java or ActiveX
See also: DEVOPTS

Syntax

AUTOSIZE=ON | OFF | DEFAULT

ON
   changes the cell size in order to preserve the number of rows and columns.

OFF
   preserves the device's original cell size and temporarily changes the number of rows and columns.

DEFAULT
   uses the default setting (ON or OFF) that is controlled by DEVOPTS bit 50 (see "DEVOPTS" on page 350).

Details

AUTOSIZE is useful when you change the size of the graphics display area using one or more of the options PAPERSIZE, XPIXELS, YPIXELS, XMAX, or YMAX. It lets you control image text size without using PROC GDEVICE. Typically, AUTOSIZE is on for most image drivers and off for all other types of drivers.

Note: If you use HSIZE of VSIZE, the character cell size changes regardless of the AUTOSIZE setting. △

BINDING

Specifies which edge of the document is the binding edge.

Used in: GOPTIONS statement OPTIONS statement

Default: DEFAULTEDGE

Restrictions: PostScript and PCL printers only. PostScript printers require a PPD file. Not supported by Java or ActiveX.

See also: DUPLEX, PPDFILE

Syntax

BINDING=DEFAULTEDGE | LONGEDGE | SHORTEDGE

Details

BINDING controls how the page is flipped when DUPLEX is in effect. It does not change the orientation of the graph. DEFAULTEDGE refers to the harware's factory-default setting. LONGEDGE and SHORTEDGE refer to the paper's long and short edges.

For PostScript printers, a PPD file must also be specified, using the PPDFILE= option. The PPD file contains the command that SAS/GRAPH needs to request the appropriate binding method on the printer being used. If a PPD file is not specified, the BINDING= option is ignored because SAS/GRAPH will lack the command needed to request the binding method.

BORDER

Specifies whether to draw a border around the graphics output area.

Used in: GOPTIONS statement

Default: NOBORDER

Syntax

BORDER | NOBORDER

Details

The placement of the border on the display is defined by the HSIZE= and VSIZE= graphics options, if used. Otherwise the placement is defined by the XMAX and YMAX device parameters.

CBACK

Specifies the background color of the graphics output.

Used in: GOPTIONS statement; GDEVICE procedure; GDEVICE Gcolors window
Default: as specified in the Gcolors window

Syntax

CBACK=*background-color*

background-color
specifies any SAS/GRAPH color name. See Chapter 12, "SAS/GRAPH Colors and Images," on page 167 for information about specifying colors.

Details

The CBACK= option is valid on all devices but can be ignored by some (for example, plotters). Specify the default in the Gcolors window of the device entry.

Note: This option overrides the Background and Foreground style attributes in the graph styles. For more information on graph styles, refer to the TEMPLATE procedure documentation in *SAS Output Delivery System: User's Guide.* △

If you explicitly specify a background color with the CBACK= option, the background color you select should contrast with the foreground colors.

If the IBACK= option is in effect, an image will appear in the background in place of the color specified with the CBACK= option.

CBY

Selects the color of the By lines that appear in the graphics output.

Used in: GOPTIONS statement
Default: (1) CTEXT= graphics option, if used; (2) first color in current color list
Restriction: not supported by Java or ActiveX

Syntax

CBY=*By line-color*

By line-color
> specifies any SAS/GRAPH color name. See Chapter 12, "SAS/GRAPH Colors and Images," on page 167 for information about specifying colors.

Details

When you use a BY statement with a SAS/GRAPH procedure to process a data set in subgroups, each graph produced by that procedure is headed by a By line that displays the BY variables and their values that define the current subgroup.

CELL

Controls whether to use cell alignment.

Used in: GOPTIONS statement; GDEVICE procedure; GDEVICE Parameters window
Default: device-dependent
Restriction: not supported by Java or ActiveX

Syntax

GOPTIONS: CELL | NOCELL

GDEVICE: CELL=Y | N

CELL
CELL=Y
> causes the device to use cell alignment. In that case SAS/GRAPH attempts to place hardware (or simulated hardware) characters inside character cells. This restriction on the location of characters means that in some cases the SAS/GRAPH procedure can generate axes that do not occupy the entire procedure output area or might be unable to create the requested graph. A blank **Cell** field in the Parameters window is the same as CELL=Y.

NOCELL
CELL=N
> suppresses cell alignment, causing the procedure to use the entire procedure output area and place axis and tick mark labels without regard to cell alignment.

Details

Specify N in the device entry or use NOCELL in a GOPTIONS statement if you want to preview a graph on a cell-aligned display but intend to produce the final graph on a device that is not cell-aligned, such as a pen plotter.

CHARACTERS

Specifies whether the device—resident font is used when no font or FONT=NONE is specified in a SAS statement.

Used in: GOPTIONS statement; GDEVICE procedure; GDEVICE Parameters window

Defaults: GOPTIONS: CHARACTERS; GDEVICE: CHARACTERS=Y

Restriction: not supported by Java or ActiveX

Syntax

GOPTIONS: CHARACTERS | NOCHARACTERS

GDEVICE: CHARACTERS=Y | N

CHARACTERS
CHARACTERS=Y
    causes SAS/GRAPH to use the device-resident font when you do not specify a font in a SAS program. A blank **Characters** field in the Parameters window is the same as CHARACTERS=Y.

NOCHARACTERS
CHARACTERS=N
    causes SAS/GRAPH to draw the characters using the SIMULATE font and suppresses the use of *all* device—resident fonts, regardless of values you specify in other SAS statements.

Details

The device—resident font is not used if you changed the HPOS= and VPOS= graphics options from the default, or if you used the HEIGHT= option in a SAS statement *and* the device does not have scalable characters.

CHARREC

Specifies a device-resident font by associating a CHARTYPE number with a device-resident font. Also defines a default size to use with that font.

Used in: GDEVICE procedure

Default: device-dependent

Syntax

CHARREC=(*charrec-list(s)*)

charrec-list
: a list of values that correspond to the fields in the Chartype window. *Charrec-list* has this form:

  type, rows, cols, 'font', 'Y' | 'N'

type	is the CHARTYPE number and can be an integer from 0 to 9999. (See "CHARTYPE" on page 338 for more information.)
rows	is the number of rows of text in the font that will fit on the display. (See "ROWS" on page 419 for more information.)
cols	is the number of columns of text in the font that will fit on the display. (See "COLS" on page 342 for more information.)
font	is a character string enclosed in quotation marks that contains the name of the corresponding device-resident font. (See "FONT NAME" on page 361 for more information.)
Y	represents a scalable font. A scalable font can be displayed at any size. (See "SCALABLE" on page 419 for more information.)
N	represents a nonscalable font. A nonscalable font can be displayed only at a fixed size. (See "SCALABLE" on page 419 for more information.)

  For example, these values assign the device's Helvetica font to be the first device-resident font in the CHARTYPE window of the driver entry:

  ```
  charrec=(1, 100, 75, 'helvetica', 'y')
  ```

CHARTYPE

Selects the number of the default hardware character set.

Used in: GOPTIONS statement; GDEVICE procedure; GDEVICE Parameters window

Default: device-dependent

Restriction: not supported by Java or ActiveX

Syntax

CHARTYPE=*hardware-font-chartype*

hardware-font-chartype
: is a nonnegative integer from 0 to 999. *hardware-font-chartype* refers to the actual number for the device-resident font you want to use as listed in the Chartype window of the device entry for the selected device driver. By default, CHARTYPE is 0, which is the default device-resident font for the device.

CIRCLEARC

Specifies whether SAS/GRAPH should use the device's hardware circle-drawing capability, if available.

Used in: GOPTIONS statement; GDEVICE procedure; GDEVICE Parameters window
Default: device-dependent
Restriction: not supported by Java or ActiveX

Syntax

GOPTIONS: CIRCLEARC | NOCIRCLEARC

GDEVICE: CIRCLEARC=Y | N

CIRCLEARC
CIRCLEARC=Y
  causes SAS/GRAPH to use the built-in hardware circle- and arc-drawing capability of the device. A blank `Circlearc` field in the Parameters window is the same as CIRCLEARC=Y.

  hardware drawing is faster, but not all devices have the capability. SAS/GRAPH device drivers do not try to use the capability if the device does not have it.

NOCIRCLEARC
CIRCLEARC=N
  causes SAS/GRAPH to use software move and draw commands to draw circles and arcs.

CMAP

Specifies a color map for the device.

Used in: GDEVICE procedure; GDEVICE Colormap window

Syntax

CMAP=('from-color : to-color' <...,'from-color-n : to-color-n'>)

from-color
  specifies the name you want to assign to the color designated by the *color* value. In the Colormap window, enter this value in the **From** field.

to-color
  specifies any SAS/GRAPH color name up to eight characters long. In the Colormap window, enter this value in the **To** field. See Chapter 12, "SAS/GRAPH Colors and Images," on page 167 for information on specifying colors.

Details

Once you have defined the color mapping, you use the new color name in any color option. For example, if your device entry maps the color name DAFFODIL to the SAS color value PAOY, you can specify the following:

```
pattern1 color=daffodil;
```

and the driver will map this to the color value PAOY.

COLLATE

Specifies whether to collate the output, if collation is supported by the device.

Used in: GOPTIONS statement; OPTIONS statement
Default: NOCOLLATE
Restriction: hardware–dependent, PostScript printers require a PPD file; not supported by Java or ActiveX
See also: GPROLOG, PPDFILE

Syntax

COLLATE | NOCOLLATE

Details

A limited number of printers can *collate* output, which means to separate each copy of printed output when you print multiple copies of output.

For PostScript printers, if a device's PPD file has Collate defined as "True", the COLLATE option is supported.

For PCL printers that support collation, use the GPROLOG= option to specify a Printer Job Language (PJL) command to enable the collation. For information on the appropriate PJL command, consult the Printer Commands section of your printer's user manual.

COLORS

Specifies the foreground colors used to produce your graphics output if you do not specify colors explicitly in program statements.

Used in: GOPTIONS statement; GDEVICE procedure; GDEVICE Gcolors window
Default: device-dependent

Syntax

GOPTIONS: COLORS=<(*colors-list* | NONE)>

GDEVICE: COLORS=(<*colors-list*>)

colors-list
: specifies one or more SAS color names. If you specify more than one color, separate each name with a blank. See Chapter 12, "SAS/GRAPH Colors and Images," on page 167 for information on specifying colors and using a color list.

    To change some of the colors in the color list and retain others, you can use a null value for colors you do not want to change. For example, to change COLORS=(RED GREEN BLUE) to COLORS=(WHITE GREEN BROWN), you can specify COLORS=(WHITE,,BROWN).

NONE
: tells SAS/GRAPH to use only the colors that you explicitly specify in program statements and to ignore the device's default color list.

    Note: If you specify COLORS=(NONE) and omit a color specification for a graphics element, such as patterns, SAS/GRAPH selects at random one of the colors already specified in your program. △

Featured in: "Example 3. Rotating Plot Symbols Through the Color List" on page 299

Details

The order of the colors in the list is important when you use default colors. For example, the colors used for titles, axes, and surfaces in the G3D procedure are assigned by default according to their position in the color list.

Note: Colors can be assigned to graph elements in different orders by different devices such as Java and ActiveX. △

If you omit or reset COLORS=, SAS/GRAPH uses the default color list for the current device. To explicitly reset the color list to the device default, specify either

```
goptions colors=;
goptions colors=();
```

If you use default patterns with a color list specified by COLORS= option, the patterns rotate through every color in the list. If the color list contains only one color, for example COLORS=(BLUE), the solid pattern is skipped and the patterns rotate through only the appropriate default hatch patterns for the graph.

Note: By default, if black is the first color in a device's color list, default pattern rotation skips black as a pattern color, but uses black as the area-outline color. Thus, the outline color is never the same as an area's fill color. Using COLORS= to change the color list changes this default pattern behavior. When COLORS= is used, all colors in the specified color list are used in color rotation, and the outline color is the first color in the specified color list. Thus, the outline color will match any area using the first color as its fill. △

See "PATTERN Statement" on page 240 for more information on pattern rotation.

COLORTBL

An eight-character field in the Gcolors window that is not currently implemented. SAS/GRAPH ignores any value entered into this field.

COLORTYPE

Specifies the color space used by the user-written part of the Metagraphics device driver.

Used in: GDEVICE procedure; GDEVICE Metagraphics window
Default: NAME

Syntax

COLORTYPE=NAME | RGB | HLS | GRAY | CMY | CMYK | HSV | HSB

NAME	SAS predefined color names.
RGB	red-green-blue (RGB) color specifications.
HLS	hue-lightness-saturation (HLS) color specifications.
GRAY	gray-scale level.
CMY	cyan-magenta-yellow color specifications.
CMYK	cyan-magenta-yellow-black color specifications.
HSV | HSB	hue-saturation-value color specifications. These specifications are also referred to as hue-saturation-brightness (HSB).

See Chapter 12, "SAS/GRAPH Colors and Images," on page 167 for a description of these color types.

Details

Use the COLORTYPE device parameter also to specify the color-naming scheme that is used for devices that support more than one color-naming scheme.

For information about Metagraphics drivers, contact Technical Support.

COLS

Sets the number of columns that the device-resident font uses.

Used in: GDEVICE Chartype window; GDEVICE procedure; CHARREC= option
Default: 0

See also: CHARREC

Syntax

See "CHARREC" on page 337 for syntax.

Details

If you are using a device driver from SASHELP.DEVICES, this parameter is already set for device-resident fonts that have been defined for your installation. If you are adding to or modifying the device-resident fonts available for a particular device driver, specify a positive value for the COLS device parameter. If COLS is greater than 0, it overrides the values of the LCOLS and PCOLS device parameters. For scalable fonts, you can specify 1 for COLS, and the actual number of columns will be computed based on the current text width.

CPATTERN

Selects the default color for PATTERN definitions when a color has not been specified.

Used in: GOPTIONS statement

Default: first color in current color list

Restriction: not supported by Java or ActiveX

Syntax

CPATTERN=*pattern-color*

pattern-color
   specifies any SAS/GRAPH color name. See Chapter 12, "SAS/GRAPH Colors and Images," on page 167 for information about specifying colors.

Details

CPATTERN= is overridden by any color specification in a PATTERN statement. For details on how CPATTERN= affects the PATTERN statement, see "The Effect of the CPATTERN= Graphics Option" on page 250.

If you specify CPATTERN=, the solid pattern is skipped and the patterns rotate through only the appropriate default hatch patterns for the graph. See "PATTERN Statement" on page 240 for more information on pattern rotation.

CSYMBOL

Specifies the default color for SYMBOL definitions when a color has not been specified.

Used in: GOPTIONS statement

Default: first color in current color list

Restriction: not supported by Java or ActiveX

Syntax

CSYMBOL=*symbol-color*

symbol-color
: specifies any SAS/GRAPH color name. See Chapter 12, "SAS/GRAPH Colors and Images," on page 167 for information about specifying colors.

Details

CSYMBOL= is overridden by any color specification in a SYMBOL statement. See "SYMBOL Statement" on page 252.

CTEXT

Selects the default color for all text and the border.

Used in: GOPTIONS statement

Default: black for Java and ActiveX devices; for other devices, the first color in current color list

See also: CTITLE

Restriction: partially supported by Java

Syntax

CTEXT=*text-color*

text-color
: specifies any SAS/GRAPH color name. See Chapter 12, "SAS/GRAPH Colors and Images," on page 167 for information about specifying colors.

Details

The CTITLE= graphics option overrides CTEXT= for all titles, notes, and footnotes, as well as the border. Any other color specifications for text in SAS statements also override the value of the CTEXT= graphics option.

Note: When you use ODS to send graphics to an HTML destination, and titles and footnotes are rendered as part of the HTML body file instead of the graphic image, you must specify the ODS USEGOPT statement for this option to work. See "Using Graphics Options with ODS (USEGOPT)" on page 195 for more information. △

CTITLE

Selects the default color for all titles, footnotes, and notes, and the border.

Used in: GOPTIONS statement

Default: (1) color specified by CTEXT=, if used; (2) black for Java and ActiveX devices; for other devices, the first color in current color list

See also: CTEXT

Syntax

CTITLE=*title-color*

title-color
> specifies any SAS/GRAPH color name. See Chapter 12, "SAS/GRAPH Colors and Images," on page 167 for information about specifying colors.

Details

Any color specification in a TITLE, FOOTNOTE, or NOTE statement overrides the value of the CTITLE= graphics option for the text. The border, however, still uses the color specified in the CTITLE= graphics option.

Note: When you use ODS to send graphics to an HTML destination, and titles and footnotes are rendered as part of the HTML body file instead of the graphic image, you must specify the ODS USEGOPT statement for this option to work. See "Using Graphics Options with ODS (USEGOPT)" on page 195 for more information. △

DASH

Specifies whether to use the device's hardware dashed-line capability, if available.

Used in: GOPTIONS statement; GDEVICE procedure; GDEVICE Parameters window

Default: device- dependent

Restriction: not supported by Java or ActiveX

See also: DASHLINE

Syntax

GOPTIONS: DASH | NODASH

GDEVICE: DASH=Y | N

DASH
DASH=Y
  causes SAS/GRAPH to use the built-in hardware dashed-line drawing capability of the device when generating graphics output. A blank **Dash** field in the Parameters window is the same as DASH=Y.

  hardware drawing is faster, but not all devices have the capability. SAS/GRAPH device drivers do not try to use the capability if the device does not have it.

NODASH
DASH=N
  causes SAS/GRAPH to draw the dashed lines.

DASHLINE

Specifies which dashed lines should be generated by hardware means if possible.

Used in: GDEVICE procedure; GDEVICE Parameters window
Default: device-dependent
See also: DASH

Syntax

DASHLINE='*dashed-line-hex-string*'X

dashed-line-hex-string
  is a hexadecimal string 16 characters long that must be completely filled. Each bit in the string corresponds to a line type. See Figure 14.22 on page 277 for line types that correspond to each bit.

  To use line type 1, turn on bit 1; to use line type 2, turn on bit 2; and so on. For example, in the following option the first byte is '1000'; only bit 1 is on and only line type 1 is selected:

  ```
  dashline='8000000000000000'x
  ```

  To turn on both bits 1 and 2, specify

  ```
  dashline='c000000000000000'x
  ```

  Bit 1 should always be on because it corresponds to a solid line.

Details

If the DASH device parameter is N in the device entry or if NODASH is used in a GOPTIONS statement, SAS/GRAPH ignores the hexadecimal string in the DASHLINE device parameter.

DASHSCALE

Scales the lengths of the dashes in a dashed line.

Used in: GOPTIONS statement

Default: DASHSCALE=1

Restriction: not supported by Java or ActiveX

Syntax

DASHSCALE=*scaling-factor*

scaling-factor
    can be any number greater than 0. For example, GOPTIONS DASHSCALE=.5 reduces any existing dash length by one-half.

Details

Only dashes or spaces with lengths greater than one pixel are scaled. Dots are not scaled because their length is effectively zero. DASHSCALE= always uses system line styles instead of the device's dashed line capabilities.

DELAY

Controls the amount of time between graphs in the animation sequence.

Used in: GOPTIONS statement

Default: 0

Restriction: GIFANIM driver only; not supported by all browsers

Syntax

DELAY=*delay-time*

delay-time
    specifies the length of time between graphs in units of 0.01 seconds. For example, to specify a delay of .03 seconds, specify DELAY=3.

Details

SAS/GRAPH puts the DELAY= value into the image file. Based on this value, the browser determines how to display the series of graphs.

DESCRIPTION

Provides a description of the device entry.

Alias: DES
Used in: GDEVICE procedure GDEVICE Detail window
Default: none

Syntax

DESCRIPTION='*text-string*'

text-string
    is a string up to 256 characters long. This is a comment field and does not affect the graphics output.

DEVADDR

Specifies the location of the device to which the output of device drivers is sent.

Used in: GOPTIONS statement
Default: host dependent
Restriction: IBM mainframe systems only

Syntax

DEVADDR=*device-address*

DEVICE

Specifies the device driver to which SAS/GRAPH sends the procedure output. The device driver controls the format of graphics output.

Alias: DEV
Used in: GOPTIONS statement OPTIONS statement
Default: device–dependent

Syntax

DEVICE=*device-entry*

device-entry
    specifies the name of a device entry that is stored in a device catalog.

Details

A device driver can direct graphics output to a hardware device, such as a terminal or a printer, or can create an external file in another graphics file format, such as TIF, GIF, or PostScript. Some device drivers also generate both graphics files and HTML files that can be viewed with a Web browser.

Usually a device driver is assigned by default. If a default driver is not assigned or if you specify RESET=ALL in a GOPTIONS statement, and you do not specify a device driver, SAS/GRAPH prompts you to enter a driver name when you execute a procedure that produces graphics output. If you are producing a graph to the screen and the Graph window is active, SAS/GRAPH selects the display driver for you automatically.

For a description of device drivers and for more information on selecting a device entry and changing device parameters, see Chapter 6, "Using Graphics Devices," on page 67.

For information on using device drivers to display and print graphics output, see Chapter 7, "SAS/GRAPH Output," on page 87.

For information on using device drivers to export graphics output to external files, see "Specifying the Graphics Output File Type for Your Graph" on page 91. For information on using device drivers to create output for the Web, see "Generating Web Presentations" on page 451.

DEVMAP

Specifies the device map to be used when device-resident fonts are used.

Used in: GOPTIONS statement; GDEVICE procedure; GDEVICE Host File Options window

Default: device-dependent

Restriction: not supported by Java or ActiveX

Syntax

DEVMAP=*device-map-name* | NONE

device-map-name
    is a string up to eight characters long that is the name of the device map entry.

NONE
    specifies that you do not want to use a device map. This can cause text to be displayed incorrectly or not at all.

Details

Device maps usually are used only when national characters appear in the text and you want them to display properly.

DEVOPTS

Specifies the hardware capabilities of the device.

Used in: GDEVICE procedure; GDEVICE Parameters window
Default: device-dependent

Syntax

DEVOPTS='*hardware-capabilities-hex-string*'X

hardware-capabilities-hex-string
is a hexadecimal string 16 characters long that must be completely filled. The following table lists the hardware capabilities of each bit:

Table 15.1 Device Capabilities Represented in the DEVOPTS String

Bit On	Capability
0	hardware circle generation
1	hardware pie fill supported
2	scalable hardware characters
3	device is a CRT-type (See TYPE device parameter)
4	translate table needed for non-ASCII hosts
5	hardware polygon fill available
6	hardware characters cell-aligned
7	user-definable colors supported
8	hardware polygons with multiple boundaries supported
9	not used
10	not used
11	adjustable hardware line width
12	double-byte font (non-US) supported
13	hardware repaint supported
14	hardware characters supported
15	no hard limit on x coordinate
16	no hard limit on y coordinate
17	not used
18	ability to justify proportional text
19	driver can produce dependent catalog entries
20	device cannot draw in default background color
21	flush device buffer when filled
22	colors defined using HLS

Bit On	Capability
23	colors defined using RGB
24	not used
25	polyline supported
26	polymarker supported
27	graphics clipping supported
28	not used
29	linkable device driver
30	pick CHARTYPE by name in CHARREC entries
31	device-dependent pattern support
32	treat SCALABLE=Y CHARREC as metric
33	size CHARTYPE as HW from CHARREC entries
34	device supports rotated arcs
35	device supports target fonts
36	device supports drawing images
37	device supports multiple color maps
38	image rotation direction
39	device requires sublib for image rotation
40	device is a 24 bit truecolor machine
41	device supports setting font attributes
42	use scan line font rendering
43	device can scale images
44	text clipping supported
45	static color device
46	driver does prolog processing
47	driver does epilog processing
48	driver output only uses a file
49	driver output requires a directory or PDS
50	autosize text to fit rows and columns
51	default binding is SHORTEDGE
52	driver supports duplex printing
53	device does right edge binding
54	ActiveX device
55	Java device
56	device uses a universal printer driver

Details

Each capability in the table corresponds to a bit in the value of the DEVOPTS device parameter. For example, if your device can generate hardware pie fills, the second bit in the first byte of the DEVOPTS string should be turned on if you want the driver to use that capability. If your device is capable of generating only hardware circles and pie fills, specify a value of 'C000000000000000'X as your DEVOPTS value (the first byte is '1100' so the first 2 bits of the first byte are set to 1). Many of the hardware capabilities specified in the DEVOPTS string are overridden by graphics options or other device parameters.

CAUTION:
Do not modify the DEVOPTS device parameter unless you are building a Metagraphics driver. If you want to prevent an SAS-supplied driver from using certain hardware capabilities, change the specific device parameter or use the corresponding graphics option. △

If the DEVOPTS string indicates that a capability is available, the driver uses it unless it is explicitly disabled by another device parameter or graphics option. If the DEVOPTS string indicates that the capability is not available, it is not used by the driver, even if the corresponding device parameter or graphics option indicates that it should be used. For example, if the DEVOPTS value indicates that the device can do a hardware pie fill, the driver uses the hardware pie fill capability unless the PIEFILL device parameter is set to N or NOPIEFILL has been specified in a GOPTIONS statement. However, if the DEVOPTS device parameter indicates that the device cannot do a hardware pie fill, the driver does not attempt to use one, even if the PIEFILL device parameter is set to Y or PIEFILL is used in a GOPTIONS statement.

DEVTYPE

Specifies the information required by SAS/GRAPH routines to determine the nature of the output device.

Used in: GDEVICE procedure; GDEVICE Host File Options window

Default: device-dependent

Syntax

DEVTYPE=*device-type*

device-type
: is a string eight characters long containing either blanks or some token name that is interpreted by the host. *Device-type* can be:

GTERM
: indicates that the output device is a graphics device that will be receiving graphics data; most device drivers use this value.

G3270
: indicates that the output device is an IBM 3270 graphics data stream. If your device is an IBM 3270 type of device, DEVTYPE= must be G3270.

Note: GTERM and G3270 are SAS/GRAPH device types. Other valid values depend on your operating environment. DEVTYPE supports any of the device-type values supported on the FILENAME statement. Refer to the SAS Help facility for the device types the FILENAME statement supports in your operating environment. In most cases, this field should not be changed. △

DISPLAY

Specifies whether output is displayed on the graphics device but does not affect whether a graph is placed in a catalog.

Used in: GOPTIONS statement
Default: DISPLAY
Restriction: not supported by Java or ActiveX

Syntax

DISPLAY | NODISPLAY

Details

In most cases, NODISPLAY suppresses *all* output except the catalog entry written to the catalog selected in the GOUT= option. Therefore, you usually specify NODISPLAY when you want to generate a graph in a catalog but do not want to display the graph on your monitor or terminal while the catalog entry is being produced.

DISPOSAL

Specifies what happens to the graphic after it is displayed.

Used in: GOPTIONS statement
Default: NONE
Restriction: GIFANIM driver only

Syntax

DISPOSAL=NONE | BACKGROUND | PREVIOUS | UNSPECIFIED

NONE
 causes the graphic to be left in place after displaying. This is the default.

BACKGROUND
 causes the background color to be returned and the graph erased after displaying.

PREVIOUS
 causes the graphic area to be restored with what was displayed in the area previously.

UNSPECIFIED
 indicates that no action is necessary.

Details

In Version 6, the ERASE | NOERASE graphics option performed this function for the GIFANIM driver.

DRVINIT

Specifies host commands to be executed before driver initialization.

Used in: GOPTIONS statement; GDEVICE procedure; GDEVICE Host Commands window

Restriction: not supported by Java or ActiveX

Syntax

DRVINIT1='*system-command(s)*'

DRVINIT2='*system-command(s)*'

system-command(s)
specifies a character string that is a valid system command and can be in upper- or lowercase letters. You can include more than one command in the string if you separate the commands with a command delimiter, which is host-specific; for example, some operating environments use a semicolon. The length of the entire string cannot exceed 72 characters.

Details

The DRVINIT command is executed before the driver is initialized. DRVINIT is typically used with FILECLOSE=DRIVERTERM to allocate a host file needed by the device driver.

DRVQRY

Specifies whether the device can be queried for information about the current device configuration.

Used in: GDEVICE procedure GDEVICE Detail window

Default: device-dependent

Syntax

DRVQRY | NODRVQRY

Details

Generally, this setting is device-dependent and you should not change it.

DRVTERM

Specifies host commands to be executed after the driver terminates.

Used in: GOPTIONS statement; GDEVICE procedure; GDEVICE Host Commands window

Restriction: not supported by Java or ActiveX

Syntax

DRVTERM1='*system-command(s)*'

DRVTERM2='*system-command(s)*'

system-command(s)
> specifies a character string that is a valid system command and can be in upper- or lowercase letters. You can include more than one command in the string if you separate the commands with a command delimiter, which is host-specific; for example, some operating environments use a semicolon. The length of the entire string cannot exceed 72 characters.

Details

The DRVTERM command is executed after the driver terminates. DRVTERM is typically used with FILECLOSE=DRIVERTERM to de-allocate a host file and execute utility programs that send the data to the graphics device. For example, DRVTERM might specify commands to send the file to a host print queue.

DUPLEX

Specifies whether to use duplex printing if available on the device.

Used in: GOPTIONS statement; OPTIONS statement
Default: NODUPLEX
Restriction: duplex printers only
See also: BINDING, GSFMODE, PPDFILE

Syntax

DUPLEX | NODUPLEX

Details

When DUPLEX is on, the driver sets up the printer for duplex operation. Before producing the first graph, set GSFMODE=REPLACE on the GOPTIONS statement, and DUPLEX on an OPTIONS or GOPTIONS statement. You can also use the BINDING= option in conjunction with DUPLEX. Before producing the second graph, set

GSFMODE=APPEND on the GOPTIONS statement so that the driver knows to place succeeding graphs on the next available side of paper.

If DUPLEX is in effect, the page's inside (binding) margin is set equal to the current HORIGIN setting, and the outside margin is set equal to

XMAX − HSIZE − HORIGIN

In terms of even- and odd-numbered pages, this means the following:

odd-numbered pages	HORIGIN determines the left margin, and XMAX-HSIZE-HORIGIN determines the right margin
even-numbered pages	XMAX-HSIZE-HORIGIN determines the left margin, and HORIGIN determines the right margin

For PostScript printers, if you do not use the PPDFILE= option to specify a PPD (PostScript Printer Description) file, a generic PostScript Level 1 duplex command is added to the driver output. If PPDFILE= is used, the duplex command is obtained from the PPD file.

ERASE

Specifies whether to erase graph after display.

Used in: GOPTIONS statement; GDEVICE procedure; GDEVICE Parameters window
Defaults: GOPTIONS: NOERASE; GDEVICE: ERASE=N
Restriction: not supported by Java or ActiveX

Syntax

GOPTIONS: ERASE | NOERASE

GDEVICE: ERASE=Y | N

ERASE
ERASE=Y
   causes the graph to be erased when you press RETURN after the graph has been displayed.

NOERASE
ERASE=N
   causes the graph to remain on the display when you press RETURN after the graph has been displayed. A blank **Erase** field in the Parameters window is the same as ERASE=N.

Details

ERASE is useful for those devices that overlay the graphics area and the message area – that is, those devices that have separate dialog box and graphics areas. On other devices, the graph is erased.

EXTENSION

Specifies the file extension for an external graphics file.

Used in: GOPTIONS statement
Default: device-dependent
Restriction: not supported by Java or ActiveX
See also: GACCESS, GSFNAME

Syntax

EXTENSION='*file-type*'

file-type
    a string up to eight characters long that is a file extension, such as GIF or CGM, that you want to append to an external file.

Details

The extension specified on EXTENSION= is used when the output destination is a storage location. The extension is ignored when the output destination is a file. To specify the output destination, you can use a FILENAME statement, or the graphics options GACCESS= or GSFNAME=.

Assuming that the output destination is a storage location,

- if EXTENSION='.', no extension is added to the filename
- if EXTENSION=' 'or EXTENSION= is not used, the driver's default extension is added to the filename
- if the driver has no default extension, SAS/GRAPH uses the default extension .GSF.

FASTTEXT

Specifies whether to use integer-based font processing for faster font rendering.

Used in: GOPTIONS statement
Default: FASTTEXT
Restriction: not supported by Java or ActiveX

Syntax

FASTTEXT | NOFASTTEXT

FBY

Selects the font for By lines.

Used in: GOPTIONS statement
Default: (1) font specified by FTEXT=, if used; (2) device–resident font (3) simulate font
Restriction: not supported by Java or ActiveX
See also: "BY Statement" on page 216

Syntax

FBY=*By line-font*

By line-font
    specifies the font for all By lines on the graphics output. See Chapter 11, "Specifying Fonts in SAS/GRAPH Programs," on page 155 for information about specifying fonts.

Details

When you use a BY statement with a SAS/GRAPH procedure to process a data set in subgroups, each graph produced by that procedure is headed by a By line that displays the BY variables and their values that define the current subgroup.

FCACHE

Specifies the number of system fonts to keep open at one time.

Used in: GOPTIONS statement
Default: FCACHE=3
Restriction: not supported by Java or ActiveX

Syntax

FCACHE=*number-fonts-open*

number-fonts-open
    specifies the number of system fonts to keep open. *Number-fonts-open* must be greater than or equal to zero.

Details

Each font requires from 4K to 10K memory. Graphs that use many fonts can run faster if you set the value of *number-fonts-open* to a higher number. However, graphs that use multiple fonts might require too much memory on some computer systems if all the fonts are kept open. In such cases, set the value of *number-fonts-open* to a lower number to conserve memory.

FILECLOSE

Controls when the graphics stream file (GSF) is closed when you are using the device driver to send graphics output to a hard copy device.

Used in: GOPTIONS statement; GDEVICE procedure; GDEVICE Host File Options window

Default: DRIVERTERM (if a device is specified)

Restriction: not supported by Java or ActiveX

See also: "Specifying the Graphics Output File Type for Your Graph" on page 91

Syntax

FILECLOSE=DRIVERTERM | GRAPHEND

DRIVERTERM
DRIVER
closes the GSF and makes it available to the device after all graphs have been produced and the procedure or driver terminates. A host command might be needed to actually send the GSF to the device. Host commands can be specified with the DRVINIT or DRVTERM parameters or entered in the Host File Options window of the device entry.

If multiple graphs are produced by a procedure, this specification creates one large file. Specifying DRIVERTERM is appropriate for batch processing because it is slightly more efficient to allocate the file only once.

GRAPHEND
GRAPH
closes the GSF after each separate graph is produced and releases it to the device before sending another. This method creates smaller files if multiple graphs are produced by a procedure. You can specify a command that sends the graph to the device with the POSTGRAPH parameter or use the Host File Options window.

Specifying GRAPHEND is appropriate for drivers that are used interactively, or for devices that require only one graph per physical file.

FILEONLY

Specifies whether a file or a storage location is the default destination for graphics output.

Used in: GOPTIONS statement

Default: device-dependent

Restriction: FILEONLY ignored if the device requires the output destination to be a storage location; not supported by Java or ActiveX

See also: DEVOPTS, GSFNAME

Syntax

FILEONLY | NOFILEONLY

FILEONLY
  specifies that a file rather than a storage location is the default destination for graphics output.

NOFILEONLY
  specifies that a storage location is the default destination for graphics output, unless a file of the same name exists.

Details

Most devices use FILEONLY as the default. However, devices that require the output destination to be a storage location use NOFILEONLY as the default. For example, the HTML device requires a storage location because it produces two types of output (HTML files and GIF image files) that cannot be written to the same file.

To determine what the default is for a particular device, look at the settings for DEVOPTS bits 48 and 49.

For more information, see "Specifying the Graphics Output File Type for Your Graph" on page 91.

FILL

Specifies whether to use the device's hardware rectangle-fill capability.

Used in: GOPTIONS statement; GDEVICE procedure; GDEVICE Parameters window
Restriction: not supported by Java or ActiveX
Default: device–dependent

Syntax

GOPTIONS: FILL | NOFILL

GDEVICE: FILL=Y | N

FILL
FILL=Y
  causes SAS/GRAPH to use the built-in hardware rectangle-filling capability of the device. A blank `Fill` field in the Parameters window is the same as FILL=Y.
  hardware drawing is faster, but not all devices have the capability. SAS/GRAPH does not try to use the capability if your device does not support it.

NOFILL
FILL=N
    causes SAS/GRAPH to use software fills to fill rectangles.

FILLINC

Specifies the number of pixels to move before drawing the next line in a software fill of a solid area.

Used in: GOPTIONS statement; GDEVICE procedure; GDEVICE Parameters window
Default: device-dependent
Restriction: not supported by Java or ActiveX
See also: FILL, PIEFILL, POLYGONFILL

Syntax

FILLINC= 0...9999

Details

In order for FILLINC to have any effect, a software fill must be used. To force a software fill, use the options NOFILL, NOPIEFILL, and NOPOLYGONFILL in a GOPTIONS statement.

If FILLINC is set to 0 or 1, adjacent lines are used (solid fill with no gaps). If FILLINC is set to 2, a pixel-width line is skipped before drawing the next line of a fill.

This option can be useful for keeping plotters from over saturating a solid area and for speeding the plotting. Some inks spread on paper. The type of paper used can also affect ink spread.

FONT NAME

Specifies the device-resident font associated with CHARTYPE.

Used in: GDEVICE Chartype window; GDEVICE procedure; CHARREC= option
Required   if adding or modifying a CHARREC
See also: CHARREC

Syntax

See "CHARREC" on page 337 for syntax.

Details

Use FONT NAME if you are adding to or modifying the device-resident fonts available for a particular device driver. The fonts that you specify must be valid for the output device. If you are using an SAS-supplied device entry, this parameter already is set for most available device-resident fonts.

FONTRES

Controls the resolution of Bitstream fonts.

Used in: GOPTIONS statement
Default: NORMAL
Restriction: not supported by Java or ActiveX
See also: FASTTEXT, FCACHE, RENDER, RENDERLIB, SWFONTRENDER

Syntax

FONTRES=NORMAL | PRESENTATION

NORMAL
renders fonts in memory using integer rendering routines, which improves character drawing speed for most host systems. NORMAL has the same effect as specifying the default values for these graphics options.

```
render=memory
renderlib=saswork
fasttext
fcache=0
```

PRESENTATION
disables the storage or use of rendered versions of Bitstream fonts, but produces the fonts at their highest resolution. FONTRES=PRESENTATION has the same effect as specifying these graphics options:

```
render=none
renderlib=saswork
nofasttext
fcache=3
```

FORMAT

Sets the file format of the metacode file produced by the SAS-supplied part of the Metagraphics device driver.

Used in: GDEVICE procedure; GDEVICE Metagraphics window
Default: CHARACTER
Restriction: Used only with user-supplied Metagraphics drivers.

Syntax

FORMAT=CHARACTER | BINARY

FTEXT

Sets the default font for all text.

Used in: GOPTIONS statement
Default: Default device–resident font (except the first title)
Restriction: partially supported by Java or ActiveX
See also: FTITLE

Syntax

FTEXT=*text-font*

text-font
  specifies the font for all text on the graphics output. See Chapter 11, "Specifying Fonts in SAS/GRAPH Programs," on page 155 for information about specifying fonts.

Details

The FTITLE= graphics option overrides FTEXT= for the *first* title. Not all fonts are supported by the ActiveX and Java devices.

Note: When you use ODS to send graphics to an HTML destination, and titles and footnotes are rendered as part of the HTML body file instead of the graphic image, you must specify the ODS USEGOPT statement for this option to work. See "Using Graphics Options with ODS (USEGOPT)" on page 195 for more information. △

FTITLE

Selects the default font for the first TITLE line.

Used in: GOPTIONS statement
Default: (1) font specified by FTEXT=, if used; (2) value of the style variable (3)device-resident font (4)simulate font
See also: FTEXT

Syntax

FTITLE=*title-font*

title-font
> specifies the font for the TITLE1 statement. See Chapter 11, "Specifying Fonts in SAS/GRAPH Programs," on page 155 for information about specifying fonts.

Details

Note: When you use ODS to send graphics to an HTML destination, and titles and footnotes are rendered as part of the HTML body file instead of the graphic image, you must specify the ODS USEGOPT statement for this option to work. See "Using Graphics Options with ODS (USEGOPT)" on page 195 for more information. △

FTRACK

Controls the amount of space between letters in the SAS-supplied Bitstream fonts (Brush, Century, Swiss, and Zapf).

Used in: GOPTIONS statement

Default: TIGHT

Restriction: not supported by Java or ActiveX

Syntax

FTRACK=LOOSE | NONE | NORMAL | TIGHT | TOUCH | V5

LOOSE
> leaves the most visible space between characters and produces a longer string.

NONE
> spacing depends on the size of the font. NONE might produce a shorter or longer string than LOOSE for the same font at different point sizes, because some sizes add space between the characters while others remove it.

NORMAL
> is the recommended setting.

TIGHT
> reduces the space between characters.

TOUCH
> leaves the least visible space between characters.

V5
> places a fixed amount of space between the characters and does not adjust for the shape of the character; that is, it does not support kerning. This spacing is compatible with Version 5 Bitstream fonts.

Details

The spacing you specify with FTRACK= affects all Bitstream text in a graph. For example, you cannot produce TIGHT Century type and LOOSE Zapf type simultaneously. This option has no effect on other font types.

Because the value of FTRACK= is stored with the graph, the spacing that you specify when the graph is created is always used when the graph is replayed.

GACCESS

Specifies the format or the destination or both of graphics data written to a device or graphics stream file (GSF).

Used in: GOPTIONS statement; GDEVICE procedure; GDEVICE Host File Options window

Default: device-dependent

Restriction: not supported by Java, ActiveX, or shortcut devices. See Chapter 6, "Using Graphics Devices," on page 67 for more information about devices.

Syntax

GACCESS=*output-format* | *'output-format destination'*

output-format
 specifies the format or the destination (the SAS log or a fileref) of the graphics data. *Output-format* varies according to the operating environment. These values can be specified in all operating environments:

 SASGASTD
  specifies that a continuous stream of data is written. SASGASTD is the default for most devices and is typically appropriate when the output file will be sent directly to a device. If you specify GACCESS=SASGASTD, use the GSFNAME= and GSFMODE= graphics options or device parameters to direct your graphics output to a GSF.

 SASGAEDT
  specifies that the file be host-specific edit format. Some hosts allow editing by inserting characters at the end of each record. SASGAEDT is typically used when the output file is to be edited later. If you specify GACCESS=SASGAEDT, use the GSFNAME= and GSFMODE= graphics options or device parameters to direct your graphics output to a GSF.

 SASGAFIX
  specifies that fixed-length records be written. (The record length is controlled by the value of the GSFLEN= graphics option or device parameter or the sixth byte of the PROMPTCHARS value.) The records are padded with blanks where necessary. SASGAFIX is typically used when the output file will be transferred to a computer that requires fixed-length records. If you specify GACCESS=SASGAFIX, use the GSFNAME= and GSFMODE= graphics options or device parameters to direct your graphics output to a GSF.

  Note: The value of the GPROTOCOL= graphics option or device parameter can greatly affect the length of the records; for example, if GPROTOCOL=SASGPLCL, the length of the records is doubled. △

 SASGALOG
  specifies that records are to be written to the SAS log.

GSASFILE
specifies that the records are to be written to the destination whose fileref is GSASFILE. The fileref can point to a specific external file or to an aggregate file location. See "FILENAME Statement" on page 36 for more information on specifying a fileref.

'output-format destination'
specifies the destination in addition to one of these output format values: SASGASTD, SASGAEDT, or SASGAFIX. *Destination* is the physical name of an external file or aggregate file location, or of a device. For details on specifying the physical name of a destination, see the SAS documentation for your operating environment.

This form is not available in all operating environments. See "Specifying the Graphics Output File Type for Your Graph" on page 91 for more information on creating graphics stream files.

Note: In the **Gaccess** field of the Host File Options window, you can specify a destination without an output format. In that case the format defaults to SASGASTD. When you specify a value in the **Gaccess** field, you do not need to quote it. △

Operating Environment Information: Depending on your operating environment, you might be able to specify other values for GACCESS=. See the SAS companion for your operating environment for additional values. △

GCLASS

Specifies the output class for IBM printers

Used in: GOPTIONS statement
Default: GCLASS=G
Restriction: used only with IBM3287 and IBM3268 device drivers on z/OS systems only

Syntax

GCLASS=*SYSOUT-class*

Details

Specifies the SYSOUT class to which the IBM3287 and IBM3268 device driver output is written.

GCOPIES

Sets the current and maximum number of copies to print.

Used in: GOPTIONS statements; GDEVICE Parameters window; GDEVICE procedure; OPTIONS statement
Defaults: GOPTIONS: GCOPIES=(0,20) GDEVICE: GCOPIES=0
Restriction: not supported by Java or ActiveX

Syntax

GOPTIONS: GCOPIES=(<*current-copies*><,*max-copies*>)
GDEVICE: GCOPIES=*current-copies*

current-copies
> is a nonnegative integer ranging from 0 through 255, but it cannot exceed the *max-copies* value specified. A value of 0 or 1 produces a single copy.

max-copies
> is a nonnegative integer ranging from 1 through 255.

If you do not specify GCOPIES, a default number of copies is searched for in this order:
1. the number of copies specified on an OPTIONS COPIES setting
2. 0 current copies, and 20 maximum copies.

Details

Not all devices have the capability to print multiple copies. See the **Gcopies** field in the Parameters window for your device to determine its capabilities.

GDDMCOPY

Instructs the driver to issue either an FSCOPY or GSCOPY call to GDDM when AUTOCOPY is in effect.

Used in: GOPTIONS statement
Default: FSCOPY
Restriction: GDDM device drivers on IBM mainframe systems only
See also: AUTOCOPY

Syntax

GDDMCOPY=FSCOPY | GSCOPY

FSCOPY
> used when sending output to an IEEE attached plotter.

GSCOPY
> used when creating an ADMPRINT file for output on 3287-type printers.

GDDMNICKNAME

Selects a GDDM nickname for the device to which output is sent.

Alias: GDDMN
Used in: GOPTIONS statement
Restriction: GDDM device drivers on IBM mainframe systems only

Syntax

GDDMNICKNAME=*nickname*

Details

Refer to the SAS Help facility for details on using GDDM drivers and options.

GDDMTOKEN

Selects a GDDM token for the device to which output is sent.

Alias: GDDMT
Used in: GOPTIONS statement
Restriction: GDDM device drivers on IBM mainframe systems only

Syntax

GDDMTOKEN=*token*

Details

Refer to the SAS Help facility for details on using GDDM drivers and options.

GDEST

Specifies the JES SYSOUT destination for IBM printers.

Used in: GOPTIONS statement
Default: LOCAL
Restriction: used only with IBM3287 and IBM3268 device drivers on z/OS systems

Syntax

GDEST=*destination*

GEND

Appends an ASCII string to every graphics data record that is sent to a device or file.

Used in: GOPTIONS statement; GDEVICE procedure; GDEVICE Gend window
Restriction: not supported by Java or ActiveX
See also: GSTART

Syntax

GEND='*string*' <...'*string-n*'>

'*string*'
  can be either of the following:
    '*hex-string*'X
    '*character-string*'
  In a GOPTIONS statement or in the GDEVICE procedure ADD or MODIFY statement, you can specify multiple strings with the GEND= option. In this case, you can mix the formats, specifying some as ASCII hexadecimal strings and some as character strings. Multiple strings are concatenated automatically.
  In the GEND window, enter the hexadecimal string without either quotation marks or a trailing x. Note, however, that the string must be entered as a hexadecimal string.
  PROC GOPTIONS always reports the value as a hexadecimal string.

Details

GEND is useful if you are creating a file and want to insert a carriage return at the end of every record. You can also use GEND in conjunction with the GSTART= graphics option or device parameter.

If you must specify the long and complicated initialization strings required by some devices (for example, PostScript printers), it is easier to use the GOPTIONS GEND= option rather than the GDEVICE Gend window because it is easier to code the string as text with GEND= than it is to convert the string to its ASCII representation, which is required to enter the string in the GDEVICE Gend window.

Note: On non-ASCII hosts, only ASCII hexadecimal strings produce consistent results in all instances because of the way the character strings are translated. In addition, the only way to specify a value for GEND that can be used by all hosts is to use an ASCII hexadecimal string; therefore, using an ASCII hexadecimal string to specify a value for GEND is the recommended method. △

GEPILOG

Sends a string to a device or file after all graphics commands are sent.

Used in: GOPTIONS statement; GDEVICE procedure; GDEVICE Gepilog window
Restriction: not supported by Java or ActiveX
See also: PREGEPILOG, POSTGEPILOG

Syntax

GEPILOG='*string*' <...'*string-n*'>

'string'
: can be either of the following:

  *'hex-string'*X

  'character-string'
  In a GOPTIONS statement or in the GDEVICE procedure ADD or MODIFY statement, you can specify multiple strings with the GEPILOG= option. In this case, you can mix the formats, specifying some as ASCII hexadecimal strings and some as character strings. Multiple strings are concatenated automatically.

  In the Gepilog window, enter the hexadecimal string without either quotation marks or a trailing x. Note, however, that the string must be entered as a hexadecimal string.

  PROC GOPTIONS always reports the value as a hexadecimal string.

Details

GEPILOG can be used in conjunction with the GPROLOG= graphics option or device parameter.

If you must specify the long and complicated initialization strings required by some devices (for example, PostScript printers), it is easier to use the GOPTIONS GEPILOG= option rather than the Gepilog window because it is easier to code the string as text with GEPILOG= than it is to convert the string to its ASCII representation, which is required to enter the string in the Gepilog window.

Note: On non-ASCII hosts, only ASCII hexadecimal strings produce consistent results in all instances because of the way the character strings are translated. In addition, the only way to specify a value for GEPILOG that can be used by all hosts is to use an ASCII hexadecimal string; therefore, using an ASCII hexadecimal string to specify a value for GEPILOG is the recommended method. △

GFORMS

Specifies the JES form name for IBM printers.

Used in: GOPTIONS statement
Default: STD
Restriction: used only with IBM3287 and IBM3268 device drivers on z/OS systems only

Syntax

GFORMS='*forms-code*'

GOUTMODE

Appends to or replaces the graphics output catalog.

Used in: GOPTIONS statement

Default: APPEND

Restriction: not supported by Java or ActiveX

Syntax

GOUTMODE=APPEND | REPLACE

APPEND
adds each new graph to the end of the current catalog.

REPLACE
replaces the contents of the catalog with the graph or graphs produced by a single procedure.

CAUTION:

If you specify REPLACE, the *entire contents* of the catalog are replaced, not just graphs of the same name. Graphs are added to the catalog for the duration of the procedure, but when the procedure ends and a new procedure begins, the contents of the catalog are deleted and the new graph or graphs are added. △

GPROLOG

Sends a string to device or file before graphics commands are sent.

Used in: GOPTIONS statement; GDEVICE procedure; GDEVICE Gprolog window

Restriction: not supported by Java or ActiveX

See also: PREGPROLOG, POSTGPROLOG

Syntax

GPROLOG='*string*' <...'*string-n*'>

'string'
can be either of the following:

*'hex-string'*X

'character-string'
In a GOPTIONS statement or in the GDEVICE procedure ADD or MODIFY statement, you can specify multiple strings with the GPROLOG= option. In this case, you can mix the formats, specifying some as ASCII hexadecimal strings and some as character strings. Multiple strings are concatenated automatically.

In the GPROLOG window, enter the hexadecimal string without either quotation marks or a trailing x. Note, however, that the string must be entered as a hexadecimal string.

PROC GOPTIONS always reports the value as a hexadecimal string.

Details

GPROLOG can be used in conjunction with the GEPILOG= graphics option or device parameter.

If you must specify the long and complicated initialization strings required by some devices (for example, PostScript printers), it is easier to use the GOPTIONS GPROLOG= option rather than the GDEVICE Gprolog window because it is easier to code the string as text with GPROLOG= than it is to convert the string to its ASCII representation, which is required to enter the string in the GDEVICE Gprolog window.

Note: On non-ASCII hosts, only ASCII hexadecimal strings produce consistent results in all instances because of the way the character strings are translated. In addition, the only way to specify a value for GEND that can be used by all hosts is to use an ASCII hexadecimal string; therefore, using an ASCII hexadecimal string to specify a value for GEND is the recommended method. △

GPROTOCOL

Specifies the protocol module to use when routing output directly to a printer or creating a graphics stream file (GSF) to send to a device attached to your host by a protocol converter.

Used in: GOPTIONS statement; GDEVICE procedure; GDEVICE Host File Options window

Restriction: not supported by Java or ActiveX

Default: host dependent

Syntax

GPROTOCOL=*module-name*

module-name **can be one of these**
   SASGPADE*

   SASGPAGL*

   SASGPASC

   SASGPAXI*

   SASGPCAB*

   SASGPCHK*

   SASGPDAT*

   SASGPDCA*

   SASGPHEX

   SASGPHYD*

   SASGPIDA*

   SASGPIDX*

   SASGPIMP*

SASGPIOC*

SASGPISI*

SASGPI24*

SASGPLCL*

SASGPNET*

SASGPMIC*

SASGPRTM*

SASGPSCS*

SASGPSTD

SASGPSTE*

SASGPTCX*

SASGPVAT*

SASGP497*

SASGP71

*Valid only for IBM mainframe systems.

Details

GPROTOCOL= specifies whether the graphics data generated by the SAS/GRAPH device driver should be altered and how the data should be altered. Unless you are using a protocol converter on an IBM mainframe, most devices do not require that the data be altered, and ordinarily, you do not have to change the default of GPROTOCOL.

On IBM hosts, the protocol module converts the graphics output to a format that can be processed by protocol converters. On other hosts, it can be used to produce a file in ASCII hexadecimal format.

Refer to the SAS Help facility for descriptions of these protocol modules.

Operating Environment Information: GPROTOCOL is valid only in certain operating environments. △

GRAPHRC

Specifies whether to return a step code at graphics procedure termination.

Used in: GOPTIONS statement

Restriction: not supported by Java or ActiveX

Default: GRAPHRC

Syntax

GRAPHRC | NOGRAPHRC

GRAPHRC
allows a return code at procedure termination. If the return code is not 0, the entire job might terminate.

NOGRAPHRC
always returns a step code of 0, even if the SAS/GRAPH program produced errors. As a result, the entire job's return code is unaffected by errors in any graphics procedure. NOGRAPHRC also overrides the ERRABEND system option.

Details

You typically use this option when you are running multiple jobs in a batch environment. It is useful primarily in an z/OS batch environment.

GSFLEN

Controls the length of records written to the graphics stream file (GSF).

Used in: GOPTIONS statement; GDEVICE procedure; GDEVICE Host File Options window

Default: device-dependent

Restriction: not supported by Java or ActiveX

See also: PROMPTCHARS

Syntax

GSFLEN=*record-length*

record-length
must be a nonnegative integer up to five digits long (0...99999). GSFLEN= specifies the length of the records written by the driver to a GSF or to the device.

If GSFLEN is 0, SAS/GRAPH uses the sixth byte of the PROMPTCHARS string to determine the length of the records. If the sixth byte of the PROMPTCHARS string is 00, the device driver sets the record length.

If you specify GACCESS=SASGAFIX and omit GSFLEN=, SAS/GRAPH uses the default length for the device.

Some values of the GPROTOCOL device parameter cause each byte in the data stream to be expanded to two bytes. This expansion is done after the length of the record is set by GSFLEN. If you are specifying a value for GPROTOCOL that does this (for example, SASGPHEX, SASGPLCL, or SASGPAGL), specify a value for GSFLEN that is half of the actual record length desired. For example, a value of 64 produces a 128-byte record after expansion by the GPROTOCOL module.

GSFMODE

Specifies the disposition of records written to a graphics stream file (GSF) or to a device or communications port by the device driver.

Used in: GOPTIONS statement; GDEVICE procedure; GDEVICE Host File Options window

Default: REPLACE

Restriction: not supported by Java or ActiveX

See also: GACCESS, GSFNAME

Syntax

GSFMODE=APPEND | PORT | REPLACE

APPEND
adds the records to the end of a GSF designated by the GACCESS= or GSFNAME= graphics option or device parameter. If the file does not already exist, it is created.
The destination can be either a specific file or an aggregate file storage location.
If the destination of the GSF is a specific file and you specify APPEND, SAS/GRAPH will add the new records to an existing GSF of the same name.
If the destination of the GSF is a file location and not a specific file, SAS/GRAPH will add the records to an external file whose name matches the name of the newly created catalog entry. For more information on how SAS/GRAPH names catalog entries, see "Specifying the Graphics Output File Type for Your Graph" on page 91.

Note: Some viewers of bitmapped output can view only one graph, even though multiple graphs are stored in the file. Therefore it might appear that a file contains only one graph when in fact it contains multiple graphs. △

PORT
sends the records to a device or communications port. The GACCESS= graphics option or device parameter should point to the desired port or device.

REPLACE
replaces the existing contents of a GSF designated by the GACCESS= or GSFNAME= graphics option or device parameter. If the file does not exist, it is created. REPLACE is always the default, regardless of the destination of the GSF.
If the destination of the GSF is a specific file and you specify REPLACE, SAS/GRAPH will replace an existing GSF with the contents of a newly created GSF of the same name.
If the destination of the GSF is a file location and not a specific file, SAS/GRAPH will replace an external file whose name matches the name of the newly created catalog entry. For more information on how SAS/GRAPH names catalog entries, see "Specifying the Graphics Output File Type for Your Graph" on page 91.

Details

When you create a GSF, the GSFNAME= or GACCESS= graphics option or device parameter controls where the output goes, and GSFMODE= controls how the driver writes graphics output records. If the output is to go to a file, specify APPEND or REPLACE. If the output is to go directly to a device or to a communications port, specify PORT. See "Specifying the Graphics Output File Type for Your Graph" on page 91 for more information on creating a graphics stream file.

GSFNAME

Specifies the fileref of the file or aggregate file location to which graphics stream file records are written.

Used in: GOPTIONS statement; GDEVICE procedure; GDEVICE Host File Options window

Restriction: Not valid for IBM32*xx*, linkable, Metagraphics, Java, or ActiveX drivers.

See also: GACCESS, GSFMODE

Syntax

GSFNAME=*fileref*

fileref
: specifies a fileref that points to the destination for the graphics stream file (GSF) output. *Fileref* must be a valid SAS fileref up to eight characters long and must be assigned with a FILENAME statement before running a SAS/GRAPH procedure that uses that fileref. The destination specified by the FILENAME statement can be either a specific file or an aggregate file location. See "FILENAME Statement" on page 36 for additional information on the FILENAME statement.

Details

Whether the resulting graphs are stored as one file or many files depends on both the type of destination and the setting of the GSFMODE= option.

If you specify a fileref with GSFNAME= and forget the FILENAME statement that defines the fileref, and if a destination is specified by the GACCESS= graphics option or device parameter, SAS/GRAPH assigns that destination to the fileref and sends the graphics output there. See also "GACCESS" on page 365.

See "Specifying the Graphics Output File Type for Your Graph" on page 91 for more information on creating graphics stream files.

GSFPROMPT

Specifies whether to write prompt messages to the graphics stream file (GSF).

Used in: GOPTIONS statement

Default: NOGSFPROMPT

Restriction: not supported by Java or ActiveX

Syntax

GSFPROMPT | NOGSFPROMPT

Details

When the GSF is processed by another program, that program can display the prompt messages. The default, NOGSFPROMPT, is compatible with Release 6.06.

Although the prompt messages appear if the graphics device is in eavesdrop mode, they do not wait for user response. If GSFPROMPT is on, the prompt messages are sent with the GSF to the device, regardless of the status of the graphics options PROMPT, GACCESS=, GSFMODE=, or GSFNAME=.

GSIZE

Sets the number of lines of display used for graphics for devices whose displays can be divided into graphics and text areas.

Used in: GOPTIONS statement; GDEVICE procedure; GDEVICE Parameters window

Restriction: not supported by Java or ActiveX

Default: device-dependent

Syntax

GSIZE=*lines*

lines
    specifies the number of lines to be used for graphics. *Lines* is a nonnegative integer up to three digits long (0...999), and can be larger or smaller than the total number of lines that can be displayed at one time. If the number is larger, scroll the graph to see it all. If GSIZE is 0, all lines are used for text.

GSTART

Prefixes every record of graphics data sent to a device or file with a string of characters.

Used in: GOPTIONS statement; GDEVICE procedure; GDEVICE Gstart window

Default: none

Restriction: not supported by Java or ActiveX

See also: GEND

Syntax

GSTART='*string* <...'*string-n*'>

'string'
    can be either of the following:

      *'hex-string'*X

      'character-string'

    In a GOPTIONS statement or in the GDEVICE procedure ADD or MODIFY statement, you can specify multiple strings with the GSTART= option. In this case, you can mix the formats, specifying some as ASCII hexadecimal strings and some as character strings. Multiple strings are concatenated automatically.

    In the GSTART window, enter the hexadecimal string without either quotation marks or a trailing x. Note, however, that the string must be entered as a hexadecimal string.

    PROC GOPTIONS always reports the value as a hexadecimal string.

Details

GSTART is useful when sending a file to a device that requires each record be prefixed with some character. You can use GSTART= in conjunction with the GEND= graphics option or device parameter.

If you must specify the long and complicated initialization strings required by some devices (for example, PostScript printers), it is easier to use the GOPTIONS GSTART= option rather than the GDEVICE Gstart window because it is easier to code the string as text with GSTART= than it is to convert the string to its ASCII representation, which is required to enter the string in the GDEVICE Gstart window.

Note: On non-ASCII hosts, only ASCII hexadecimal strings produce consistent results in all instances because of the way the character strings are translated. In addition, the only way to specify a value for GEND that can be used by all hosts is to use an ASCII hexadecimal string; therefore, using an ASCII hexadecimal string to specify a value for GEND is the recommended method. △

GUNIT

Specifies the default unit of measure to use with height specifications.

Used in: GOPTIONS statement

Default: CELLS

Restriction: partially supported by Java or ActiveX

Syntax

GUNIT=*units*

  units must be one of

CELLS	character cells
CM	centimeters
IN	inches
PCT	percentage of the graphics output area

PT   points (there are approximately 72 points in an inch).

Details

Used with options in the AXIS, FOOTNOTE, LEGEND, NOTE, SYMBOL, and TITLE statements and in some graphics options. If you specify a value but do not specify an explicit unit, the value of the GUNIT= graphics option is used. If the HSIZE= and VSIZE= options are specified then GUNIT is ignored and inches will be used.

GWAIT

Specifies the time between each graph displayed in a series.

Used in: GOPTIONS statement
Default: GWAIT=0
Restriction: not supported by Java or ActiveX

Syntax

GWAIT=*seconds*

seconds
  specifies the number of seconds between graphs. *Seconds* can be any reasonable positive integer. By default, GWAIT=0, which means that you must press the RETURN key between each display in a series of graphs.

Details

GWAIT= enables you to view a series of graphs without having to press the ENTER key (or the RETURN or END key, depending on your device) between each display. For example, if you specify GWAIT=5, five seconds elapse between the display of each graph in a series. If you use the NOPROMPT graphics option, the GWAIT= graphics option is disabled.

GWRITER

Specifies the name of the external writer used with IBM printers.

Used in: GOPTIONS statement
Default: SASWTR
Restriction: Used only with IBM3287 and IBM3268 device drivers on z/OS systems

Syntax

GWRITER='*writer-name*'

HANDSHAKE

Specifies the type of flow control used to regulate the flow of data to a hard copy device.

Used in: GOPTIONS statement; GDEVICE procedure; GDEVICE Parameters window
Default: host dependent
Restriction: not supported by Java or ActiveX

Syntax

HANDSHAKE=HARDWARE | NONE | SOFTWARE | XONXOFF

HARDWARE
HARD
   specifies that SAS/GRAPH instruct the device to use the hardware CTS and RTS signals. (This is not appropriate for some devices.)

NONE
   specifies that SAS/GRAPH send data without providing flow control. Specify NONE only if the hardware or interface program you are using provides its own flow control.

SOFTWARE
SOFT
   specifies that SAS/GRAPH use programmed flow control with plotters in eavesdrop mode.

XONXOFF
X
   specifies that SAS/GRAPH instruct the device to use ASCII characters DC1 and DC3. (This is not appropriate for some devices.)

Details

HANDSHAKE regulates flow of control by specifying how and if a device can signal to the host to temporarily halt transmission and then resume it. Flow control is important because it is possible to send commands to a hard copy device faster than they can be executed.

HANDSHAKE can be used when you are using a protocol converter, interface program, or host computer that can perform XONXOFF or hardware handshaking. You can also use this option if you are routing output through flow-control programs of your own, as in a multiple-machine personal computer environment where the graphics plotter is a shared resource. SAS/GRAPH software sends output to a server (the file transfer does not require flow control). The server queues incoming graphs and sends them to the plotter. The server, rather than SAS/GRAPH software, is responsible for handling flow control. An interface program is usually invoked by the line printer daemon and provides formatting or control signals for a system destination. The interface program typically includes port configuration options, such as baud, parity, and special character processing requirements (raw or cooked mode) for that destination.

If you do not use HANDSHAKE, the value in the driver entry is used.

If you use HANDSHAKE=XONXOFF or HANDSHAKE=HARDWARE, SAS/GRAPH does not actually do the handshaking. It tells the device which type of handshake is being used. The protocol converter, interface program, or host computer actually does the handshake.

Note: If you are creating a graphics stream file using a driver for a plotter and you specify HANDSHAKE=SOFTWARE, the software that you use to send the file to the plotter must be able to perform a software handshake. You will probably want to specify one of the alternative values if you route output to a file. △

HBY

Specifies the height of By lines generated when you use BY-group processing.

Used in: GOPTIONS statement

Default: One cell unless the HTEXT= option is used

Restriction: not supported by Java or ActiveX

See also: "BY Statement" on page 216

Syntax

HBY=*By line-height* <*units*>

By line-height <*units*>
: specifies the height of By line text; by default *By line-height* is 1. If you specify HBY=0, the BY headings are suppressed. For a description of *units*, see "Specifying Units of Measurement" on page 328.

    Note: If a value for *units* is not specified, the current units associated with the GUNIT graphics option are used. △

Details

When you use a BY statement with a SAS/GRAPH procedure to process a data set in subgroups, each graph produced by that procedure is headed by a By line that displays the BY variables and their values that define the current subgroup.

HEADER

Specifies the command that executes a user-supplied program to create HEADER records for the driver.

Used in: GDEVICE procedure; GDEVICE Metagraphics window

Restriction: Used only with user-supplied Metagraphics drivers.

See also: HEADERFILE

Syntax

HEADER='*command*'

command
: specifies a command that runs a user-written program that creates the file of HEADER records. *Command* is a string up to 40 characters long.

Details

For information about Metagraphics drivers, contact Technical Support.

HEADERFILE

Specifies the fileref for the file from which the Metagraphics driver reads HEADER records.

Used in: GDEVICE procedure; GDEVICE Metagraphics window
Restriction: Used only with user-supplied Metagraphics drivers.
See also: HEADER

Syntax

HEADERFILE=*fileref*

fileref
: specifies a valid SAS fileref up to eight characters long. *Fileref* must have been previously assigned with a FILENAME statement or a host command before running the Metagraphics driver. See "FILENAME Statement" on page 36 for details.

Details

For information about Metagraphics drivers, contact Technical Support.

HORIGIN

Sets the horizontal offset from the lower-left corner of the display area to the lower-left corner of the graph.

Used in: GOPTIONS statement; GDEVICE procedure; GDEVICE Detail window
Restriction: not supported by Java or ActiveX
See also: VORIGIN

Syntax

HORIGIN=*horizontal-offset* <IN | CM | PT>

horizontal-offset <IN | CM | PT>
> must be a nonnegative number and can be followed by a unit specification, either IN for inches (default), or CM for centimeters, or PT for points. If you do not specify HORIGIN, a default offset is searched for in this order:
> 1 the left margin specification on an OPTIONS LEFTMARGIN setting
> 2 HORIGIN setting in the device catalog.

Details

The display area is defined by the XMAX and YMAX device parameters. By default, the origin of the graphics output area is the lower-left corner of the display area; the graphics output is offset from the lower-left corner of the display area by the values of HORIGIN and VORIGIN. HORIGIN + HSIZE cannot exceed XMAX.

Note: When sending output to the PRINTER destination (ODS PRINTER), if you specify the VSIZE= option without specifying the HSIZE= option, the default origin of the graphics output area changes. The default placement of the graph changes from the lower-left corner of the display area to the top-center of the graphics output area. Likewise, if you specify the HSIZE= option without specifying the VSIZE= option, the graph is positioned at the top-center of the graphics output area by default. △

See "The Graphics Output and Device Display Areas" on page 59 for details.

HOSTSPEC

Stores FILENAME statement options in the device entry.

Used in: GDEVICE procedure; GDEVICE Host File Options window

Syntax

HOSTSPEC='*text-string*'

text-string
> specifies FILENAME statement options that are valid for the operating environment. *Text-string* accepts characters in upper or lower case. See the SAS documentation for your operating environment for details.

Details

HOSTSPEC can be used when the driver dynamically allocates a graphics stream file or spool file. It can specify the attributes of the file, such as record format or record length. It cannot be used with Metagraphics drivers.

HPOS

Specifies the number of columns in the graphics output area.

Used in: GOPTIONS statement

Default: device-dependent: the value of the LCOLS or PCOLS device parameter

Restriction: not supported by Java or ActiveX

See also: PCOLS, LCOLS, VPOS

Syntax

HPOS=*columns*

columns
: specifies the number of columns in the graphics output area, which is equivalent to the number of hardware characters that can be displayed horizontally. Specifying HPOS=0 causes the device driver to use the default hardware character cell width for the device.

Details

The HPOS= graphics option overrides the values of the LCOLS or PCOLS device parameters and temporarily sets the number of columns in the graphics output area. HPOS= does not affect the width of the graphics output area but merely divides it into columns. Therefore, you can use HPOS= to control cell width.

The values specified in the HPOS= and VPOS= graphics options determine the size of a character cell for the graphics output area and consequently the size of many graphics elements, such as device–resident text. The larger the size of the HPOS= and VPOS= values, the smaller the size of each character cell.

See "Overview" on page 59 for more information.

HSIZE

Sets the horizontal size of the graphics output area.

Used in: GOPTIONS statement; GDEVICE procedure; GDEVICE Detail window

Restriction: partially supported by Java or ActiveX

See also: VSIZE, XMAX

Syntax

HSIZE=*horizontal-size* <IN | CM | PT>

horizontal-size <**IN | CM | PT**>
: specifies the width of the graphics output area; *horizontal-size* must be a positive number and can be followed by a unit specification, either IN for inches (default), or CM for centimeters, or PT for points.

If you do not specify HSIZE=, a default size is searched for in this order:
1 the horizontal size is calculated as

   XMAX − LEFTMARGIN − RIGHTMARGIN

   Note that LEFTMARGIN and RIGHTMARGIN are used in the OPTIONS statement.
2 HSIZE setting in the device catalog.

HTEXT

Specifies the default height of the text in the graphics output.

Used in: GOPTIONS statement
Default: One cell
Restriction: partially supported by Java

Syntax

HTEXT=*text-height* <*units*>

text-height <*units*>
   specifies the height of the text; by default *text-height* is 1. For a description of *units*, see "Specifying Units of Measurement" on page 328.

   Note: If a value for *units* is not specified, the current units associated with the GUNIT graphics option are used. △

Details

HTEXT= is overridden by the HTITLE= graphics option for the *first* TITLE line.

Note: When you use ODS to send graphics to an HTML destination, and titles and footnotes are rendered as part of the HTML body file instead of the graphic image, you must specify the ODS USEGOPT statement for this option to work. See "Using Graphics Options with ODS (USEGOPT)" on page 195 for more information. △

HTITLE

Selects the default height used for the first TITLE line.

Used in: GOPTIONS statement
Default: Two cells unless HTEXT= is used

Syntax

HTITLE=*title-height* <*units*>

title-height <units>
specifies the height of the text in the TITLE1 statement. By default, *title-height* is 2. For a description of *units*, see "Specifying Units of Measurement" on page 328.

Note: If a value for *units* is not specified, the current units associated with the GUNIT graphics option are used. △

Details

If you omit the HTITLE= option, TITLE1 uses the height specified by the HTEXT= graphics option, if used.

Note: When you use ODS to send graphics to an HTML destination, and titles and footnotes are rendered as part of the HTML body file instead of the graphic image, you must specify the ODS USEGOPT statement for this option to work. See "Using Graphics Options with ODS (USEGOPT)" on page 195 for more information. △

IBACK

Specifies an image file to display in a graph's background area.

Restriction: partially supported by Java
See also: CBACK, IMAGESTYLE

Syntax

IBACK=*fileref* | '*external-file*' | '*URL*' | " "

fileref
specifies a fileref that points to the image file you want to use. *Fileref* must be a valid SAS fileref up to eight characters long and must have been previously assigned with a FILENAME statement.

external-file
specifies the complete filename of the image file you want to use. The format of external-file varies across operating environments.

URL
specifies the URL of the image file that you want to use.

Details

The image can be used with any procedures that produce a picture or support the CBACK= option. The IBACK option is supported by the Graph applet and the Map applet, but it is not supported by the Contour applet. See Chapter 16, "Introducing SAS/GRAPH Output for the Web," on page 439 for information about these applets.

This option overrides the BackGroundImage and Image styles attribute in the graph styles. To suppress a background image that is defined in a style or to reset the value of the IBACK= option, specify a blank space:

IBACK=" "

For more information on graph styles, refer to the TEMPLATE procedure documentation in *SAS Output Delivery System: User's Guide*.

For a list of the file types that you use, see "Image File Types Supported by SAS/GRAPH" on page 181.

ID

Specifies the description string used by the Metagraphics driver.

Used in: GDEVICE procedure; GDEVICE Metagraphics window

Restriction: Used only with user-supplied Metagraphics drivers.

Syntax

ID='*description*'

description
> is a character string up to 70 characters long. If this field is blank, the name and description of the graph as specified in the PROC GREPLAY window of the GREPLAY procedure are used.

Details

For information about Metagraphics drivers, contact Technical Support.

IMAGEPRINT

Enables or disables image output

Used in: GOPTIONS statement

Default: IMAGEPRINT

Restriction: not supported by Java or ActiveX

Syntax

IMAGEPRINT | NOIMAGEPRINT

IMAGEPRINT
> default value specifies that any images are to be included in graphics output.

NOIMAGEPRINT
> specifies that images are to be withheld from graphics output.

IMAGESTYLE

Specifies the way to display the image file that is specified on the IBACK= option.

Default: TILE

Restriction: not supported by Java

Syntax

IMAGESTYLE= TILE | FIT

TILE
   tile the image within the specified area. This copies the images as many times as needed to fit the area.

FIT
   fit the image within the background area. This stretches the image, if necessary.

Details

Note: This option overrides the BackGroundImage and Image styles attribute in the graph styles. For more information on graph styles, refer to the TEMPLATE procedure documentation in *SAS Output Delivery System: User's Guide*. △

INTERACTIVE

Sets level of interactivity for Metagraphics driver.

Used in: GDEVICE procedure; GDEVICE Metagraphics window

Default: USER

Restriction: Used only with user-supplied Metagraphics drivers.

Syntax

INTERACTIVE=USER | GRAPH | PROC

USER
   specifies that the user-written part of the driver be executed outside of SAS/GRAPH.

PROC
   specifies that the user-written part of the Metagraphics driver be invoked after the procedure is complete.

GRAPH
   specifies that the user-written part be invoked for each graph.

Details

For information about Metagraphics drivers, contact Technical Support.

INTERLACED

Specifies whether images are to be displayed as they are received in the browser.

Used in: GOPTIONS statement
Default: NONINTERLACED
Restriction: driver-dependent, GIF series of drivers only

Syntax

INTERLACED | NONINTERLACED

Details

With interlacing it is possible to get a rough picture of what a large image will look like before it is completely drawn in your browser. Your browser might allow you to set an option that will determine how images are displayed.

INTERPOL

Sets the default interpolation value for the SYMBOL statement.

Used in: GOPTIONS statement
Restriction: not supported by Java or ActiveX

Syntax

INTERPOL=*interpolation-method*

interpolation-method
  specifies the default interpolation to be used when the INTERPOL= option is not specified in the SYMBOL statement. See "SYMBOL Statement" on page 252 for the complete syntax of all interpolation methods.

ITERATION

Specifies the number of times to repeat the animation loop.

Used in: GOPTIONS statement

Default: 0

Restriction: GIFANIM driver only

Syntax

ITERATION=*iteration-count*

iteration-count
 specifies the number of times that your complete GIF animation loop is repeated. It is assumed that the animation is always played once; this option specifies how many times the animation is repeated. *Iteration-count* can be a number from 0...65535. A value of 0 causes the animation to loop continuously.

Details

In Version 6, the GCOPIES graphics option controlled iteration for the GIFANIM driver.

KEYMAP

Selects the keymap to use.

Used in: GOPTIONS statement

Default: installation dependent

Restriction: not supported by Java or ActiveX

Syntax

KEYMAP=*key-map-name* | NONE

key-map-name
 specifies the name of a keymap.

NONE
 suppresses the keymap assigned by default to a non-U.S. keyboard. If you specify KEYMAP=NONE, text might display incorrectly or not at all.

Details

Non-default key maps usually are used only with non-U.S. Keyboards.

LCOLS

Sets the number of columns in the graphics output area for landscape orientation.

Used in: GDEVICE procedure; GDEVICE Detail window
Default: device-dependent
See also: HPOS, LROWS, PCOLS

Syntax

LCOLS=*landscape-columns*

landscape-columns
   must be a nonnegative integer up to three digits long (0...999).

Details

Either the LROWS and LCOLS pair of device parameters or the PROWS and PCOLS pair of device parameters are required and must be nonzero.
   The HPOS= graphics option overrides the value of LCOLS.
   See "Overview" on page 59 for more information.

LFACTOR

Selects the default hardware line thickness.

Used in: GOPTIONS statement; GDEVICE procedure; GDEVICE Parameters window
Default: device-dependent
Restriction: Used only with devices that can draw hardware lines of varying thicknesses. Not supported by Java or ActiveX.

Syntax

LFACTOR=*line-thickness-factor*

line-thickness-factor
   can range from 0 through 9999. A value of 0 for LFACTOR is the same as a factor of 1. Lines are drawn *line-thickness-factor* times as thick as normal.

Details

LFACTOR is useful when you are printing graphics output on a plotter. Depending on the orientation and type of device, some plotters might require LFACTOR=10 to get the same thickness of lines as on the display of some devices.

LROWS

Sets the number of rows in the graphics output area for landscape orientation.

Used in: GDEVICE procedure; GDEVICE Detail window
Default: device-dependent
See also: LCOLS, PROWS, VPOS

Syntax

LROWS=*landscape-rows*

landscape-rows
   is a nonnegative integer up to three digits long (0...999).

Details

Either the LROWS and LCOLS pair of device parameters or the PROWS and PCOLS pair of device parameters are required and must be nonzero.
   The VPOS= graphics option overrides the value of LROWS.
   See "Overview" on page 59 for more information.

MAXCOLORS

Sets the total number of colors that can be displayed at once.

Used in: GDEVICE procedure; GDEVICE Parameters window
Default: device-dependent
See also: PENMOUNTS

Syntax

MAXCOLORS=*number-of-colors*

number-of-colors
   must be an integer in the range 2 through 256. The total number of colors includes the foreground colors plus the background color.

Details

The PENMOUNTS= graphics option overrides the value of MAXCOLORS.

MAXPOLY

Sets the maximum number of vertices for hardware-drawn polygons.

Used in: GDEVICE procedure; GDEVICE Parameters window

Default: device-dependent

Syntax

MAXPOLY=*number-of-vertices*

number-of-vertices
: is a nonnegative integer up to four digits long. A value of 0 means that there is no limit to the number of vertices that can be specified in the hardware's polygon-drawing command. The maximum value of MAXPOLY depends on the number of vertices your device can process.

MODEL

Specifies the model number of the output device.

Used in: GDEVICE procedure; GDEVICE Detail window
Default: device-dependent

Syntax

MODEL=*model-number*

model-number
: is a nonnegative integer up to five digits long that is the SAS-designated model number for the corresponding device. It is not the same as a manufacturer's model number.

Details

Do not change this field in SAS-supplied drivers or in drivers that you copy from SAS-supplied drivers.

MODULE

Specifies the name of the corresponding executable driver module for the device.

Used in: GDEVICE procedure; GDEVICE Detail window
Default: device-dependent

Syntax

MODULE=*driver-module*

driver-module
   is a literal string up to eight characters long. All standard driver modules begin with the characters SASGD.

Details

Do not change this field in SAS-supplied drivers or in drivers that you copy from SAS-supplied drivers.

NAK

Specifies the negative response for software handshaking for Metagraphics drivers.

Used in: GDEVICE procedure; GDEVICE Metagraphics window

Restriction: Used only with user-supplied Metagraphics drivers.

Syntax

NAK='*negative-handshake-response*'X

negative-handshake-response
   is a hexadecimal string up to 16 characters long.

Details

For information about Metagraphics drivers, contact Technical Support.

OFFSHADOW

Controls the width and depth of the drop shadow in legend frames.

Used in: GOPTIONS statement

Default: (0.0625, −0.0625) IN

Restriction: not supported by Java or ActiveX

Syntax

OFFSHADOW=(*x* <*units*>, *y* <*units*>) | (*x,y*) <*units*>

x,y
   specify the width (x) and depth (y) of the drop shadow generated by the LEGEND statement.

If a value for *units* is not specified, the current units associated with the GUNIT graphics option are used. For a description of *units*, see "Specifying Units of Measurement" on page 328.

Details

The values specified by OFFSHADOW= are used with the CSHADOW= and CBLOCK= options in a LEGEND statement. For details, see "LEGEND Statement" on page 225.

PAPERDEST

Specifies which output bin the printer should use if multiple bins are available on the device.

Used in: GOPTIONS statement; OPTIONS statement
Default: 1 (the upper output bin)
Restrictions: hardware-dependent, PostScript printers require a PPD file; not supported by Java or ActiveX
See also: PAPERSOURCE, PPDFILE

Syntax

PAPERDEST=*bin*

bin
: specifies the name or number of the output bin. Values for *bin* depend on the type of printer and can be one of the following:

  bin
  : the name or number of the output bin – for example, PAPERDEST=4, PAPERDEST=BIN2, PAPERDEST=SIDE

  '*long bin name*'
  : a character string that is the name of the output bin – for example, PAPERDEST='Top Output Bin'. Names with blanks or special characters must be quoted.

  For PostScript printers, the value for *bin* must correspond to an OutputBin value in the PPD file.

  For PCL printers, consult the printer's documentation for valid bin values. If a numeric value exceeds the maximum bin value allowed for the printer, a warning message is issued . For string values, the string is checked against a list of strings that are valid for the driver (for example, 'UPPER', 'LOWER', or 'OPTIONALOUTBIN*n*', where *n* is the bin number). If the string is not valid for the driver, a warning message is issued.

PAPERFEED

Specifies the increment of paper that is ejected when a graph is completed.

Used in: GOPTIONS statement; GDEVICE procedure; GDEVICE Detail window

Default: PAPERFEED=0.0 IN

Restriction: device-dependent; not supported by Java or ActiveX

Syntax

PAPERFEED=*feed-increment* <IN | CM>

feed-increment <IN | CM>
> must be a nonnegative number and can be followed by a unit specification, either IN for inches (default) or CM for centimeters.

Details

PAPERFEED does not control the total length of the ejection. If you specify PAPERFEED=1, the driver ejects paper in 1 inch increments until the total amount of paper ejected is at least half an inch greater than the size of the graph last printed. If you specify PAPERFEED=8.5 IN, the paper is ejected in increments of 8.5 inches, measuring from the origin of the first graph.

PAPERFEED is provided mainly for plotters that use fanfold or roll paper. If you are using fanfold paper, specify a value for PAPERFEED that is equal to the distance between the perforations.

PAPERLIMIT

Sets the width of the paper used with plotters.

Used in: GOPTIONS statement

Default: maximum dimensions specified in the device driver

Restriction: ZETA plotters and KMW rasterizers

Syntax

PAPERLIMIT=*width* <IN | CM>

width <IN | CM>
> specifies the paper width in IN for inches (default) or CM for centimeters. If PAPERLIMIT= is not specified, the maximum dimensions of the graph are restricted by the hardware limits of the graphics device.

Details

If you want to use a driver with a device that has a larger plotting area than the device for which the driver is intended (for example, using the ZETA887 driver with a ZETA 836 plotter), the PAPERLIMIT= graphics option can be used to override the size limit of the driver.

PAPERSIZE

Specifies the name of a paper size.

Used in: GOPTIONS statement; OPTIONS statement

Default: device-dependent

Restriction: hardware- dependent, PostScript printers require a PPD file; not supported by Java or ActiveX

See also: PAPERSOURCE, PPDFILE

Syntax

PAPERSIZE='*size-name*'

size-name
: specifies the name of a paper size, such as LETTER, LEGAL, or A4.

    If you do not specify the PAPERSIZE= option, the PAPERSIZE= option setting on an OPTIONS statement is used. If no OPTIONS statement sets a paper size, the value for paper size is device-dependent:

    □ The universal printing devices use the size specified in the Page Setup dialog box.

    □ All other printer devices use the LETTER paper size.

Details

Typically, you might use the PAPERSIZE= option with the Output Delivery System (ODS). For some printers, the PAPERSIZE= option overrides the PAPERSOURCE= option selection.

For PostScript devices, the name must match the name of a paper size in the PPD file. Refer to the PPD file for a list of valid names. *Size-name* is case-insensitive and can contain a subset of the full name. For example, if the name in the PPD file is *PageSize A4/A4, you can specify PAPERSIZE='A4'. If a PPD file is not specified, the PAPERSIZE= option is ignored.

For PCL devices, the device driver searches the SAS Registry for supported paper size values. To see the supported list of sizes, submit the following statements:

```
proc registry listhelp
    startat='options\papersize';
run;
```

For more information about the SAS Registry, refer to the SAS Help facility.

PAPERSOURCE

Specifies which paper tray the printer should use if multiple trays are available on the device.

Used in: GOPTIONS statement; OPTIONS statement

Default: device-dependent

Restriction: hardware- dependent, PostScript printers require a PPD file; not supported by Java or ActiveX

See also: PAPERDEST, PAPERSIZE, PPDFILE

Syntax

PAPERSOURCE=*tray*

tray
> specifies the name or number of the paper tray. Values for *tray* depend on the type of printer and can be one of the following:
>
> *tray*   the name or number of the paper tray, for example, PAPERSOURCE=3, PAPERSOURCE=TRAY3, PAPERSOURCE=Upper
>
> '*long tray name*'   a character string that is the name of the paper tray, for example, PAPERSOURCE='Optional Output Tray'. Names with blanks or special characters must be quoted.

Details

On some printers, if the PAPERSIZE= option is also specified, it overrides the setting on the PAPERSOURCE= option.

For PostScript printers, a tray number, such as PAPERSOURCE='tray3', must correspond to an InputSlot value in the PPD file.

For PCL printers, consult the printer's documentation for valid tray values. If a numeric value exceeds the maximum tray value allowed for the printer, a warning message is issued. For string values, the string is checked against a list of strings that are valid for the driver:

- 'AUTO'
- 'HCI' or 'HCI*n*', where *n* is a number from 2 to 21
- 'MANUAL'
- 'MANUAL_ENVELOPE'
- 'TRAY*n*', where *n* is 1, 2, or 3.

If the string is not valid for the driver, a warning message is issued.

PAPERTYPE

Specifies the name of a paper type.

Used in: GOPTIONS statement; OPTIONS statement

Default: PLAIN

Restriction: hardware- dependent, PostScript printers require a PPD file; not supported by Java or ActiveX

See also: PPDFILE

Syntax

PAPERTYPE='*type-name*'

type-name
> specifies the name of a paper type. Valid values depend on the type of printer.
> For PostScript devices, *type-name* must match the name of a paper type in the PPD file, such as TRANSPARENCY or PLAIN. Refer to the PPD file for a list of valid names. *Type-name* is case-insensitive and can contain a subset of the full name. For example, if the name in the PPD file is *MediaType Plain/Paper you can specify PAPERTYPE='PLAIN/PAPER'.
> For PCL devices, *type-name* specifies the name of a paper type that is available on the current printer, such as GLOSSY, PLAIN, SPECIAL, or TRANSPARENCY. Consult your printer's user manual for the complete list of available paper types on your printer.

Details

For PostScript devices, if a PPD file is not specified, the PAPERTYPE= option is ignored.

PATH

Sets the increment of the angle for device-resident text rotation.

Used in: GDEVICE procedure; GDEVICE Metagraphics window
Default: PATH=0
Restriction: Used only with user-supplied Metagraphics drivers.

Syntax

PATH=*angle-increment*

angle-increment
> is an integer in the range 0 to 360 that specifies the angle at which to rotate the text baseline. A value of 0 means that the device uses its default orientation. Specify 0 if your device does not perform string angling in hardware.

Details

For information about Metagraphics drivers, contact Technical Support.

PCLIP

Specifies whether a clipped polygon is stored in its clipped or unclipped form.

Used in: GOPTIONS statement

Default: NOPCLIP

Restriction: not supported by Java or ActiveX

See also: POLYGONCLIP

Syntax

PCLIP | NOPCLIP

PCLIP
   stores clipped polygons with the graph in the default catalog WORK.GSEG, or in the catalog you specify.

NOPCLIP
   stores the unclipped form of the polygon and causes the polygon to be clipped when replayed.

Details

The effects of this option are seen only when you use the graphics editor to edit a graph.

When a procedure produces a graph with intersecting polygons or blanking areas, it clips portions of the polygons to prevent the ones behind from showing through. When the graph is created and stored in a catalog, if PCLIP is in effect, the clipped form of the polygon is stored with it. If NOPCLIP is specified, the complete polygon is stored in the catalog and the graph is clipped each time it is replayed.

For example, suppose you create a block map like the one in Figure 15.3 on page 400.

Figure 15.3 Intersecting Polygons

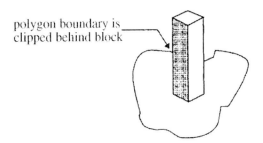

The block clips the boundary of the map area polygon. If you specify PCLIP, the map area polygon is stored in its clipped form, as shown in Figure 15.4 on page 400.

Figure 15.4 Clipped Polygon with PCLIP Option

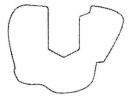

NOPCLIP stores the map area in its unclipped form, as shown in Figure 15.5 on page 401.

Figure 15.5 Polygon with NOPCLIP Option

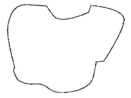

In this case, when the graph is recalled from the catalog, the map area polygon must be clipped before it is displayed with the block. If you plan to edit the graph with the graphics editor, specify NOPCLIP so polygons retain their original form.

PCOLS

Sets the number of columns in the graphics output area for portrait orientation.

Used in: GDEVICE procedure; GDEVICE Detail window
Default: device–dependent
See also: HPOS, LCOLS, PROWS

Syntax

PCOLS=*portrait-columns*

portrait-columns
   must be a nonnegative integer up to three digits long (0...999).

Details

Either the LROWS and LCOLS pair of device parameters or the PROWS and PCOLS pair of device parameters are required and must be nonzero.
   The HPOS= graphics option overrides the value of PCOLS.
   See "Overview" on page 59 for more information.

PENMOUNTS

Specifies the number of active pens or colors.

Used in: GOPTIONS statement
Default: device-dependent
Restriction: not supported by Java or ActiveX

See also: MAXCOLORS

Syntax

PENMOUNTS=*active-pen-mounts*

active-pen-mounts
: specifies the number of pens for a plotter with multiple pens. After the specified number of pens have been used, you are prompted to change the pens.

Details

For devices that are not pen plotters, PENMOUNTS= can be used to indicate the number of colors that can be displayed at one time. In this case, PENMOUNTS= performs the same function as the MAXCOLORS device parameter except that the value specified for MAXCOLORS includes the background color and PENMOUNTS only refers to foreground colors. Thus, PENMOUNTS=4 implies MAXCOLORS=5.

PENMOUNTS= overrides the value of the MAXCOLORS device parameter. You can specify MAXCOLORS= in a GOPTIONS statement as a synonym for PENMOUNTS=.

PENSORT

Specifies whether plotters draw graphics elements in order of color.

Used in: GOPTIONS statement; GDEVICE procedure; GDEVICE Parameters window
Restriction: not supported by Java or ActiveX
Default: device-dependent

Syntax

GOPTIONS: PENSORT | NOPENSORT

GDEVICE: PENSORT=Y | N

PENSORT
PENSORT=Y
> causes the plotter to draw all graphics elements of one color at one time. For example, it draws all the red elements in the output, then all the blue elements, and so on. This specification is compatible with previous releases. Use it for plotters with real pens.

NOPENSORT
PENSORT=N
> causes the plotter to draw each element as it is encountered, regardless of its color. For example, the plotter might draw a red circle, then a blue line, and then a red line, and so on. This method is best for electrostatic printers implemented with Metagraphics drivers of TYPE=PLOTTER. In addition, NOPENSORT enables you to specify non-standard color names.

PIEFILL

Specifies whether to use the device's hardware pie-fill capability.

Used in: GOPTIONS statement; GDEVICE procedure; GDEVICE Parameters window
Default: device-dependent
Restriction: not supported by Java or ActiveX

Syntax

GOPTIONS: PIEFILL | NOPIEFILL

GDEVICE: PIEFILL=Y | N

PIEFILL
PIEFILL=Y
> causes SAS/GRAPH to use the built-in hardware capability of the device, if available, to fill pies and pie sections. A blank `Piefill` field in the Parameters window is the same as PIEFILL=Y.
>
> Hardware drawing is faster, but not all devices have the capability. SAS/GRAPH does not try to use the capability if your device does not support it.

NOPIEFILL
PIEFILL=N
> causes SAS/GRAPH to fill pies and pie sections using software pie fills.

POLYGONCLIP

Specifies the type of clipping used when two polygons overlap.

Used in: GOPTIONS statement

Default: device-dependent
Restriction: not supported by Java or ActiveX
See also: PCLIP

Syntax

POLYGONCLIP | NOPOLYGONCLIP

POLYGONCLIP
    specifies polygon clipping, which enables a clipped polygon to be filled with a hardware pattern. POLYGONCLIP affects only graphs that have blanking areas or intersecting polygons.

NOPOLYGONCLIP
    specifies line clipping; a polygon that has been line-clipped cannot use a hardware pattern.

Details

Clipping is the process of removing part of one polygon when two polygons intersect. For example, in a block map, a block might overlap the boundary of its map area. In this case, the polygon that makes up the map area is clipped so that you do not see the boundary line behind the block. (See Figure 15.3 on page 400 for an illustration of a clipped polygon.) The type of clipping used by a graph affects whether a clipped area can use hardware patterns.

POLYGONCLIP is affected by the PCLIP graphics option:

POLYGONCLIP with PCLIP or NOPCLIP
    all areas can use hardware patterns

NOPOLYGONCLIP with NOPCLIP
    all areas use only software patterns

NOPOLYGONCLIP with PCLIP
    areas can use either hardware or software patterns depending on the nature of the clipped polygons.

Under some conditions the polygons might not be clipped correctly. Specifying both POLYGONCLIP and NOPCLIP will produce the correct graph.

POLYGONFILL

Specifies whether to use the hardware polygon-fill capability.

Used in: GOPTIONS statement; GDEVICE procedure; GDEVICE Parameters window
Default: device-dependent
Restriction: not supported by Java or ActiveX

Syntax

GOPTIONS: POLYGONFILL | NOPOLYGONFILL

GDEVICE: POLYFILL=Y | N

POLYGONFILL
POLYFILL=Y
    causes SAS/GRAPH to use the built-in hardware capability of the device to fill polygons. A blank `Polyfill` field in the Parameters window is the same as POLYGONFILL.
    hardware drawing is faster, but not all devices have the capability. SAS/GRAPH does not try to use the capability if your device does not support it.

NOPOLYGONFILL
POLYFILL=N
    causes SAS/GRAPH to use software fills to fill polygons.

POSTGEPILOG

Specifies data to send immediately after the data that is stored in the Gepilog field of the device entry is sent.

Used in: GOPTIONS statement
Default: Null string
Restriction: not supported by Java or ActiveX
See also: GEPILOG, PREGEPILOG

Syntax

POSTGEPILOG='*string*'

'*string*'
    can be either of the following:
    '*hex-string*'X
    '*character-string*'
        PROC GOPTIONS always reports the value as a hexadecimal string.

POSTGPROLOG

Specifies the data to send immediately after the data that is stored in the Gprolog field of the device entry is sent.

Used in: GOPTIONS statement
Default: Null string
Restriction: not supported by Java or ActiveX
See also: GPROLOG, PREGPROLOG

Syntax

POSTGPROLOG=*'string'*

'string'
   can be either of the following:
   *'hex-string'*X
   'character-string'
   PROC GOPTIONS always reports the value as a hexadecimal string.

POSTGRAPH

Specifies host commands to be executed after the graph is produced.

Used in: GOPTIONS statement; GDEVICE procedure; GDEVICE Host Commands window

Restriction: not supported by Java or ActiveX

See also: FILECLOSE

Syntax

POSTGRAPH1=*'system-command(s)'*

POSTGRAPH2=*'system-command(s)'*

system-command(s)
   specifies one or more valid system commands. The string can contain upper- or lowercase characters. Separate multiple commands with a command delimiter, which is host-specific; for example, some operating environments use a semicolon. The total length of the string cannot exceed 72 characters. The commands are executed right after the graph is produced.

Details

If you want to use a host command to send output to the device after each graph executes, use the POSTGRAPH parameter with FILECLOSE=GRAPHEND.

PPDFILE

Specifies the location of an external file containing PostScript Printer Description (PPD) information.

Used in: GOPTIONS statement

Restriction: PostScript printers only

See also: BINDING, COLLATE, DUPLEX, PAPERDEST, PAPERSIZE, PAPERSOURCE, PAPERTYPE, REVERSE

Syntax

PPDFILE=*fileref* | '*external-file*'

fileref
: specifies a fileref that points to the PPD file you want to use. *Fileref* must be a valid SAS fileref up to eight characters long and must have been previously assigned with a FILENAME statement.

external-file
: specifies the complete filename of the PPD file you want to use. The format of *external-file* varies across operating environments. For details, see the SAS documentation for your operating environment.

Details

A PostScript Printer Description (PPD) file is a text file that contains commands required to access features of the device. These files are available from Adobe. Also, many printer manufacturers provide the appropriate PPD file for their PostScript printers.

PREGEPILOG

Specifies data to send immediately before the data that is stored in the Gepilog field of the device entry is sent.

Used in: GOPTIONS statement
Default: Null string
Restriction: not supported by Java or ActiveX
See also: GEPILOG, POSTGEPILOG

Syntax

PREGEPILOG='*string*'

'*string*'
: can be either of the following:
   '*hex-string*'X
   '*character-string*'
   PROC GOPTIONS always reports the value as a hexadecimal string.

PREGPROLOG

Specifies the data to send immediately before the data that is stored in the Gprolog field of the device entry is sent.

Used in: GOPTIONS statement
Default: Null string
Restriction: not supported by Java or ActiveX
See also: GPROLOG, POSTGPROLOG

Syntax

PREGPROLOG='*string*'

'*string*'
    can be either of the following:
        '*hex-string*'X
        '*character-string*'
    PROC GOPTIONS always reports the value as a hexadecimal string.

PREGRAPH

Specifies host commands to be executed before the graph is produced.

Used in: GOPTIONS statement; GDEVICE procedure; GDEVICE Host Commands window
Restriction: not supported by Java or ActiveX
See also: FILECLOSE

Syntax

PREGRAPH1='*system-command(s)*'

PREGRAPH2='*system-command(s)*'

system-command(s)
    specifies one or more valid system commands. The string can contain upper- or lowercase characters. Separate multiple commands with a command delimiter, which is host-specific; for example, some operating environments use a semicolon. The total length of the string cannot exceed 72 characters. The commands are executed immediately before the graph is produced.

Details

The PREGRAPH parameter should be used with FILECLOSE=GRAPHEND.

PROCESS

Specifies the command that translates the metafile into commands for the device.

Used in: GDEVICE procedure; GDEVICE Metagraphics window
Restriction: Used only with user-supplied Metagraphics drivers.
See also: INTERACTIVE

Syntax

PROCESS='*command*'

command
: specifies the command that translates the metafile produced by the Metagraphics driver into commands for the device. The command runs your program to produce the output. *Command* is a string up to 40 characters long.

Details

PROCESS is required if the value of the INTERACTIVE device parameter is PROC or GRAPH.

For information about Metagraphics drivers, contact Technical Support.

PROCESSINPUT

Specifies the fileref for the file that contains input for the user-written part of the Metagraphics driver.

Used in: GDEVICE procedure; GDEVICE Metagraphics window
Restriction: Used only with user-supplied Metagraphics drivers.

Syntax

PROCESSINPUT=*fileref*

fileref
: specifies a valid SAS fileref up to eight characters long. *Fileref* must be assigned with a FILENAME statement or a host command before running the Metagraphics driver. See "FILENAME Statement" on page 36 *SAS/GRAPH: Reference* for additional information.

Details

For information about Metagraphics drivers, contact Technical Support.

PROCESSOUTPUT

Specifies the fileref for the file that receives output from the user-written part of the Metagraphics driver.

Used in: GDEVICE procedure; GDEVICE Metagraphics window
Restriction: Used only with user-supplied Metagraphics drivers.

Syntax

PROCESSOUTPUT=*fileref*

fileref
   specifies a valid SAS fileref up to eight characters long. *Fileref* must be assigned with a FILENAME statement or a host command before running the Metagraphics driver. See "FILENAME Statement" on page 36 for additional information.

Details

For information about Metagraphics drivers, contact Technical Support.

PROMPT

Specifies whether prompts are issued.

Used in: GOPTIONS statement; GDEVICE procedure; GDEVICE Parameters window
Restriction: not supported by Java or ActiveX
Default: device-dependent

Syntax

GOPTIONS: PROMPT | NOPROMPT

GDEVICE: PROMPT=0...7

PROMPT
   causes all prompts to be displayed.

NOPROMPT
   suppresses all prompts. NOPROMPT overrides the GWAIT= graphics option.

PROMPT=0...7
   in the GDEVICE procedure, specifies the level of prompting:

0	provides no prompting
1	issues startup messages only. Startup messages are messages such as PLEASE PRESS RETURN TO CONTINUE.
2	signals end of graph if device is a video display or sends message to change paper if device is a plotter.
3	combines the effects of 1 and 2.
4	sends a message to mount pens if the device is a plotter.

5	combines the effects of 4 and 1.
6	combines the effects of 4 and 2.
7	sends all prom

Note: If you specify either 0 for the PROMPT device parameter or NOPROMPT in a GOPTIONS statement for a display device, the display clears immediately after the graph is drawn. △

In the GDEVICE Parameters window, the PROMPT parameter consists of four fields that describe the type of prompt:

start up
  issues a message to turn the device on (if the device is a hardcopy device) or the message PLEASE PRESS RETURN AFTER EACH BELL TO CONTINUE.

end of graph
  signals, usually by a bell, when the graph is complete (valid for video displays only).

mount pens
  issues a message to mount pens in a certain order and (for certain devices only) to ask for pen priming strokes for plotters.

change paper
  prompts the user to change the paper (valid for plotters only).

Enter an X for each prompt that you want to be given. If no Xs appear in these fields, no prompt messages are issued, and the device does not wait for you to respond between graphs.

PROMPTCHARS

Selects the prompt characters to be used by SAS/GRAPH device drivers.

Used in: GOPTIONS statement; GDEVICE procedure; GDEVICE Parameters window
Default: host dependent
Restriction: not supported by Java or ActiveX
See also: GSFLEN, HANDSHAKE

Syntax

PROMPTCHARS='*prompt-chars-hex-string*'X

prompt-chars-hex-string
  is an 8-byte hexadecimal string that is specified as 16 hexadecimal characters. In GDEVICE procedure statements, enclose the string in single quotation marks, followed by an X. In the Parameters window, enter the hexadecimal string without either quotation marks or a trailing X.

Note: Bytes 1, 4, and 5 are the safest for you to change because you are most likely to know the correct value for them. Check with Technical Support before changing any of the other bytes. △

The following list describes each byte in the string:

byte 1
is the ASCII code of the system prompt character (for software handshaking). The system prompt character is the last character that the host sends before waiting for a response from the plotter. For example, 11 means the host sends an XON or DC1 character as a prompt. If the host does not send a special character for a prompt, set this byte to 00.

byte 2
is the ASCII code of the echo-terminator character (for software handshaking). This character is sent at the beginning of each record.

byte 3
prevents splitting commands across records if the value is 01. If you are creating a graphics stream file to send to a device at a later time, and there is the possibility that extra characters will be added between records during transmission, setting the third byte to 01 reduces the likelihood that the extra characters will be interpreted as graphics commands and cause stray lines or other device characters. If the third byte is set to 00, the driver makes the records as long as possible and splits device commands across records if necessary. Setting the third byte to 00 is more efficient but is more likely to result in device errors if output is written to a file and later transmitted to the device.

byte 4
is the line-end character (for software handshaking). It indicates that more data can be sent. This character is almost always a carriage-return character, 0D.

byte 5
specifies turnaround delay in tenths of a second (for software handshaking). The turnaround delay is the amount of time the device waits after receiving the prompt character before sending the line-end character. For example, a value of 05 represents a half-second delay.

byte 6
sets default record length using a hexadecimal value 00–FF. This byte sets the length of the records sent to the device or to a file. If this byte is set to 00 (the default), SAS/GRAPH uses the longest record length possible for the device. To specify an alternate length, set the sixth byte to the hexadecimal value for the desired length. For example, to generate records of length 80, specify 50 for the sixth byte. If the GSFLEN device parameter or graphics option is specified, its value overrides the value of the sixth prompt character.

Some values of the GPROTOCOL device parameter cause each byte in the data stream to be expanded to two bytes. This expansion is done after the length of the record is set by PROMPTCHARS. If you are specifying a value for GPROTOCOL that does this (for example, SASGPHEX, SASGPLCL, or SASGPAGL), specify a value for the sixth byte of PROMPTCHARS that is half of the actual record length desired. For example, a hexadecimal value of 40 (64 decimal) produces a 128-byte record after expansion by the GPROTOCOL module.

bytes 7 and 8
are unused and should be set to 0000.

Details

PROMPTCHARS is most commonly used to specify parameters used in software handshaking (see "HANDSHAKE" on page 380), but it can also be used to control the length of records written by most drivers. You can also use the GSFLEN= graphics option for this purpose.

PROWS

Sets the number of rows in the graphics output area for portrait orientation.

Used in: GDEVICE procedure; GDEVICE Detail window
Default: device–dependent
See also: LROWS, PCOLS, VPOS

Syntax

PROWS=*portrait-rows*

portrait-rows
is a nonnegative integer up to three digits long (0...999).

Details

Either the LROWS and LCOLS pair of device parameters or the PROWS and PCOLS pair of device parameters are required and must be nonzero.
 The VPOS= graphics option overrides the value of PROWS.
 See "Overview" on page 59 for more information.

QMSG

Specifies whether log messages are held until after the graphics output is displayed.

Used in: GDEVICE procedure; GDEVICE Detail window
Default: device–dependent

Syntax

GOPTIONS: QMSG | NOQMSG

GDEVICE: QMSG=Y | N

QMSG QMSG=Y
 queues driver messages while the device is in graphics mode (default for video devices).

NOQMSG QMSG=N
prevents the queuing of messages (default for plotters, cameras, and printers).

Details

Message queuing is desirable on display devices that do not have a separate dialog box and graphics area. If messages are not queued, they are written to the log as the graphics output is being generated. This behavior can cause problems on some devices.

A blank `Queued messages` field in the Parameters window can mean either Y or N, depending on the device.

RECTFILL

Specifies which rectangle fills should be performed by hardware.

Used in: GDEVICE procedure; GDEVICE Parameters window
Default: device-dependent
See also: FILL

Syntax

RECTFILL='*rectangle-fill-hex-string*'X

rectangle-fill-hex-string
  is a hexadecimal string that is 16 characters long. In GDEVICE procedure statements, enclose the string in single quotation marks, followed by an X. In the Parameters window, enter the hexadecimal string without either quotation marks or a trailing X.

  The following table shows which bit position (left-to-right) within the hexadecimal string controls each fill pattern.

Bit	Fill pattern	Bit	Fill pattern
1	R1	9	L4
2	R2	10	L5
3	R3	11	X1
4	R4	12	X2
5	R5	13	X3
6	L1	14	X4
7	L2	15	X5
8	L3	16	S

  For example, if you want the driver to use only the L1 and R1 fills in hardware, the first and sixth bits of the first byte of the hexadecimal string should be turned on,

which corresponds to a value of '8400000000000000'X ('84'X is equivalent to '1 0 0 0 0 1 0 0' in binary). If a particular hardware rectangle fill is not available or not to be used (as indicated by the value of RECTFILL), the fill is generated by the software.
See "PATTERN Statement" on page 240 for an illustration of the fill patterns.

Details

Note: Not all devices support this capability. If FILL=N is specified or the NOFILL option is used in a GOPTIONS statement, RECTFILL is ignored. △

RENDER

Controls the creation and disposition of rendered Bitstream fonts.

Used in: GOPTIONS statement

Default: MEMORY

Restriction: not supported by Java or ActiveX

See also: RENDERLIB

Syntax

RENDER=APPEND | DISK | MEMORY | NONE | READ

APPEND
creates files to store rendered versions of Bitstream fonts if the files do not already exist, reads previously rendered characters from the font files, and appends rendered versions of new characters to the font files when the SAS/GRAPH procedure terminates.

DISK
creates files to store rendered versions of Bitstream fonts if the files do not already exist, reads previously rendered characters from the font files, and appends rendered versions of new characters to the font files as they are encountered. This method is slower on some hosts, but it can work in memory-constrained conditions where the other rendering methods fail.

MEMORY
renders all fonts in memory without creating any font files on disk. Font files are not used even if they already exist. New characters are not written to existing font files when SAS/GRAPH procedures terminate.
  This is the default and should be the fastest method on hosts that support virtual memory.

NONE
disables the font rendering features.

READ
reads existing rendered font files but does not create new font files or write new characters to existing font files. This is useful only when font files already exist in the rendered font library.

Details

The memory capacity and input/output characteristics of your host system determine which value for the RENDER= option provides the best performance.

RENDERLIB

Specifies the SAS library in which rendered font files are stored.

Used in: GOPTIONS statement
Default: WORK
Restriction: not supported by Java or ActiveX
See also: RENDER

Syntax

RENDERLIB=*libref*

libref
  specifies a previously defined libref that identifies the SAS library. The default library is WORK. See "LIBNAME Statement" on page 36 for more information on assigning a libref.

REPAINT

Specifies how many times to redraw the graph.

Used in: GOPTIONS statement; GDEVICE procedure; GDEVICE Parameters window
Default: device–dependent
Restriction: not supported by Java or ActiveX

Syntax

REPAINT=*redraw-factor*

redraw-factor
  is a nonnegative integer up to three digits long (0...999).

Details

Use this option with printers that produce light images after only one pass. This option also is useful for producing transparencies; multiple passes make the colors more solid or more intense.

  Not all devices have this capability.

RESET

Resets graphics options to their defaults and/or cancels global statements.

Used in: GOPTIONS statement

Syntax

RESET=ALL | GLOBAL | *statement-name* | (*statement-name(s)*)

ALL
   sets all graphics options to defaults and cancels all global statements.

GLOBAL
   cancels all global statements (AXIS, FOOTNOTE, LEGEND, PATTERN, SYMBOL, and TITLE). Options in the GOPTIONS statement are unaffected.

statement-name
   resets or cancels only the specified global statements. For example, RESET=PATTERN cancels all PATTERN statements only. To cancel several statements at one time, enclose the statement names in parentheses. For example, RESET=(TITLE FOOTNOTE AXIS).

   Note: RESET=GOPTIONS sets all graphics options to defaults but does not cancel any global statements. △

Featured in: "Example 10. Creating a Bar Chart with Drill-Down Functionality for the Web" on page 321

Details

RESET=ALL or RESET=GOPTIONS must be the first option specified in the GOPTIONS statement; otherwise, the graphics options that precede the RESET= option in the GOPTIONS statement are reset. Other options can follow the RESET= graphics option in the statement.

REVERSE

Specifies whether to print the output in reverse order, if reverse printing is supported by the device.

Used in: GOPTIONS statement
Default: NOREVERSE
Restrictions: hardware-dependent, PostScript printers require a PPD file; not supported by Java or ActiveX
See also: PPDFILE

Syntax

REVERSE | NOREVERSE

Details

The purpose of REVERSE is to control the stacking order of printer output, depending on how the printer stacks paper. On some printers, reverse implies using the alternate output bin (back of the printer).

For PCL devices, REVERSE sends output to the LOWER out bin, which is the face-up output bin.

For PostScript devices, if the PPD file has an "OutputOrder" entry and one of its entries is "Reverse," the device supports reverse order printing and the appropriate PostScript code to activate reverse will be used. If the PPD file does not have an "OutputOrder" entry but does have a "PageStackOrder" entry and corresponding OutputBin value, then reverse order printing is supported indirectly, using the PPD file's PageStackOrder/OutputBin entries.

Note: Some PostScript devices implement Reverse as the default output mode for one of the output bins. In this case, selecting either the "reverse" output bin or specifying REVERSE mode produces identical results. △

ROTATE

Specifies whether and how to rotate the graph.

Used in: GOPTIONS statement; GDEVICE procedure; GDEVICE Detail window
Restriction: not supported by Java or ActiveX

Syntax

GOPTIONS: ROTATE=LANDSCAPE | PORTRAIT
GOPTIONS: ROTATE | NOROTATE
GDEVICE: ROTATE=LANDSCAPE | PORTRAIT

ROTATE | NOROTATE
  specifies whether to rotate the graph 90 degrees from its default orientation.

ROTATE=LANDSCAPE
  specifies landscape orientation (the graph is wider than it is high).

ROTATE=PORTRAIT
  specifies portrait orientation (the graph is higher than it is wide).

If you do not specify a rotation, a default is searched for in this order:
1 the ORIENTATION setting on an OPTIONS statement
2 device-dependent default.

ROTATION

Sets the increment of the angle by which the device can rotate any given letter in a string of text in a Metagraphics driver.

Used in: GDEVICE procedure; GDEVICE Metagraphics window
Default: ROTATION=0
Restriction: Used only with user-supplied Metagraphics drivers.

Syntax

ROTATION=*angle-increment*

angle-increment
>specifies the increment of the angle at which to rotate individual characters, for example, every 5 degrees, every 45 degrees, and so on. *Angle-increment* is an integer in the range 0 to 360. A value of 0 means that the device uses its default character rotation. Specify 0 if your device does not perform hardware character rotation.

Details

For information about Metagraphics drivers, contact Technical Support.

ROWS

Specifies the number of rows the device-resident font uses in graphics output.

Used in: GDEVICE Chartype window; GDEVICE procedure; CHARREC= option
Default: 0
See also: CHARREC

Syntax

See "CHARREC" on page 337 for syntax.

Details

If you are using a device driver from SASHELP.DEVICES, this parameter already is set for device-resident fonts that have been defined for your installation. For scalable fonts, you can specify 1 for ROWS, and the actual number of rows will be computed based on the current text width. If you are adding to or modifying device-resident fonts available for a particular device driver, specify a positive value for the ROWS device parameter. If ROWS is greater than 0, it overrides the values of the LROWS and PROWS device parameters.

SCALABLE

Specifies whether a font is scalable.

Used in: GDEVICE Chartype window; GDEVICE procedure; CHARREC= option

Default: device–dependent

See also: CHARTYPE

Syntax

See "CHARREC" on page 337 for syntax.

Details

A device-resident font is scalable if it can be used with any combination of rows and columns. Use the SCALABLE device parameter if you are adding to or modifying the fonts available for a particular device driver. If you are using a device driver from SASHELP.DEVICES, this parameter already is set for device-resident fonts that have been defined for your installation.

SIMFONT

Specifies a SAS/GRAPH font to use if the default device-resident font cannot be used.

Used in: GOPTIONS statement

Default: SIMULATE

Restriction: not supported by Java or ActiveX

Syntax

SIMFONT=*SAS/GRAPH-font*

SAS/GRAPH-font
  specifies a SAS/GRAPH font to use instead of the default device-resident font. By default, this is the SIMULATE font, which is stored in the SASHELP.FONTS catalog.

Details

SAS/GRAPH substitutes the SAS/GRAPH font specified by the SIMFONT= option for the default device-resident font in these cases:

- when you use the NOCHARACTERS option in a GOPTIONS statement
- when you specify a non-default value for the HPOS= or VPOS= graphics option and your device does not have scalable hardware characters
- when you replay a graph using a device driver other than the one used to create the graph
- when you specify an angle or rotation for your hardware text that the device is not capable of producing
- when you specify a device-resident font that is not supported by your device.

See Chapter 11, "Specifying Fonts in SAS/GRAPH Programs," on page 155 for details.

SPEED

Selects pen speed for plotters with variable speed selection.

Used in: GOPTIONS statement; GDEVICE procedure; GDEVICE Parameters window
Default: device-dependent
Restriction: not supported by Java or ActiveX

Syntax

SPEED=*pen-speed*

pen-speed
> specifies a percentage (1 through 100) of the maximum pen speed for the device. For example, SPEED=50 slows the drawing speed by half. In general, slowing the drawing speed produces better results.
> By default, the value of SPEED is the normal speed for the device.

SWAP

Specifies whether to reverse BLACK and WHITE in the graphics output.

Used in: GOPTIONS statement; GDEVICE procedure; GDEVICE Parameters window
Defaults: GOPTIONS: NOSWAP; GDEVICE: SWAP=N
Restriction: not supported by Java or ActiveX

Syntax

GOPTIONS: SWAP | NOSWAP

GDEVICE: SWAP=Y | N

SWAP
SWAP=Y
> swaps BLACK for WHITE and vice versa.

NOSWAP
SWAP=N
> does not swap the colors. A blank **Swap** field in the Parameters window is the same as SWAP=N.

Details

SWAP does not affect the background color and only affects BLACK and WHITE foreground colors specified as predefined SAS color names. SWAP ignores BLACK and WHITE specified in HLS, RGB, or gray-scale format. This option is useful when you want to preview a graph on a video device and send the final copy to a printer that uses a white background.

```
goptions reset=all cback=blue ctitle=black swap;
title1 h=8 'swap test';
title2 h=8 'another title';
proc gslide border;
run;
```

SWFONTRENDER

Specifies the method used to render system fonts.

Used in: GOPTIONS statement

Default: device-dependent

Restriction: not supported by Java or ActiveX

Syntax

SWFONTRENDER = POLYGON | SCANLINE

SWFONTRENDER = POLYGON
 uses polygon rendering

SWFONTRENDER = SCANLINE
 uses scanline rendering

Details

SWFONTRENDER determines the method used to render system text to a vector graphics file. In some graphics formats, SCANLINE rendering can produce better quality output might be distorted if the output is replayed on a device with a different resolution than the original device. If the system text is rendered as a POLYGON, resizing the graph will not distort the text.

SYMBOL

Specifies whether to use the device's symbol-drawing capability.

Used in: GOPTIONS statement; GDEVICE procedure; GDEVICE Parameters window

Default: device-dependent

Restriction: not supported by Java or ActiveX

See also: SYMBOLS

Syntax

GOPTIONS: SYMBOL | NOSYMBOL

GDEVICE: SYMBOL=Y | N

SYMBOL
SYMBOL=Y
causes SAS/GRAPH to use the built-in symbol-drawing capability of the device, if available. A blank `Symbol` field in the Parameters window is the same as SYMBOL=Y.

hardware drawing is faster, but not all devices have the capability. SAS/GRAPH does not try to use the capability if your device does not support it.

NOSYMBOL
SYMBOL=N
causes SAS/GRAPH to draw the symbols using SAS/GRAPH fonts.

SYMBOLS

Specifies which symbols can be generated by hardware.

Used in: GDEVICE procedure; GDEVICE Parameters window
Default: device–dependent
See also: "SYMBOL Statement" on page 252

Syntax

SYMBOLS='*hardware-symbols-hex-string*'X

hardware-symbols-hex-string
is a hexadecimal string that is 16 characters long and must be completely filled. This table shows which bit position (left-to-right) within the hexadecimal string controls each hardware symbol.

Bit to turn on	Symbol Description	Symbol
1	PLUS	+
2	X	×
3	STAR	✶
4	SQUARE	□
5	DIAMOND	◇
6	TRIANGLE	△
7	HASH	♯
8	Y	Y
9	Z	Z
10	PAW	⁖
11	POINT	·
12	DOT	●
13	CIRCLE	○

For example, if you want the driver to do only the PLUS and X symbols in hardware, the first and second bits of the first byte of the hexadecimal string should be turned on, which would correspond to a value of 'C000000000000000'X ('C0'X is equivalent to '1 1 0 0 0 0 0 0' in binary).

Details

These are not the only symbols that can be generated for graphics output but are the symbols that can be drawn by the hardware. SAS/GRAPH can draw other symbols.

Note: Not all devices are capable of drawing every symbol. If a particular hardware symbol is not available or not to be used (as indicated by the value of SYMBOLS), the symbol is generated by the software. If the value of the SYMBOL device parameter in the device entry is N or the NOSYMBOL graphics option is used, the value of SYMBOLS is ignored. △

TARGETDEVICE

Displays the output as it would appear on a different device. Also, specifies the device driver for the PRINT command.

Alias: TARGET

Used in: GOPTIONS statement

Restriction: not supported by Java or ActiveX

Syntax

TARGETDEVICE=*target-device-entry*

target-device-entry
: specifies the name of a device entry in a catalog.

Details

Use TARGETDEVICE= to specify a device driver when you want to:

- preview graphics output on your monitor as it would appear on a different output device. For details, see "Previewing Output" on page 109.
- print output from the Graph window or the Graphics Editor window with the PRINT command. For details, see "Printing Your Graph" on page 110.
- specify a device driver for graphics output created by the ODS HTML statement.

TRAILER

Specifies the command that creates TRAILER records for the Metagraphics driver.

Used in: GDEVICE procedure; GDEVICE Metagraphics window
Restriction: Used only with user-supplied Metagraphics drivers
See also: TRAILERFILE

Syntax

TRAILER='*command*'

command
: specifies a command that runs a user-written program that creates the TRAILER file. *Command* is a string up to 40 characters long.

Details

For information about Metagraphics drivers, contact Technical Support.

TRAILERFILE

Specifies the fileref of the file from which the Metagraphics driver reads TRAILER records.

Used in: GDEVICE procedure GDEVICE Metagraphics window
Restriction: Used only with user-supplied Metagraphics drivers
See also: TRAILER

Syntax

TRAILERFILE=*fileref*

fileref
: specifies a valid SAS fileref up to eight characters long. *Fileref* must have been previously assigned with a FILENAME statement or a host command before running the Metagraphics driver. See "FILENAME Statement" on page 36 for additional information on the FILENAME statement.

Details

For information about Metagraphics drivers, contact Technical Support.

TRANSPARENCY

Specifies whether the background of the image should appear to be transparent when the image is displayed in the browser.

Used in: GOPTIONS statement
Default: NOTRANSPARENCY
Restriction: This option is supported by the ACTIVEX and ACTXIMG drivers when the output is used in a PowerPoint presentation and by the GIF series of drivers only.

Syntax

TRANSPARENCY | NOTRANSPARENCY

Details

When the image is displayed and TRANSPARENCY is in effect, the browser's background color replaces the driver's background color, causing the image to appear transparent.

Note: It is recommended that you set the background color of your GIF output to match the background color of the presentation in which you want to use the GIF image. As an alternative, consider using the UPNGT device. △

TRANTAB

Selects a translate table for your system that performs ASCII-to-EBCDIC translation.

Used in: GOPTIONS statement; GDEVICE procedure; GDEVICE Host File Options window

Default: host dependent

Restriction: not supported by Java or ActiveX

Syntax

TRANTAB=*table* | *user-defined-table*

table
: specifies a translate table stored as a SAS/GRAPH catalog entry. *Table* can be one of the following:

    SASGTAB0 (default translate table for your operating environment)

    GTABVTAM

    GTABTCAM

user-defined-table
: specifies the name of a user-created translate table.

Details

TRANTAB is set by the SAS Installation Representative and is needed when an EBCDIC host sends data to an ASCII graphics device. See the SAS/GRAPH installation instructions for details. You can also create your own translate table using the TRANTAB procedure. For a description of the TRANTAB Procedure, see *Base SAS Procedures Guide*.

TYPE

Specifies the type of output device to which graphics commands are sent.

Used in: GDEVICE procedure; GDEVICE Detail window

Default: device–dependent

Syntax

TYPE=CAMERA | CRT | EXPORT | PLOTTER | PRINTER

CAMERA
: specifies a film-recording device.

CRT
: specifies a monitor or terminal.

EXPORT
: identifies the list in which the device appears under SAS/ASSIST software. This is used for drivers that produce output to be exported to other software applications, such as CGM or HPGL.

PLOTTER
  specifies a pen plotter.

PRINTER
  specifies a printer

Details

You should not modify this value for SAS-supplied device drivers.

UCC

Sets the user-defined control characters for the device.

Used in: GOPTIONS statement; GDEVICE procedure; GDEVICE Parameters window
Restriction: device–dependent; not supported by Java or ActiveX

Syntax

UCC='*control-characters-hex-string*'X

control-characters-hex-string
  is a hexadecimal string that can be up 32 bytes (64 characters) long. You only need to specify up to the last non-zero byte; the remaining bytes will be set to zero.

Details

Not all devices support this feature, and the meaning of each byte of the string varies from device to device.

Typically the UCC byte position is indicated by a bracketed value. For example, UCC[2] refers to the second byte of the string. For assistance with determining UCC values for your specific device, please contact SAS Technical Support.

USERINPUT

Determines whether user input is enabled for the device.

Used in: GOPTIONS statement
Default: NOUSERINPUT
Restrictions: GIFANIM driver only; not supported by all browsers

Syntax

USERINPUT | NOUSERINPUT

USERINPUT
  enables user input

NOUSERINPUT
  disables user input

Details

When user input is enabled, processing of the animation is suspended until a carriage return, mouse click, or some other application-dependent event occurs. The user input feature works with the delay time setting so that processing continues when user input occurs or the delay time has elapsed, whichever comes first.

VORIGIN

Sets the vertical offset from the lower-left corner of the display area to the lower-left corner of the graph.

Used in: GOPTIONS statement; GDEVICE procedure; GDEVICE Detail window

Restriction: not supported by Java or ActiveX

See also: HORIGIN

Syntax

VORIGIN=*vertical-offset* <IN | CM | PT>

vertical-offset **<IN | CM | PT>**
  must be a nonnegative number and can be followed by a unit specification, either IN for inches (default), or CM for centimeters, or PT for points. If you do not specify VORIGIN, a default offset is searched for in this order:

  1 the bottom margin specification on an OPTIONS BOTTOMMARGIN setting
  2 VORIGIN setting in the device catalog.

Details

The display area is defined by the XMAX and YMAX device parameters. By default, the origin of the graphics output area is the lower-left corner of the display area; the graphics output is offset from the lower-left corner of the display area by the values of HORIGIN and VORIGIN. VORIGIN + VSIZE cannot exceed YMAX.

  Note: When sending output to the PRINTER destination (ODS PRINTER), if you specify the VSIZE= option without specifying the HSIZE= option, the default origin of the graphics output area changes. The default placement of the graph changes from the lower-left corner of the display area to the top-center of the graphics output area. Likewise, if you specify the HSIZE= option without specifying the VSIZE= option, the graph is positioned at the top-center of the graphics output area by default. △

See "The Graphics Output and Device Display Areas" on page 59 for details.

VPOS

Sets the number of rows in the graphics output area.

Used in: GOPTIONS statement

Default: device-dependent: the value of the LROWS or PROWS device parameter

Restriction: not supported by Java or ActiveX

See also: HPOS, LROWS, PROWS

Syntax

VPOS=*rows*

rows
> specifies the number of rows in the graphics output area, which is equivalent to the number of hardware characters that can be displayed vertically. Specifying VPOS=0 causes the device driver to use the default hardware character cell height for the device.

Details

The VPOS= graphics option overrides the values of the LROWS or PROWS device parameters and temporarily sets the number of columns in the graphics output area. VPOS= does not affect the height of the graphics output area but merely divides it into rows. Therefore, you can use VPOS= to control cell height.

The values specified in the HPOS= and VPOS= graphics options determine the size of a character cell for the graphics output area and consequently the size of many graphics elements, such as hardware text. The larger the size of the HPOS= and VPOS= values, the smaller the size of each character cell.

See "Overview" on page 59 for more information.

VSIZE

Sets the vertical size of the graphics output area.

Used in: GOPTIONS statement; GDEVICE procedure; GDEVICE Detail window

Restriction: partially supported by Java or ActiveX

See also: HSIZE, YMAX

Syntax

VSIZE=*vertical-size* <IN | CM | PT>

vertical-size <IN | CM | PT>

specifies the height of the graphics output area; *vertical-size* must be a positive number and can be followed by a unit specification, either IN for inches (default), or CM for centimeters, or PT for points. If you do not specify the VSIZE= option, a default size is searched for in this order:

1 the vertical size is calculated as

YMAX − BOTTOMMARGIN − TOPMARGIN

Note that BOTTOMMARGIN and TOPMARGIN are used in the OPTIONS statement.

2 VSIZE setting in the device catalog.

V6COMP

Allows programs that are run in the current version of SAS to run with selected Version 6 defaults.

Used in: GOPTIONS statement

Default: NOV6COMP

Restriction:
Partially supported by Java or ActiveX

Ignored unless OPTIONS NOGSTYLE is also specified

Syntax

V6COMP | NOV6COMP

V6COMP
causes SAS/GRAPH programs to use these Version 6 behaviors:

- By default, patterns are hatched patterns, not solid, and the default outline color matches the pattern color.
- By default, the GCHART and GPLOT procedures do not draw a frame around the axis area.

NOV6COMP
causes SAS/GRAPH programs to use all the features of the current SAS version.

Details

V6COMP performs the necessary conversions so that, for selected defaults, you get the same results in the current SAS version that you did in Version 6.

Note: V6COMP does not convert Version 6 catalogs to catalogs with the current SAS catalog format. △

XMAX

Specifies the width of the addressable graphics display area; affects the horizontal resolution of the device and the horizontal dimension of the graphics output area.

Used in: GOPTIONS statement; GDEVICE procedure; GDEVICE Detail window

Restriction: Ignored by default display drivers, universal printing drivers, Java, and ActiveX

See also: HSIZE, PAPERSIZE, XPIXELS

Syntax

XMAX=*width* <IN | CM | PT>

width
> is a positive number that can be followed by a unit specification, either IN for inches (default), or CM for centimeters, or PT for points. If you do not specify XMAX, a default width is searched for in this order:
>
> 1 the *width* specification on an OPTIONS PAPERSIZE setting
>
> 2 XMAX in the device entry catalog.
>
> If XMAX=0, default behavior is used. If both XMAX and PAPERSIZE have been specified on GOPTIONS, the last request is used.

Details

Like the XPIXELS device parameter, XMAX controls the width of the display area, but the width is in inches, centimeters, or points rather than pixels. Typically, you might use XMAX to change the width of the display area for a hardcopy device.

SAS/GRAPH uses the value of XMAX in calculating the horizontal resolution of the device:

x-resolution = XPIXELS / XMAX

However, changing XMAX does not necessarily change the resolution:

- If you use the GOPTIONS statement to change only the value of XMAX= and do not change XPIXELS=, SAS/GRAPH retains the default resolution of the device and recalculates XPIXELS, temporarily changing the width.

- If you specify values for both XMAX= and XPIXELS=, SAS/GRAPH recalculates the resolution of the device using both of the specified values. The new resolution might be different. For example, both of these pairs of values produce the same resolution, 300dpi:

   XPIXELS=1500 and XMAX=5

   XPIXELS=1800 and XMAX=6

XMAX also affects the value of HSIZE, which controls the horizontal dimension of the graphics output area.

- If you change the value of XMAX and do not change HSIZE=, SAS/GRAPH calculates a new value for HSIZE=, using this formula:

   HSIZE = XMAX − *margins*

Note: The *margins* quantity, here, is not a device parameter. It represents the value of the left margin plus the right margin. The left margin is the value of HORIGIN. The right margin is whatever is left over when you subtract HSIZE and HORIGIN from XMAX. The value of *margins* is always based on the original XMAX and HSIZE values that are stored in the device entry. △

□ If you specify values for both XMAX= and HSIZE=, SAS/GRAPH uses the specified values plus the value of device parameter HORIGIN. Anything left over is added to the right margin. For example, if XMAX=6IN and HSIZE=4IN and HORIGIN=.5IN, the right margin will be 1.5in. If HSIZE= is larger than XMAX=, HSIZE= is ignored.

To permanently change the value of the XMAX device parameter in the device entry, use the GDEVICE procedure. This can change the resolution.

To temporarily change the size of the display and the resolution of the device for the current graph or for the duration of your SAS session, use XMAX= and XPIXELS= in the GOPTIONS statement.

To reset the value of XMAX to the default, specify XMAX=0. To return to the default resolution for the device, specify both XMAX=0 and XPIXELS=0.

See "Overview" on page 59 for more information.

XPIXELS

Specifies the width of the addressable display area in pixels and in conjunction with XMAX determines the horizontal resolution for the device.

Used in: GOPTIONS statement; GDEVICE procedure; GDEVICE Detail window

Default: device–dependent

See also: XMAX

Restriction: Partially supported by Java and ActiveX

Syntax

XPIXELS=*width-in-pixels*

width-in-pixels
   is a positive integer up to eight digits long (0...99999999).

Details

Like the XMAX device parameter, XPIXELS controls the width of the display area, but the width is in pixels rather than inches, centimeters, or points. Typically, you might use XPIXELS to change the width of the display area for an image format device.

Note: This option overrides the OutputWidth style attribute in the graph styles. For more information on graph styles, refer to the TEMPLATE procedure documentation in *SAS Output Delivery System: User's Guide.* △

The value of XPIXELS is used in calculating the resolution of the device:

x-resolution = XPIXELS / XMAX

However, changing XPIXELS does not necessarily change the device resolution:

- If you use the GOPTIONS statement to change only the value of XPIXELS= and do not change XMAX=, SAS/GRAPH retains the default resolution of the device and recalculates XMAX, temporarily changing the width of the display. If HSIZE= is also not specified, SAS/GRAPH uses the new XMAX value to calculate a new HSIZE value, using this formula:

    HSIZE = XMAX − *margins*

    Note: Margins are not device parameters, but represent the value of HORIGIN (the left margin) plus the right margin. The right margin is whatever is left over when you subtract HSIZE and HORIGIN from XMAX. The values of *margins* is always based on the original XMAX and HSIZE values that are stored in the device entry. △

    If HSIZE= is specified and its value is larger than XMAX, HSIZE= is ignored.

- If you use the GDEVICE procedure to permanently change the value of the XPIXELS device parameter in the device entry, SAS/GRAPH automatically recalculates the resolution of the device is using the value of XMAX device parameter.

- If you change the values of both XMAX= and XPIXELS=, SAS/GRAPH recalculates the resolution of the device using both of the specified values.

Note: When SAS/GRAPH recalculates the resolution, the resolution does not necessarily change. For example, both of these pairs of values produce the same resolution, 300dpi:

```
XPIXELS=1500 and XMAX=5
XPIXELS=1800 and XMAX=6
```

△

To reset the value of XPIXELS to the default, specify XPIXELS=0. To return to the default resolution for the device, specify both XPIXELS=0 and XMAX=0.

YMAX

Specifies the height of the addressable graphics display area; affects the vertical resolution of the device and the vertical dimension of the graphics output area.

Used in: GOPTIONS statement; GDEVICE procedure; GDEVICE Detail window

Restriction: ignored by default display drivers and universal printing drivers; not supported by Java or ActiveX

See also: PAPERSIZE, VSIZE, YPIXELS

Syntax

YMAX=*height* <IN | CM | PT>

height
: is a positive number that can be followed by a unit specification, either IN for inches (default), or CM for centimeters, or PT for points. If you do not specify YMAX, a default height is searched for in this order:

1 the *height* specification on an OPTIONS PAPERSIZE setting
2 YMAX in the device entry catalog.

If YMAX=0, default behavior is used. If both YMAX and PAPERSIZE have been specified on GOPTIONS, the last request is used.

Details

See "XMAX" on page 432.

YPIXELS

Specifies the height of the addressable display area in pixels and in conjunction with YMAX determines the horizontal resolution for the device.

Used in: GOPTIONS statement; GDEVICE procedure; GDEVICE Detail window
Default: device-dependent
See also: YMAX
Restriction: Partially supported by Java and ActiveX

Syntax

YPIXELS=*height-in-pixels*

height-in-pixels
is a positive integer up to eight digits long (0...99999999).

Details

See "XPIXELS" on page 433.

Note: This option overrides the OutputHeight style attribute in the graph styles. For more information on graph styles, refer to the TEMPLATE procedure documentation in *SAS Output Delivery System: User's Guide.* △

PART 2

Bringing SAS/GRAPH Output to the Web

Chapter 16 .... Introducing SAS/GRAPH Output for the Web *439*

Chapter 17 .... Creating Interactive Output for ActiveX *453*

Chapter 18 .... Creating Interactive Output for Java *469*

Chapter 19 .... Attributes and Parameters for Java and ActiveX *485*

Chapter 20 .... Generating Static Graphics *503*

Chapter 21 .... Generating Web Animation with GIFANIM *519*

Chapter 22 .... Generating Interactive Metagraphics Output *531*

Chapter 23 .... Generating Web Output with the Annotate Facility *539*

Chapter 24 .... Creating Interactive Treeview Diagrams *543*

Chapter 25 .... Creating Interactive Constellation Diagrams *553*

Chapter 26 .... Macro Arguments for the DS2CONST and DS2TREE Macros *569*

Chapter 27 .... Enhancing Web Presentations with Chart Descriptions, Data Tips, and Drill-Down Functionality *595*

Chapter 28 .... Troubleshooting Web Output *633*

CPSIA information can be obtained at www.ICGtesting.com
Printed in the USA
LVOW072223070612

285023LV00001B/5/P